THE CONTINUING RELEVANCE OF JOHN DEWEY

VIBS

Volume 223

a volume in
Central European Value Studies
CEVS
Emil Višňovský, Editor

THE CONTINUING RELEVANCE OF JOHN DEWEY

Reflections on Aesthetics, Morality, Science, and Society

Edited by
Larry A. Hickman
Matthew Caleb Flamm
Krzysztof Piotr Skowroński
and
Jennifer A. Rea

Rodopi

Amsterdam - New York, NY 2011

Cover Image: University of Chicago Laboratory Schools, founded by John Dewey in 1896.

Cover Design: Studio Pollmann

The paper on which this book is printed meets the requirements of "ISO 9706:1994, Information and documentation - Paper for documents - Requirements for permanence".

ISBN: 978-90-420-3232-3
E-Book ISBN: 978-90-420-3233-0

Printed in the Netherlands

CONTENTS

SCIENCE AND LOGIC

SOCIETY

INTRODUCTION

Larry Hickman

October 20, 2009 marked the 150th anniversary of the birth of John Dewey. During the early decades of the 20th century Dewey was a major figure in philosophy and education in America. On the occasion of his 90th birthday, the *New York Times* hailed him as "America's Philosopher." Beyond the shores of his native country, Dewey's work was also widely read and appreciated. His influence reached even beyond the places where he was invited to lecture–Japan, China, the Soviet Union, Mexico, South Africa, and many countries in Europe. By the middle of the twentieth century, however, due in part to the emergence of new movements such as Existentialism, Anglo-American analytic philosophy, Critical Theory, and Vienna Circle Positivism, his influence began to wane.

During the past two decades, however, Dewey's star has once again been on the rise. In addition to the Center for Dewey Studies at Southern Illinois University Carbondale, there are now Dewey research centers in China, Germany, Hungary, Italy, Japan, and Poland. In other countries such as South Korea, Russia, Croatia, Finland, and Vietnam philosophers, teacher-trainers, public administrators, and others come together to discuss the relevance of Dewey's ideas to the challenges of their respective professions.

The essays in this volume grew out of just such a venture. They are revised versions of some of the papers presented at the conference on "John Dewey in the Context of American and European Values" that was convened in Poland at the University of Opole during June 2009. Taken together, they present the views of internationally known scholars on some of the most important aspects of Dewey's work–in aesthetics, ethics, philosophy of science and logic, and social philosophy.

In the first section, *Aesthetics*, **John Ryder's** essay "Experience, Knowledge and Art" addresses the theme of Dewey's aesthetics from the standpoint of its relation to cognition. Ryder argues that the experience of art has an intellectual dimension, that is, that knowledge is derived from the art experience. But he is also quick to deny that such knowledge is of the same type that is generated and exhibited in the sciences. Ryder builds his essay around three questions. First, in what senses do art and aesthetic expcriences generate knowledge? Second, what is the relation of art as cognitive activity to other cognitive activities? Third, what are the implications of art as cognitive activity for an overall understanding of knowledge? Drawing on a wealth of examples from painting and music, he explores some possible answers to these questions. He concludes with three suggestions for further research. First, there needs to be a major revision of the analytic view of naturalist epistemology. Second, there needs to be a revision of Dewey's theory of inquiry that will flesh out the hermeneutic links between art and science. Finally, there needs to be further investigation of Susan Langer's claim that discursive reason and art constitute different forms.

In his essay "Aesthetics as Social Philosophy," **James Campbell** examines the roles that art and aesthetic experience play within Dewey's larger philosophy. More specifically, he argues that despite the fact that Dewey's *Art as Experience* (1934) was not published until late in his career, art and aesthetic experience had been central to the treatments of democracy morals, and education that he had published much earlier. Campbell emphasizes Dewey's view that without an adequate appreciation for the aesthetic there cannot be an adequate understanding of experience. He constructs his account around the following themes: Dewey's insistence that the aesthetic must be reconnected with the "ordinary;" his examination of the power of aesthetic imagination to establish the connections that allow for the reconstruction of experience; his detailed argument that art and the aesthetic must play a role in any viable society; and the importance of aesthetic experience in the formation of the individual, and thus the community.

In his complex and richly detailed essay "The 'Quality' of Philosophy: On the Aesthetic Matrix of Dewey's Pragmatism." **Robert Innis** explores the connections between the semiotic work of Charles S. Peirce and Dewey's aesthetics. Innis argues that Dewey's discussion of the pervasive quality of a "situation" is closely linked to Peirce's category of Firstness. On both accounts, he suggests, there is an "explosion of the claim of uniqueness and exclusive power of propositional-linguocentric-symbolization, and that iconic symbolization is sui generis." Innis places Dewey's aesthetics in the context of accounts by Lyotard, Heidegger, Langer and others as he makes his case.

In his wide-ranging and erudite essay "Social and Political Powers in John Dewey's Aesthetics," **Krzysztof (Chris) Pitor Skowroński** discusses Dewey's aesthetics in the context of the debates between essentialist and non-essentialist aestheticians; finds relations between Dewey's work and *avant guard* artists such as Pablo Picasso and Marcel Duchamp and suggests that Dewey would agree that the arts can serve as a litmus test for the scope of liberty in a given stage of development of a given culture or society. It is the question of the relation of aesthetics to power, however, that is Skowroński's paramount concern. He discusses institutional power as it affects the aesthetic qualities of museums, advertising, and even schools and universities. The importance of a work of art, he argues, can be measured by its influence on social and political powers, even though *as* a work of art it may not have been politically motivated.

Richard Anthony Spurgeon Hall begins his essay “The Deweyan Aesthetic of Charles Ives” by noting that numerous philosophers have influenced artists, and then asking where among artists, if anywhere, Dewey's influence reached. He then proposes and defends the thesis that the music of Charles Ives is Deweyan in character, “a practical enactment, as it were, of Dewey’s aesthetics.” More specifically, he suggests that there is in the work of both men a rejection of dualism: the dualism of art and life, the dualism of “aesthetic experience versus “ordinary” experience and the dualism of popular, or vernacular music, and classical, or cultivated music. Both Dewey and Ives thought that art failed when it was remote from daily social living; that art at its best is integrated with organic life. In a series of examples drawn from the works of Emerson and Thoreau, Hall stresses the experimental aspects of Ives’ work, and places that aspect of his work, too, within the Deweyan vein.

Sebastian Stankiewicz’s essay “Qualitative Thought, Thinking Through the Body, and Embodied Thinking: Dewey and his Successors” is a thoughtful and detailed exploration of the epigram “think with the sense–feel with the mind.” Stankiewicz demonstrates that Dewey’s pragmatist aesthetics provides a picture of thinking that appreciates all human abilities, not just the linguistic. He first reconstructs Dewey’s conception of qualitative thinking, then explores the notions of “thinking through the body” as it has been advanced by Richard Shusterman, and “embodied thinking” as it has been articulated by Mark Johnson. Finally, he considers some of the consequences of these ideas for a philosophy of art. In all this Stankiewicz pays homage to Dewey’s contention that aesthetic experience is the exemplar of other modes of experiencing.

In the section on *Ethics*, **Ramón del Castillo’s** essay “John Dewey and the Ethics of Recognition” presents a sophisticated and detailed view of Dewey’s political philosophy as neither the type of communicative action that Juergen Habermas has advanced, nor the communitarian social philosophy that has been a part of the debates within the United States. Castillo utilizes the work of Axel Honneth as an aid to reconnecting Dewey's work to contemporary politics. In his view, Dewey stresses that communication within democratic societies does not consist simply in the expression of interests and claims, but rather in the interplay “in which social actors overcome barriers and become transformed and enriched.” This distinguishes Dewey's view–which emphasizes growth–from versions of participatory democracy that ground themselves in participatory practices. Castillo argues that it is recognition–recognition of differences–that is intrinsically connected with growth.

Ángel M. Faerna’s essay “Dewey’s Value Theory and the Analytic Tradition of Moral Philosophy” takes up two central tenets of moral philosophy as it has been developed within Anglo-American analytical philosophy. The first involves the putative split between facts and values that he believes led analytic moral philosophy to avoid involvement in normative commitments. The second, which he treats as a consequence of that debilitating split, is the tendency within analytic philosophy for ethics to be absorbed into a metaethical preoccupation with the language of ethics. As a part of his deft analysis of *Dewey's Theory of Valuation* and other texts, Faerna provides a rich account of Dewey's responses to ethicists working in the analytic tradition. He concludes that Dewey's empirical ethics is not so much interested in preconceived notions of what good is, but rather an investigation of the possibilities of more meaningful experiences.

In "Pragmatic Moralism and the Politicization of Philosophy," **Matthew Caleb Flamm** revisits the interchange between Dewey and George Santayana regarding their respective versions of naturalism. Recalling the verdict of Santayana, that Dewey's naturalism was all "foreground," and Dewey's charge that Santayana's naturalism was "broken-backed," Flamm nevertheless points to a significant affinity between the two thinkers: both diagnosed the mistake of traditional philosophy with respect to its depictions of nature as committing an idealistic fallacy. Flamm further argues that Santayana's proposal for setting matters right was better than Dewey's, since it involved the development of a type of cosmology that Dewey's position denied. Flamm suggests that the "moralism" Santayana diagnosed in Dewey's philosophy may have opened the gates for the politicizing neo-pragmatism of some contemporary philosophers writing in the shadow and influence of Richard Rorty.

"Dewey's Theory of Values," by **Hugh McDonald**, provides a detailed account of the material indicated by its title, as well as of some common misreadings and misconceptions of Dewey's treatment of the subject. McDonald takes special care to distinguish the terms "value," "valuing," "valuation," and „valuable." He points out that although the conative and the emotional are aspects of the process of valuation for Dewey, he refused to identify values with psychological states, as do some of his critics. Dewey consequently ruled out any private or introspective theory of values. In a helpful contribution to the discussion, McDonald also parses the terms "instrumental," "inherent," and "intrinsic." His essay concludes with a discussion of Dewey's rejection of the traditional split between facts and values, that is, between the "is"and the "ought."

In the section on *Science*, **Larry Hickman's** essay "Scientific Concepts in Dewey's Pedagogy" explores the process of formation of what Dewey terms "genuine" or "scientific" concepts. He distinguishes scientific concepts from those that are merely nominal and those that are merely analytic. Nominal concepts are generated by enumeration and memorization, a technique that is widely used by educators who "teach to a test." Analytic concepts are generated by definitional analysis that involves acquisition, categorization, and reference determination. Concept-analysis is closely associated with Anglo-American analytic philosophy. Although there are nominal (by enumeration and memorization) and analytic (by definitional analysis) components in the process of genuine concept-formation, it is in fact much more complex than either of its alternatives. Scientific concept-formation starts not from enumeration, but from identification of particular experiences, supplementing those experiences with others of the same type based on behavioral responses, and locating the refined concept in a larger context, perhaps legal or economic. A consequence is the development of conceptual tools that can be of service in the future control of facts. Two examples suggest the applicability of scientific concept-formation to social and political issues: adoption of children by same-sex couples and the search for a more scientific concept of democracy.

In "Dewey's Science: A Transactive Model of Research Processes," **Philipp Dorstewitz** develops a model of Dewey's process of inquiry that identifies the main modes of scientific research (hypothesis formation, reasoning, experiment) as concurrent and mutually dependent practices which center on the production of warranted beliefs and values. He contends that Dewey's "rhythm of situations" builds the temporal context of a process-model of inquiry. This model views an inquiring community as a transactive process in unfolding problematic situations. After a review of some pertinent debates within the philosophy of science, such as that between Kuhn and Lakatos, Dorstewitz defends the notion that Dewey's theory of experience and inquiry should no longer be ignored by mainstream philosophy of science.

Ahti-Veikko Pietarinen's essay "A Hedgehog Who Thought He Was a Fox: Dewey Betwixt the One and Many-World Philosophies" traces the ways in which Dewey came to accept a universalist theory of meaning, which involved "discouraging the falsification of philosophical claims, advocating a naturalistic interpretation of habits, and evaluating the identity of problematic situations with respect to the one, actual world." Pietarinen's central argument is that even though Dewey followed his teacher Charles Peirce in appreciating the value of "metasystematic approaches to scientific inquiry," in the final analysis he was even more influenced by William James in the sense that he ultimately opted for a "one world" theory of meaning. After presenting a series of tightly constructed arguments, Pietarinen points out that although a philosophical fox (who thinks of many possible worlds) may occasionally pretend to be a hedgehog (who thinks of one big world), it would not be possible for a hedgehog to be an occasional fox. He leaves his reader with a question: given what appears to be an inner conflict in Dewey's position, might there be in Dewey's work some deeper metaphysical point that his readers have missed?

In the section on *Society*, **Walter Feinberg's** essay "Teaching Religion in Public Schools: A Critical Appraisal of Dewey's Ideas on Religion and Education," takes on a subject that is both difficult and highly controversial. After informing his readers about how he will use the terms "post-modern," "society," and "liberal," he characterizes the current educational system as tending to a split between schools that teach one religion and those who teach none at all. His own suggestion, which builds on the work of Dewey, is that we need schools in which dialogue across and among religious orientations is encouraged. His presentation is supported with a brief historical sketch of the problem, followed by a discussion of Dewey's major work on religion that is supplemented by other aspects of his philosophy, and finally by a presentation of the results of his study of classroom teachers who are helping their students develop the skills that will be required for dialogues across religious boundaries.

In her essay "John Dewey and Josiah Royce in Dialogue on the Individual and Community," **Jacquelyn Ann K. Kegley** highlights the concerns and philosophical positions that the two men held in common, as well as some significant issues on which they differed, or at least on which they placed different emphases. She points out that Dewey and Royce shared the belief that their time was one of insecurity and the "lost individual," that this condition was the result of over emphasis on self-interest and economic values, that "individualism" needed to be redefined, and that a true individuality could be developed only in the context of community of shared interests and values. But Kegley thinks that Dewey failed to develop the notion of communication to the extent that Royce did. Further, she suggests that whereas Dewey viewed scientific method as a model for revitalizing democracy, Royce thought that the scientific community provided but one paradigm of a community of interpretation. She presents Dewey as more interested in social, political, and economic issues than Royce, and Royce as more interested than Dewey in the prospects of freeing individuals and communities from blindness to the needs of others.

Maciej Kassner's essay "John Dewey and Friedrich von Hayek on Individualism and Freedom" presents a detailed account of the contrasting positions of the two thinkers regarding the role of individuality within social institutions. Kassner finds that C. B. McPherson's distinction between possessive individualism and developmental individualism provides an excellent tool with which parse the opposing positions. After an extended discussion of Dewey's attacks on the notion of "rugged individualism" during the 1930s, he turns to a discussion of Hayek and his influence. Finally, he responds to Hayek from a Deweyan perspective, utilizing arguments advanced by Dewey himself and also by Karl Polanyi. Perhaps most telling, he finds in Hayek's argument a deep contradiction: the maintenance of *laissez faire* individualism will depend on individuals who "submit to conventions . . . whose justification in the particular instance may not be recognizable and which to him will often appear unintelligible and irrational." Kassner concludes that these issues are still very much alive, and that a Deweyan inspired "developmental" individualism is more coherent and sustainable than one that is "possessive."

In the final essay in this collection, "Dewey's Ethical-Political Philosophy as Resource in Today's Global Crises and as a Guide to a Post-Ideological Politics for the 21st Century," **Gregory Pappas** argues that Dewey offers a way out of the ideological oppositions in socio-political discourse. There is no reason, he suggests, for Americans to get stuck between libertarianism and communitarianism. Dewey's own democratic contextualism holds that each case of whether market or social forces should prevail should be considered on its own merits. Pappas suggests that the philosophy of governance exhibited by the Obama administration in fact honors some of Dewey's basic insights. Pappas concludes that Dewey's ethical-political thought can guide us toward a post-ideological politics by refusing to pre-define "what works" ahead of careful consideration of problematic situations. He responds to critics who claim that Dewey's ethics is too thin, too relativistic, by pointing out that although he rejected moral absolutes, he nevertheless insisted that good judgment and deliberation depends on the cultivation of good habits.

AESTHETICS

One

EXPERIENCE, KNOWLEDGE AND ART

John Ryder

1. Introduction

For many people, myself included, one comes away from an artistic experience with a sense that something intellectually significant has occurred. I am thinking here of experience with art as a spectator rather than as an artist. I would not want to speak for all those who make or perform something, but speaking only for myself, when I play a piece of music I do not have the sense that I have accomplished something intellectual. When I listen to a performance, especially a live performance, I often do have just such a sense, as I do during a visit to an exhibition. I do not here want to make anything of the distinction between making or performing and observing or listening because I am not sure how far to generalize my own experience. I mean to point out simply that there is an intellectual dimension to the experience of art. And I would not want to say that aesthetic experience in this general sense must have an intellectual dimension, only that it may.

To be a bit more precise, this intellectual experience through art is, for me, the sense that I have learned something, or that I now know something I did not previously. It is not always a simple matter to specify what it is I think I have learned, or what I now know, though in that respect the same can sometimes be said for the intellectual experience of listening to a lecture or reading a non-fiction book. Despite the complexity of specifying what one has learned, it is clear to me that knowledge is derived in and through aesthetic experience, or more modestly that knowledge can be derived in and through aesthetic experience. To the extent that this claim is true, it is then fair to say that art has a cognitive dimension. Again, this is not to say that all art is cognitive, or that all art must lead to enhanced or increased knowledge, merely that it may. I am not offering a theory of art or an aesthetic theory in which it is a defining characteristic of art that it is cognitive, or that the many other things that art is and does may be subsumed under or reducible to the cognitive. I also do not intend to take a position on the question whether the cognitive dimension of art is among its aesthetic characteristics. The point is simply that a cognitive dimension is at least a contingent feature of art.

I have so far equivocated between art and the experience of art in the sense that I have attributed an intellectual or cognitive dimension to both. Art and the experience of it, or aesthetic experience in a general sense, are not the same thing, and it is quite possible that one might attribute a cognitive dimension to one, for example to the experience of art, and not to the other. Though this distinction is genuine, I will continue to talk about art and aesthetic experience interchangeably. One reason to do so is that because I am not here attempting to develop a theory of art it does not seem to me to be a critical question whether it is among the defining traits of art that it be or can be cognitive, as I have pointed out above. That question may remain open as far as this discussion is concerned. What is relevant is whether in our engagement with art we may have a cognitive experience. Whether that experience is properly to be attributed to a cognitive dimension of art or to a cognitive dimension of aesthetic experience is a question that can wait for another occasion, and the answer to it does not have any critical bearing on the question if and in what senses we may enhance our understanding and insight through our engagement with art. So for ease of formulation I will refer to art and aesthetic experience interchangeably.

Dewey briefly addresses the question of art and knowledge in *Art as Experience* in the chapter titled The Challenge to Philosophy.[1] His concern there, and he is right to have it, is that among the numerous attempts to develop a philosophy of art some have tended to reduce art to a cognitive exercise. He ascribes this view, or at least the danger of this view, to the romantic impulse to understand art as a uniquely powerful entre into the nature of things. The problem with this approach, he points out, is twofold: 1) there is a tendency to ignore the many other characteristics of art by overemphasizing the cognitive; and 2) without a more careful analysis of knowledge, to say that art is primarily cognitive is to make it more or less equivalent in function to science and other disciplines, but in that case art does not fare well.

Dewey is clear in this short section that he has no problem with acknowledging a cognitive dimension of art, as long as we are careful about what we take that to mean. He also indicates that even this modest assertion does a disservice to art unless there is a more general understanding of art in which its potential cognitive dimension can be situated and understood. The aesthetic theory he develops in *Art as Experience* is an effort to develop an adequate theory of art. Whether his general theory of knowledge is sufficient to the task is a question to which we will return.

That art has a cognitive dimension suggests three general problems: 1) In what sense or senses do art and aesthetic experience generate knowledge; 2) What is the relation of art as a cognitive activity to other cognitive activities such as science, mathematics or philosophy; and 3) What might be the implications for our understanding of knowledge of the fact that art has a cognitive dimension?

2. What Knowledge Does Art Inspire?

The first point to be made in considering this question is that art gives rise to knowledge in many different ways, or that there are many different senses in which art is cognitive. Dewey indicates one of the ways in which art is cognitive when he says that "Tangled scenes of life are made more intelligible in esthetic experience; not, however, as reflection and science render things more intelligible by reduction to conceptual form, but by presenting their meanings as the matter of a clarified, coherent, and intensified or 'impassioned' experience."[2] To put the point another way, one sense in which we may derive knowledge from art is through the capacity of art to arrange its subject matter in such a way as to bring to the fore or highlight elements of the subject that are otherwise less clear or less pointed. This is the basis of the cliché that a picture is worth a thousand words. Descriptions of the effect of war on people can be moving and profound, of course, but the arrangement of the elements of village life that Picasso presents in *Guernica* focuses the point in a unique way. The cliché has it wrong, though, because finally wrought words, especially in fiction, can do much the same. Remarque's account of the impact of the Great War is a good example. Through these and similar works of art we better understand war in its lived dimension; they present "their meanings as the matter of a clarified, coherent, and intensified or 'impassioned' experience."

We derive knowledge from and through art in other ways as well. It has served to express ideals, as in classical Greek sculpture, in which case we come away understanding ideal forms, and the civilization that thought in terms of ideal forms, better than we had before. In other cases it serves to teach something, whether the moral lessons of medieval Christian painting or the ideological lessons of the more crass forms of socialist realism. It can suggest a critical analysis, as does some modernist art in its assessment of industrial society, for example in Leger's use of tubular forms to depict people, or it can describe social status, as frequently did privately commissioned painting in early modern Europe. It can be highly politically charged, as in a Diego Rivera mural, or it can puncture political pretensions, as do Shostakovich's parodic ballets and operas.

Literature can have similar properties. We know more about people and their psychology after reading Dostoevsky than we did before, and we better comprehend the frustration of delusional but firmly believed ideals through Cervantes, so much so that to "tilt at windmills" has entered the lexicon. Volumes have been written about what we learn and know from Hamlet, Macbeth and Lear, and surely we have a deeper understanding of the potential passion and tragedy of love after Romeo and Juliet, not to mention the disastrous consequences of tribal feuds.

Expressionist and abstract art too has a cognitive dimension. In a direct sense, we know more about colors and their relations to one another after Albers and Rothko, not to mention the Fauve, Blaue Reiter and Brücke schools. John Cage and other experimental composers stretched our understanding of sound, and by implication music itself. The greatest of the Baroque and Classical composers described what was possible within certain clearly delineated formal constraints more thoroughly than anyone could have imagined. This is part of the reason their music continues to fascinate us. Conversely, we can admire Wagner and Schönberg in their capacity to break through traditional structure to reveal what is possible beyond it. In this sense, the knowledge we derive from abstract art can concern the generation of possibilities. Kazimir Malevich may appear to have done something extremely simple with his *Black Square*, but not only did that image become an iconic expression of the Russian *avant garde*, it also engendered an explosion of possibilities, as is made clear by a recent exhibition in the Russian Museum in St. Petersburg called The Adventures of the Black Square.[3] To this day sculptors, painters and installation artists continue to generate new possibilities, and thus expand our experience and insight.

Dewey is right to be skeptical of the Romantic and generally mystical inclination to think that art penetrates to some "higher" knowledge of reality, but one can understand how the Romantics and mystics come to that point of view. Some works of art do seem to transport us, so to speak, or at least they seem to point to something deeper. The Adagio movement of Schubert's C Major Quintet is a breathtaking piece of music, and when I hear it I cannot help but think that Schubert knew something the rest of us do not. Whether he did in fact "know something" none of us can say. Perhaps it was simply that he was a genius as a composer. But I think it is fair to say that *our* knowledge is enriched as we listen to Schubert. For one thing, we have a better understanding of the capacity of human creativity; we understand ourselves better through having the sort of experience Schubert and other great artists make possible; and our understanding of the possibilities of creativity, experience, and life itself is immeasurably enhanced.

The centrality of possibilities in the cognitive dimension of art is not to be underestimated. We have so far pointed to various works of art and artists to suggest ways in which art is cognitive, but we could also point to philosophers for whom the aesthetic dimension of experience has been central to their philosophical work and hence to the understanding and wisdom their work expresses. One of the clearest examples of such a philosopher is John J. McDermott, for whom experience as aesthetic and the aesthetic in experience permeates his thinking no matter what philosophical theme he is exploring. In McDermott's hands art and the aesthetic is the very environment in which reflective individuals both live their lives and render them meaningful, surely processes that involve understanding and insight in some ways and to some degrees. There is no better example of the centrality of possibility in the cognitive dimension of art.[4]

A very young Susan Sontag wrote in her diary that reading Romain Rolland gave her a "knowledge of aliveness". In a similar vein, Orhan Pamuk has recently said that though he regrets the "utilitarian" nature of the idea, he cannot help but think that "books exist to prepare one for life."[5] Pamuk has in mind books of all kinds, but certainly including prose and poetry. Both writers are expressing an important sense in which art, in their cases literature, contributes knowledge, which is to say that it has an import for one's life, how it is understood and how it is lived. They have in mind, I imagine, the fact that literature offers one not so much information about the world, or something to which to aspire, as it exposes one to the range of experience and the possibilities inherent in it. Stories that occur in a setting with which one is familiar, one's own time and place for example, and that include characters one can recognize, thereby expand one's own experience. They often involve events in which one can place oneself and imagine one's own reactions and behavior. But literature that places the reader beyond his own time and place can be, and probably usually is, even more expansive. Perhaps there are sufficient commonalities among people across cultures and through history that we are able to expand the possibilities of our own lives by engaging with them. We can feel, which is to say we can understand, both the romanticism and the hopelessness of Don Quixote's efforts, and we know what it means to "tilt at windmills," though none of us have ever lived or will ever live in early modern Spain. We can feel the internal struggle and trauma of Sensei in Natsume Soseki's *Kokoro*, though none of us have experienced the individual and cultural confusion of Meiji Japan's struggle to confront the modern West. In such cases one learns not so much about early modern Spain or about Meiji Japan, but about the possibilities of experience; one's own experience and its possibilities expand accordingly. Something like this, I take it, is the "knowledge of aliveness," and the sense in which literature can "prepare one for life."

Even this short and hurried list of art's cognitive capacities suggests that when we speak about knowledge as enabled by art we are speaking about more or something other than "knowledge that," whether by knowledge and truth we mean justified true belief, or assertions that accord either with an independent reality or with a prior stock of beliefs, or warranted assertability. This observation leads us to the question of the relation of art as a cognitive activity to more commonly acknowledged forms of cognition such as science and mathematics, or for that matter philosophy.

3. Knowledge in General

If the cognitive dimension of art deals not so much with "knowledge that" in any standard sense, and little with "knowledge how," but more importantly, as Justus Buchler called it, "knowledge through," then when dealing with knowledge we are faced with a complex situation that we need some way to sort out systematically.[6] Fortunately we are not left to our own devices in this matter because there is wise guidance available to us. Several philosophers in the American tradition have addressed the question of the nature of knowledge in a way broad enough to encompass knowledge in and through art. Dewey is a rich source of insights and possibilities in this respect, as is Susanne Langer. Langer more than Dewey dealt directly with the issue of the cognitive capacity of art and the ways in which knowledge through art differs from propositional knowledge. The other valuable source for us is Justus Buchler in his theory of judgment and his concept of query.[7] We will return to a consideration of Dewey and Langer below. At this point, though, I would like to focus on Buchler's theory of judgment because it provides particularly useful concepts and distinctions.

To avoid unnecessary confusion let us stipulate from the start that in his use of the term "judgment" Buchler has enabled potential misunderstanding because the meaning of the term in his theory is somewhat idiosyncratic. It is worth keeping in mind that the theory of judgment as we will briefly describe it is a theory of a person's interaction with the complexes of his environment that cumulatively define his life. In this respect the theory of judgment is a component of a broader theory of human being. In order to understand what Buchler means by "judgment," and why he would use that term and not another, one has to have the patience to follow his explication and rationale. Toward this end it may be worthwhile to keep in mind the fact that when he developed the theory in the early 1950s he was explicitly attempting to improve on Dewey's conception of experience and interaction.

The term "judgment" here is to be understood not as a mental act but as any sort of manipulation of complexes toward some end. The manipulation can be mental, but it can also be physical, emotional, repetitive or creative. It can produce an idea, a book, a poem, a painting, a piece of music, an apple pie or a home run. There are three forms of judgment – assertive, exhibitive and active – though any product of judgment may participate in more than one of them.

An assertive judgment is a stating or saying of some kind; it is generally a proposition. The results of scientific inquiry for the most part consist of assertive judgments, as do mathematical propositions and the products of philosophical thinking. Assertive judgments tend to be the sort to which truth-value can be attributed; they make claims about their subject matter that can be said to be true or false, or at least they admit such evaluative attempts. As a practical matter we may not be able to say whether a particular assertion is true or false, but for the most part they can be so evaluated. Though the standard sort of assertive judgment is linguistic, it is not the case that all linguistic judgments are assertive. A poem is not, for example, and neither is an ordinary speech act such as "I now pronounce you husband and wife." A poem is an exhibitive judgment, in ways we shall soon see, and a speech act of this kind is an active judgment.

A judgment is exhibitive when it shows rather than states something; it does not consist of propositions, but of exhibitions or demonstrations. Works of art are paradigmatic cases of exhibitive judgments. They do not as a rule assert anything, and they are generally not susceptible to attributions of truth or falsity, at least not in the standard senses. Rather than assert a proposition, exhibitive judgments consist of complexes purposefully ordered in such a way as to bring to light aspects or traits of the complexes ordered or of the product as a whole that were not available to us previously. Think of Monet's series of paintings along the Thames or of the façade of the cathedral at Rouen. Nothing is asserted, but traits of the Parliament Building and of the Rouen Cathedral are available to us as they had not been before. Or perhaps it is meanings that are so rendered, as in *Guernica*. Exhibition in this sense applies more broadly than merely to pictorial expressions. Musical meaning is enriched through the Wagnerian *leitmotiv*, for example, as is the expanded capacity of words in *Finnegan's Wake* or in any good poem. Furthermore, even assertive judgments can have an exhibitive character, as through their extraordinary structure does Spinoza's *Ethics* or Hegel's grand intellectual architectonic. And there is beauty, an exhibitive characteristic, in an elegant proof, as all mathematicians know.

Active judgments are those that do something, as in a carpenter's product or a footballer's goal or a baseball pitcher's curveball. They too can function in other modes. The well coordinated skating and passing of a hockey team's forward line as it moves down the ice can be a thing of beauty and function exhibitively, and when two men are arguing and one spits at the other, the assertive point is usually clear. All of these are examples of instances in which complexes of nature are manipulated to produce an action of some kind, and to that extent are active judgments.

Some judgments, though not all, are attempts systematically to investigate a subject, and such judgments are instances of query. There are, however, importantly different kinds of query in that there are importantly different kinds of investigative judgments. Some investigations are made by putting questions to a subject either experimentally or rationally. Such investigations are cases of inquiry, and the sciences, mathematics and philosophy in their standard forms consist of judgments of inquiry. But not all systematic investigation is experimental or rational. Monet was investigating light, color, and his pond at Giverny, just as Schönberg was investigating the possibilities of twelve-tone rows in composition. These are instances of query no less than are science and mathematics, and as we have seen their results can be cognitive just as the results of inquiry can be.

When knowledge is understood as arising from judgments that query nature, and if various forms of query produce knowledge, then it becomes much less odd or threatening to either science or art to speak of art's cognitive dimension. It also suggests, however, that traditional approaches to epistemology need to be revised.

4. The Implication for Epistemology of Art's Cognitive Dimension

I use the term "epistemology" to mean any systematic attempt to understand knowledge and truth, so that it is intended to include the various forms of analytic epistemology, pragmatist approaches to knowledge, and other creative approaches such as Buchler's and Susanne Langer's. In the English-speaking world the most pervasive forms of epistemology in this broad sense have been the analytic and pragmatist approaches, so I will turn first to the implications of the cognitive dimension of art for epistemology in these senses.

An obvious problem arises when we consider what is called naturalist epistemology, which derives its name from the fact that it is a series of developments from Quine's initial call to naturalize epistemology. The problem is in part a consequence of the fact that in this tradition the natural sciences hold a privileged position with respect to the understanding of knowledge. We should distinguish two ways this occurs. One of them is that naturalists of this stripe argue for turning to the results of the sciences, primarily to psychology and other cognitive disciplines, in our efforts to understand the nature of knowledge and its acquisition. This move amounts to an abandonment of more traditional *a priori* analysis in the attempt to understand knowledge. The second way the emphasis on science appears in naturalist epistemology is in its tendency to privilege the results of the natural sciences as the paradigmatic instances of knowledge. This is significant because the naturalist needs to have something in mind as to what counts as knowledge if he is to inquire into its nature, and his inclination is to point to the propositions that result from natural scientific inquiry and say "that's what we mean by knowledge." This move can have a weak and a strong form. The weak version will say that the propositions of the sciences are the clearest instances of knowledge, though other sorts of activities may also generate knowledge in less clear and perhaps less reliable forms. The strong version is to say that the propositions of the natural sciences are the only instances of knowledge, which is to say the only way we can have knowledge of nature is through the sciences. Everything else may have some value, but it is not cognitive.[8]

Another observation worth making at this point is that naturalist epistemologists share with their non-naturalist opponents an assumption about knowledge, which is that it is always to be understood as a belief. Thus the question becomes "what kinds of belief count as knowledge, and why." Naturalists will say that empirically derived beliefs either modeled on or rooted in the sciences are the only ones that count as knowledge, and non-naturalists will allow for other, non-empirical or rationally derived beliefs to count. Either way, the analysis of knowledge is about which beliefs are cognitively acceptable.

Naturalist epistemology in any of its analytic forms cannot handle the fact that art has a cognitive dimension. Nor, we should add, can any form of epistemology that understands knowledge entirely or paradigmatically in terms of propositions and / or beliefs. We must keep in mind the senses in which we have ascribed a cognitive dimension to art. The observation that art has a cognitive capacity has been made before, though usually by aestheticians rather than epistemologists. Even there, though, the tendency has been to ascribe to art the capacity to generate knowledge that is similar to the knowledge enabled by the sciences. So, for example, the cognitive capacity of literature has been debated in terms of the truth-claims that can be found in it or derived from it. Even those who are inclined to defend art's cognitive dimension in these terms are reluctant to extend the point beyond literature to more abstract arts such as music or dance because these do not issue in propositions or truth-claims.[9] In our ascription of a cognitive dimension of art we have been careful, it should be remembered, not to limit knowledge in art to propositions or to truth-claims of the traditional sorts. In fact we have not limited knowledge in the arts to beliefs that can be propositionally expressed at all. This is the point of the distinction between query and inquiry.

If we are right in attributing to art the possibility of knowledge that is not propositional and not a matter of beliefs with truth-value, then any epistemological approach that understands knowledge as justified belief is inadequate to the cognitive capacity of art, and therefore will result in an inadequate understanding of knowledge. Naturalist epistemology suffers from the further disability of limiting from the outset its conception of knowledge to the propositions of the natural sciences, in either its weak or strong versions. With that point of departure naturalist epistemology can never generate an adequate understanding of knowledge. One might ask at this point whether it might be possible to recast naturalist epistemological approaches, or other, non-naturalist forms of analytic epistemology, by taking knowledge in art, perhaps alongside knowledge in science, as paradigmatic. The answer is that it is certainly possible, though we may find in doing so that the analytic tools characteristically used in this tradition are not up to the task. If not everything is a nail then one needs more than a hammer. But we do not know the results of such an experiment until it is tried, and it would likely be an interesting project to do so.

Another approach at this point is to say that there are alternatives with rich histories to which we might turn. The most obvious of them, for those of us who are inclined to do philosophy in the American tradition, is the broadly pragmatist or a more precisely Deweyan approach. A weak version of this can be found in Rorty's interest in literature as no less interesting and valuable than science, and in his overall rejection of the need for a theory of knowledge. The problem with Rorty, at least for our purposes, is that he nowhere discussed the arts in general in the context of our ongoing dialogue. In fact he himself said that he has little feel for the arts, and therefore did not consider them in this light.[10] We may find greater possibilities, then, in turning to Dewey. Is there in Dewey's understanding of knowledge and truth a way to accommodate the cognitive dimension of art?

There are certainly *prima facie* reasons to think that Dewey's approach to knowledge has potential for us. First, he rejected the subjectivism of traditional epistemology so that knowledge is no longer to be understood in terms of beliefs held in the mind that then need to be evaluated for their reliability in relation to an external world. Knowledge is no longer a matter of beliefs, except possibly in the more extended Peircean sense. A second and related point is that knowledge is a matter not of what we think but of what we can do in relation to our lived environment. This understanding of knowledge provides much more room for the broader sense of knowledge and cognition to which the arts point. Third, as we have seen, Dewey explicitly acknowledges the possibility that art can be cognitive. The fourth and perhaps the most important reason for hopefulness with respect to Dewey's approach is that both knowledge and art are understood within the broader category of experience. There is what we might call a hermeneutic link between them that may be immensely helpful. It appears to be possible, then, to build into Dewey's sense of logic the cognitive dimension of art in ways that are impossible in contemporary analytic epistemology, naturalist or otherwise.

What, if any, are the obstacles to so doing? There do appear to be potential stumbling blocks. First, Dewey, like the naturalist epistemologists, understood science to be the paradigmatic instance of knowledge. Of course he had a broader understanding of scientific inquiry than the Quinean naturalists have, but still his approach to epistemology, or to logic as he preferred, is undertaken in relation to inquiry. Even if inquiry in his sense is not understood to rely on representations, it is not clear that it can accommodate knowledge that results from query of the sort that characterizes the arts. This points to a second potential problem, which is that in the end for Dewey knowledge is about propositions and warranted assertability, which as we have seen is likely to be too restrictive to handle cognition in the arts simply because the arts are not for the most part about assertions, warranted or otherwise. Dewey appears to have understood this; in fact it is the very point he is making in *Art as Experience* when he warns that we do not want to construe knowledge in art in such a way that it imitates knowledge in science. This insight, however, does not appear to have been turned back on his own conception of knowledge such that knowledge can be understood more broadly than as a matter of propositions and assertions, even if their warrant is to be judged in terms of functional success.

It is instructive to note that Dewey, unlike Buchler and as we shall see Langer, was not disposed to emphasize the differences between knowledge in science and knowledge in art. On the contrary, his inclination was to emphasize their similarities. He does so, however, not by trying to argue that either is simply a form of the other, but rather that both exemplify a more general, creative process. Consider, by way of illustration, the following passages from *Experience and Nature* and "Qualitative Thought":

> …if modern tendencies are justified in putting art and creation first…It would then be seen that science is an art, that art is practice, and that the only distinction worth drawing is not between practice and theory, but between those modes of practice that are not intelligent, not inherently and immediately enjoyable, and those which are full of enjoyed meanings.[11]
>
> Science is an instrumentality of and for art because it is the intelligent factor *in* art.[12]
>
> Thinking is preeminently an art; knowledge and propositions which are the products of thinking, are works of art, as much so as statuary and symphonies.[13]
>
> The logic of artistic construction and esthetic appreciation is peculiarly significant because they exemplify in accentuated and purified form the control of selection of detail and of mode of relation, or integration, by a qualitative whole…Artistic thought is not however unique in this respect but only shows an intensification of a characteristic of all thought. In a looser way, it is a characteristic of all non-technical, non-"scientific" thought. Scientific thought is, in its turn, a specialized form of art, with its own qualitative control.[14]
>
> …the gist of the matter is that the immediate existence of quality, and of dominant and pervasive quality, is the background, the point of departure, and the regulative principle of all thinking…"Scientific" thinking, that expressed in physical science, never gets away from qualitative existence…Construction that is artistic is as much a case of genuine thought as that expressed in scientific and philosophical matters, and so is all genuine esthetic appreciation of art, since the latter must in some way, to be vital, retrace the course of the creative process.[15]

There is a sense in which Dewey in these passages can be read as proposing something like Buchler's point that science and art are both forms of query. Dewey puts it differently by emphasizing the point that both science and art exemplify a creative process in which, as he describes it, there is invariably a qualitative background in terms of which scientific and artistic products derive their meaning and import. When he says, for example, that both science and art are instances of "the control of selection of detail and of mode of relation, or integration, by a qualitative whole," he seems to be pointing to something similar to what Buchler means by judgment. Thus there is an important and valuable dimension to Dewey's approach in that he neither radically distinguishes science from art such that neither can do anything that the other does, such as have a cognitive character, nor does he reduce either to the other and thereby limit their distinctive functions. In this respect Dewey appears to enable a more satisfactory understanding of the possibility of cognition in art.

Indeed, as in the case of naturalist epistemology, it may be possible to recast Dewey's logic in such a way that it can accommodate the cognitive dimension of the arts. In fact it seems to be a much more promising project than an attempt to do so for Quinean naturalism. Whether it can be successful, however, is not obvious, and certainly not something we are entitled simply to assume.

Another source of insight into the question of the cognitive dimension of art is Susanne Langer's treatment of the issue. As will be clear from the quotations below, Langer understands art as a symbolic form that has the function of expressing human feeling. Science and other modes of discourse, by contrast, are symbolic forms that have the function of enabling thought and discursive communication. Thus science and art are both symbolic forms, though they differ in that one enables conceptual thought and discourse while the other enables the expression of the subjective feeling of experience. Each in its own way can generate knowledge.

Langer developed this analysis in *Feeling and Form* and in *Problems of Art*. The latter is a more summary expression of her views and provides a clear entre into her understanding of the issue. She makes the clear distinction there between discursive and artistic symbolic forms:

> Language is the symbolic form of rational thought…The structure of discourse expresses the forms of rational cognition; that is why we call such thinking "discursive"…To express the forms of what we might call "unlogicized" mental life…or what is usually called the "life of feeling," requires a different symbolic form…This form, I think, is characteristic of art and is, indeed, the essence and measure of art.[16]

That the expression of the "life of feeling" has or can have a cognitive dimension in Langer's view is clear enough: "even the discursive pattern has its limits of usefulness…Yet there is a great deal of experience that is knowable…yet defies discursive formulation, and therefore verbal expression: this is what we sometimes call the *subjective aspect* of experience, the direct feeling of it…"[17] The function of art is the expression of this "direct feeling" of experience, and that expression, in that it brings into focus a dimension of experience "that is knowable," is cognitive.

Thus the initially necessary ingredients for a suitable analysis of the cognitive dimension of art are to be found in Langer. First, it is clear that for her art enables us to know something, specifically the subjective aspect of experience. Second, art is distinguished from discursive, propositional knowledge in such a way that both share the critical feature of cognitive activities, i.e. both are symbolic forms, but neither is reduced to the other, which is to say that each retains its distinctive features as cognitive activities. But as for Dewey, it is not clear on the face of it whether her understanding of knowledge in general,and art in particular, is adequate to a fuller understanding of cognition in art. For one thing, it is not clear that it is appropriate to say of art that it has only one function, for example that it expresses feeling. It is one thing to understand art as a distinctive symbolic form, analogous to Buchler's understanding of it as paradigmatic of exhibitive judgment. But it is quite another thing to attribute one function to art. On the face of it art does many things and has many functions. To express subjective experience or "feeling" may or may not be one of them. But even if it is, there does not appear to be good reason to limit its function in that way.

Despite these potential shortcomings, Langer's analysis bears further scrutiny. We have suggested that it may be worthwhile in the case of naturalist epistemology to make an effort to recast it with the cognitive dimension of art in mind. We have also said that Dewey's understanding warrants further consideration because there is a great deal of promise there. We may comfortably say the same for Langer. It is likely to be a profitable exercise to make use of her concepts and insights to develop a more adequate understanding of art, one that answers to our experience of art's cognitive dimension and to its multi-functional character.

Buchler was almost certainly aware of Langer's work on art when it appeared in the 1950s, if for no other reason than that they both moved in the same circles in the New York philosophical community. Surprisingly, though, Buchler does not take up her analysis as he developed his theory of judgment and art's character as exhibitive judgment. Dewey's tendency to think of knowledge in terms of inquiry, however, is a shortcoming to which Buchler has directly pointed. Indeed it is from Buchler's point of view a major shortcoming of Dewey's conception of experience in general.[18] If this criticism has a point, then indeed we need a more adequate understanding of the various ways nature may be questioned and investigated, and by implication of the knowledge that may result. We have suggested earlier that Buchler's theory of judgment and his concept of query provide such an improved understanding. Judgments of any kind may yield knowledge. Sometimes they are assertive and propositional, in which case we may achieve knowledge that is susceptible to the evaluative categories appropriate to assertive judgments, such as truth-value. Sometimes they are exhibitive, in which cases we may achieve knowledge that is to be evaluated in other terms, for example by the deeper understanding and appreciation it enables or by the expanded possibilities it reveals. The former cases are knowledge of the sort that results from the sciences, mathematics and philosophy; the latter are instances of knowledge appropriate to the arts. Neither is reducible to the other, neither is defined in terms of the other, and taken together they enable a broader understanding of knowledge and they reflect the richness of our experience, aesthetic and otherwise.

We have suggested three possible research programs or projects that are indicated by this effort to take seriously the cognitive dimension of art. The first would be a revision of analytic naturalist epistemology by regarding knowledge through art as an additional paradigm along with the knowledge that results from the sciences. The second would be a revision of Dewey's theory of inquiry such that the hermeneutic link between knowledge and art that Dewey himself points too can be made the basis of an expanded understanding of knowledge. The third is a consideration of Langer's view that discursive reason and art are two different symbolic forms, each capable of generating knowledge. We may now suggest a fourth, which would be a more thorough development of the theory of judgment into a conception of knowledge that not only enables but also demonstrates in greater detail the cognitive dimension of art. In this effort one might also bring to bear the insights from other theories that represent similar attempts. The hermeneutic tradition from Heidegger through Gadamer to Vattimo is the most obvious possible source. In the end Rorty may or may not have been right that epistemology as it is practiced in the analytic tradition is a dead end. As we have seen with respect to analytic naturalist epistemology he was almost certainly right. But he was wrong to think that there is no good reason to continue to ask after the nature and varieties of knowledge.[19] If nothing else, doing so allows us to develop a coherent understanding of the cognitive dimension of art, and in so doing we are able to enhance our appreciation of the profound role that art and aesthetic experience play in our lives.

NOTES

1. I would like to thank the participants in the Opole conference for their comments and suggestions, especially Gregory Pappas, Robert Innis and Larry Hickman. I would like to thank Lyubov Bugaeva and Robert Howell for their reading of a draft of the paper and their suggestions.

2. John Dewey, *Art as Experience*, Later Works, Volume 10, edited by Jo Ann Boydston, Carbondale, IL: Southern Illinois University Press, 1981. See Chapter XII.

3. ibid.

4. See the catalog from this exhibition: *The Adventures of the Black Square*, The State Russian Museum, Palace Editions, St. Petersburg 2007.

5. One might look at virtually any of McDermott's writings as illustrations of the point, but a god place to begin would be "To Be Human is to Humanize: A Radically Empirical Aesthetic," in John McDermott, *The Culture of Experience*, New York: New York University Press, 1976, pp 21-62. One might also consult his most recent collection of essays, *The Drama of Possibility: Experience as Philosophy of Culture*, Douglas R. Anderson, ed., New York: Fordham University Press, 2007. For a valuable overview of the aesthetic as the seed of philosophic insight see Richard E. Hart's "Landscape and Personscape in Urban Aesthetics," in James Campbell and Richard E. Hart, eds., *Experience as Philosophy*, New York: Fordham University Press, 2006.

6. Susan Sontag, *Reborn: Journals and Notebooks, 1947-1963*, David Rieff, ed., NY: Farrar, Straus and Giroux, 2008; Orhan Pamuk, "My Turkish Library, *New York Review of Books*, December 18, 2008, pp 69 – 72.

7. Buchler uses the expression "knowledge through" to refer to poetic knowledge. See Justus Buchler, *The Main of Light*, New York: Oxford University Press, 1974. See pages 150 ff.

8. The theory of judgment and the concept of query are developed most thoroughly in Justus Buchler, *Nature and Judgment*, New York: Columbia University Press, 1955. See Chapters I and II.

9. For a clear exposition of these points from a proponent of naturalized epistemology see Hilary Kornblith, "In Defense of a Naturalized Epistemology," in John Greco and Ernest Sosa, eds., *The Blackwell Guide to Epistemology*, Oxford: Blackwell Publishers, Ltd., 1999, pp 158-69. Another useful overview, though in this case for the purpose of critical rejection, see Steven J. Wagner and Richard Warner, eds., *Naturalism*, Notre Dame, IN: University of Notre Dame Press, 1993. See especially the Introduction, pp 1-21.

10. Good examples of this can be found in Matthew Kieran, ed., *Contemporary Debates in Aesthetics and the Philosophy of Art*, Malden, MA: Blackwell Publishers, Ltd, 2006. See specifically Chapter 7, Berys Gaut, "Art and Cognition," pp115-126, and Chapter 8, Peter Lamarque, "Cognitive

Values in the Arts: Marking the Boundaries," pp 127-142.

11. Rorty makes the point about his disinclination to discuss the arts in an interview with Alexander Kremer. See Alexander Kremer and John Ryder, *Self and Society*, Rodopi Press, 2009, pp 227 - 241.

12. John Dewey, *Experience and Nature*, LW Volume 1, edited by Jo Ann Boydston, Carbondale, IL: Southern Illinois University Press, 1988, pp 268 – 269.

13. ibid., p. 276.

14. ibid., p. 283.

15. John Dewey, "Qualitative Thought," in *On Experience, Nature and Freedom*, Richard Bernstein, ed., New York, NY: The Bobbs-Merrill Company, Inc., 1960, pp 186-187.

16. ibid., p. 198.

17. Susanne K. Langer, *Problems of Art*, New York, NY: Charles Scribner's Sons, 1957, pp 124 – 125.

18. ibid., pp 21 – 22.

19. See *Nature and Judgment*, pp 104-05.

Two

AESTHETICS AS SOCIAL PHILOSOPHY

James Campbell

1. Introduction

John Dewey was one of the great stars of Pragmatism, the American philosophy that attempted to re-balance the relationship between thinking and life, and between philosophy (and academia in general) and the rest of life. In the process, Pragmatism developed a reputation for bringing a lean and serious demeanor to philosophical practice. In spite of William James's somewhat tenderer side, Pragmatism was a philosophy that sought to replace the inherited and the customary with the products of publically-tested experience. Dewey dedicated himself especially to bringing the advantages of the scientific method to our thinking about education and morality and politics.[1] It was, then, somewhat puzzling when he announced in 1931 that the theme of the initial William James Lectures at Harvard was to be "Art and the Esthetic Experience."[2]

Early in the process of preparing these lectures, which took place between February and May of 1931, Dewey admitted to a desire "to take up some new field . . . a field I haven't treated systematically."[3] The adverb is important here, because Dewey's thinking had long treated themes aesthetic.[4] This is especially true of his philosophy of education, where aesthetic considerations play a central, if under-emphasized, role in his presentations of growth and his calls for the enrichment of life. Prior to *Art as Experience* (1934), however, Dewey had not examined aesthetics in as thorough a fashion as he had recently examined metaphysics in *Experience and Nature* (1925/9) and epistemology in *The Quest for Certainty* (1929), or was shortly to examine religion in (1934) and logic in *Logic: The Theory of Inquiry* (1938).

On this 150th anniversary of Dewey's birth, I suspect that few read his writings chronologically, as his early readers had to. As a consequence, the sense of surprise at encountering his systematic aesthetic is now missing. For us, the aesthetic is an element in all of Dewey's philosophizing. As Irwin Edman writes in 1950, *Art as Experience* spells out "for a good many readers the lesson that, far from neglecting or ignoring the claims of art, art was the illustration *par excellence* for Dewey of the ultimate meaning of intelligence, morals, democracy, education and even ultimately of the nature of things."[5] Dewey's educational and moral and political philosophy is equally a philosophy of living aesthetically; and his aesthetics functions in consequence as a critique of how an-aesthetically and, worse, unquestioningly we now live.

"If artistic and esthetic quality is implicit in every normal experience," Dewey wonders, "how shall we explain how and why it so generally fails to become explicit? Why is it that to multitudes art seems to be an importation into experience from a foreign country and the esthetic to be a synonym for something artificial?"[6] In a similar fashion, he notes that aesthetics should play a more central role across our philosophical inquiries. This is the import of such comments as "[t]o esthetic experience, then, the philosopher must go to understand what experience is."[7]

This advice remains powerful at present, when philosophy as a professional discipline all too often misses the centrality of aesthetic experience to human existence. Instead, it is excessively concerned with reasoning and 'truth,' and consequently isolates aesthetics as a 'lesser' area of inquiry. Perhaps worse, philosophy at present often accepts our bifurcated economic and political situation as 'natural' or 'necessary'; and, rather than offering a critique of our sad reality, it attempts to formulate an aesthetic theory that papers it over. This is at least one way to read Stephen C. Pepper's essay in the 1939 Schilpp volume, *The Philosophy of John Dewey*. At the end of his piece, Pepper indicates his disappointment that Dewey would want to mix aesthetic and social themes: "Of the social message in *Art as Experience*, its plea to break down the separation between art and life, to realize that there is beauty in the commonest and meanest things, and on Tuesday and Wednesdays as well as on Sundays, I have said nothing." Although Pepper admits that "[p]ossibly this message is the chief intention of the book,"[8] for him Dewey's call to increase the possibilities for living esthetically holds no interest.

2. Overcoming Bifurcations

From the very beginning of *Art as Experience*, Dewey indicates that his focus will be on reconnecting aesthetic experience with 'normal' experience. The realm of art is not the private playground of the museum- and tuxedo-crowd, nor is it "the beauty parlor of civilization"[9] to which we retire for rejuvenation before our inevitable return to the dull and crippling reality of day-to-day existence. "When artistic objects are separated from both conditions of origin and operation in experience," he writes, "a wall is built around them that renders almost opaque their general significance." As a result, art is exiled "to a separate realm, where it is cut off from that association with the materials and aims of every other form of human effort, undergoing, and achievement."[10] Properly understood, however, art presents instances of the highest moments that are possible in human life, and indicates that – with effort – we can make life better. Consequently, when we study a work of art, our emphasis should not be so much upon narrowly appreciating the music or the poetry as much as upon comprehending fuller human life. Further, our goal in approaching the "intrinsic nature of esthetic experience" is not just to get a better theory of art – as important as that goal is – but to foster more aesthetic lives. For Dewey, "a philosophy of art is sterilized unless it makes us aware of the function of art in relation to other modes of experience, and unless it indicates why this function is so inadequately realized, and unless it suggests the conditions under which the office would be successfully performed."[11]

In his concern for continuity, for reconnecting 'normal' and aesthetic experience, craft-workers and artists, and producers and consumers, Dewey calls on us "to restore continuity between the refined and intensified forms of experience that are works of art and the everyday events, doings, and sufferings that are universally recognized to constitute experience."[12] Before we can achieve this, however, it will be necessary to be clearer about who we are as creatures, and why we make, enjoy and need art. This anthropological approach runs counter to the approach of those art historians and aestheticians who have accentuated the presumed perfection of buildings like the Parthenon and canvases by El Greco and Cézanne, and neglected art's human function. We need to consider how these objects functioned in the individual and collective lives of their creators and their contemporaries. Only when we can uncover what these people "had in common, as creators [of this art] and as those who were satisfied with it, with people in our own homes and on our own streets,"[13] can we hope to restore continuity between aesthetic and 'normal' experience.

What are the elements of life that we, even today, regard as 'aesthetic' – or, if that is too fancy a term – 'enjoyable'? What are, in Dewey's words, "the events and scenes that hold the attentive eye and ear of man, arousing his interest and affording him enjoyment as he looks and listens"? He mentions such particulars as emergency vehicles, construction sites and athletic events. "The sources of art in human experience will be learned," he continues, also by theindividual "who notes the delight of the housewife in tending her plants, and the intent interest of her goodman in tending the patch of green in front of the house; the zest of the spectator in poking the wood burning on the hearth and in watching the darting flames and crumbling coals." We miss the roots of the aesthetic, he writes, when our appreciation of masterpieces overwhelms our attempts to understand the overall function of art in living. "In order to understand the meaning of artistic products," he writes, "we have to forget them for a time, to turn aside from them and have recourse to the ordinary forces and conditions of experience that we do not usually regard as esthetic." We think too often in terms of the object in the frame or on the pedestal, and not enough about how it arrived there. Our lives are full of enjoyments like singing and dancing, building and hiking, and cooking and gardening; but we think that they have nothing to do with art, because art is wholly other. As Dewey puts it, we have taken things that "in their own time and place" functioned as "enhancements of the processes of everyday life" and relegated them "to the museum and gallery."[14] Works of art should be "celebrations, recognized as such, of the things of ordinary experience"; but we have set them "in a niche apart." If we return to 'normal' experience, however, to our fascination with fires and fire trucks, with sounds and language, and with gardening and decoration, we can recognize something very important about the integration of aesthetic and 'normal' experience. As he concludes, "by going back to experience of the common or mill run of things," we can "discover the esthetic quality such experience possesses."[15]

As to why the split between the aesthetic and the 'normal' components of experience occurred in the first place, Dewey suggests a series of interrelated socio-historical – i.e., non-philosophical – factors. One is the political cluster of monarchism and nationalism and imperialism, under which wealth demonstrated power and artworks functioned as loot. After the capitalists replaced the monarchs, they used art to demonstrate that they were not just tasteless plutocrats, unable to appreciate the 'finer things' of life. Later, the desire of communities and nations to build museums intensified, both to preserve cultural heritage and to 'uplift' their citizens. As what he calls the market system of "economic cosmopolitanism" developed further, art objects become commodities to be sold or auctioned as "specimens of fine art" divorced from any particular community.[16] As familiar and sensible as all of this may seem to mid-20th-century people, Dewey writes, the resultant history of aesthetics has been a denial of the rootedness of art in natural processes. Further, this compartmentalization of 'normal' and aesthetic experience has become the sub-text of contemporary life – or at least it was the subtext until the most recent economic collapse. More or less willing followers of the suggestion of Frederick Winslow Taylor to maximize 'efficiency' by 'rationalizing' the manufacturing process through 'scientific management,'[17] we surrendered our forty hours of mindless activity to fund our 'free' time. Often, we drove the wedge further into our lives by working overtime to pay for major purchases and vacations.

Into this bifurcated society comes the philosopher – the aesthetician – who offers us an understanding of art that explains why the tradition of museumization and pedestalization should be seen as the correct way to appreciate the higher, or spiritual, or perhaps even eternal aspects of art. The art object, it seems, is a unique thing produced by the extraordinary talent of this remarkable individual which it is our duty as non-artists to contemplate. Consider Edman's formulation: "Connoisseurs and critics have encouraged this notion of art by treating the aesthetic experience as if it were the exclusive potentiality of a rare class of gifted and sensitive darlings. The snobbishness of the aesthetes encouraged on the other hand the contempt of the practically or the socially active, who regard the arts and the artists as truants from important human concerns, the preoccupations of the trifling, the irresponsible and the perverse."[18]

If Dewey is correct, we must still wonder why so many philosophical thinkers stress the bifurcation of the aesthetic and the 'normal.' Why is there such willingness to abandon the natural for the supernatural, to regard any connection between "the higher and ideal things of experience with basic vital roots . . . as betrayal of their nature and denial of their value?"[19] He offers a number of reasons. From religion, there is often hatred for the physical or the animal, and praise for the spiritual and 'eternal.' From metaphysics, there is often a higher regard for the permanent and unchanging than for the temporal and developing. In economics, there is the painful reality of class society in which play is done for enjoyment, for its own sake, whereas work is done only under compulsion. All too often, what seems to be of central importance to human existence has been seen as somehow separate from and 'above' normal human existence.

There is nothing intrinsic to the art object itself, however, that demands bifurcation. As Dewey writes, "*theories* which isolate art and its appreciation by placing them in a realm of their own, disconnected from other modes of experiencing, are not inherent in the subject-matter but arise because of specifiable extraneous conditions." Still, these theories are familiar: "Embedded as they are in institutions and in habits of life, these conditions operate effectively because they work so unconsciously." For Dewey, on the other hand, our problem is to reconnect aesthetic experience with the 'normal' processes of living. "Even a crude experience," he writes, "if authentically an experience is more fitting to give a clue to the intrinsic nature of esthetic experience than is an object already set apart from any other mode of experience." He continues that, when art emphasizes "its connection with discovered qualities of ordinary experience," it will be able "to indicate the factors and forces that favor the normal development of common human activities into matters of artistic value."[20] He thus rejects the claim that to be of value, the standards of quality in art must come from outside of our experience, and asserts, on the contrary, that "the esthetic is no intruder in experience from without, whether by way of idle luxury or transcendent ideality." It is, rather, "the clarified and intensified development of traits that belong to every normally complete experience."[21]

Using this consideration of the possibilities of experience as a starting point, Dewey maintains that the familiar distinction between fine and useful arts is not itself aesthetic in origin. It comes, as we have seen, from the outside, from an economic realm that contains an already bifurcated social situation. He recognizes that our inherited systems of religion and economics live comfortably within this bifurcation, and even that these inherited ways of thinking are often troubled by the suggestion that 'normal' lives – or normal peoples' lives – deserve to be esthetic. "The hostility to association of fine art with normal processes of living is a pathetic, even a tragic, commentary on life as it is ordinarily lived," he writes. "Only because that life is usually so stunted, aborted, slack, or heavy laden, is the idea entertained that there is some inherent antagonism between the process of normal living and creation and enjoyment of works of esthetic art." For Dewey, on the contrary, "art itself is the best proof of the existence of a realized and therefore realizable, union of material and ideal."[22] While occasionally the continuity of the aesthetic and the normal thrives, most of the time it does not; and making it happen more often is not thought to be worth the reallocation of the necessary economic resources. The majority of our experiences – experiences that are bad or an-esthetic – do not reflect how life, or even work, has to be. They reflect, rather, the familiar separation of the useful and the fine.

Dewey also attacks the inherited Western bifurcation of arts into popular and fine, art for the low-brow and art for the high-brow. As with the other bifurcations he sees this distinction as based upon some inherited non-aesthetic criterion. He considers, e.g., Joshua Reynolds's claim that art should focus upon what cultured people find to be "'*generally* interesting'" – i.e., "'the great events of Greek and Roman fable and history . . . the capital events of Scripture'" – to be without foundation. Dewey wonders what Reynolds would have said "if he had been able to anticipate the ballet girls of Degas, the railway-coaches of Daumier . . . or the apples, napkins, and plates of Cézanne."[23] For Dewey, *pace* Reynolds, art can take its subject from any sector of human existence. To point to four more examples, we can consider: Carl Sandburg's volume of poems, *Good Morning, America* (1928); Edward Hopper's painting, *Room in New York* (1932); Clifford Odets's play, *Waiting for Lefty* (1935); and Virgil Thomson's ballet, *Filling Station* (1938). There is much of 'general interest' in this quartet of works – and in many others – that the tradition of 'high' aesthetics would cause us to overlook.

It is, of course, always possible to wonder how much Dewey's emphasis upon continuity helps us to understand and appreciate the most refined aspects of art. Susanne K. Langer complains that Dewey ranked "ballet and golf, and sculpture and hairdressing, all on a par," and that he did not have an adequate understanding of the uniqueness of creativity: "Why is a piece of music a creation and a shoe usually just a product? The distinction, though it is protested by some philosophers who are anxious champions of democracy, like John Dewey, is a commonly accepted one. An automobile is not created on the conveyor belt, but manufactured. We don't create bricks, aluminum pots, or toothpaste; we simply make such things. But we create works of art."[24] Dewey, as we have seen, rejects such bifurcations as harmful. His complete confidence that he is on the right track with his emphasis upon continuity is even shown in his style of presentation, which is less an argument than a metaphor: "Mountain peaks do not float unsupported; they do not even just rest upon the earth. They *are* the earth in one of its manifest operations."[25]

Dewey emphasizes that art fits into 'normal' experience and sanctifies it. The use of what we often think of as a religious term is not improper here. If we can recover the word 'sanctify' from its supernatural kidnappers, and return it to its proper natural use, it will carry a meaning of heightening or enhancing the experiences of human existence. This is what art can do. Of course, we all know the difference between our daily lives and what they might be, and few of us are able to consider our lives at present as artistic. The reasons for the separation, however, are economic and cultural rather than aesthetic.

3. Imagining Possibilities

Dewey believes that we can improve our attempts to understand the role of art in life with insights from the Darwinian perspective, and from a bit of speculative anthropology. By going far back along the evolutionary trail and studying the live creature in its environment – "not merely *in* it but because of it" – we can learn something fundamental about the role of the aesthetic in a life that, as he puts it, "falls out of step with the march of surrounding things and then recovers unison with it." Living creatures are impulsive, with needs and desires; and they find themselves in a complex process of doing and undergoing, with only a slowly developing understanding of what it all means. Life is a series of transitions, of loss and recovery. It is a career of sleep and wakefulness, hunger and satisfaction, heath and illness, activity and rest; and, when life is successful, there develops a rhythm of balance and imbalance, of harmony and tension. While this natural rhythm is central to the lives of worms and birds and antelope, with humans this process of transition from harmony to tension and back also becomes conscious. There is in our lives what Dewey calls a 'pattern of inquiry' by means of which we attempt to rectify our disharmonies; and, while we have done a great deal to minimize these imbalances – e.g., grocery stores and air conditioning and antibiotics – we cannot prevent them. We can, however, use our successful resolutions of these imbalances for growth. "Life grows," he writes, "when a temporary falling out is a transition to a more extensive balance of the energies of the organism with those of the conditions under which it lives."[26]

Dewey maintains that, if our world were without change or if there were only chaos, there would be no life, and consequently no aesthetic dimension of life. In lives such as ours, however, the aesthetic dimension should be very powerful. He writes that "the experience of a living creature is capable of esthetic quality" because of these "breaks and re-unions," and "[t]he moment of passage from disturbance into harmony is that of intensest life." Still, to sustain this aesthetic level requires more than our accidental stumbling upon pleasurable resolutions. It requires a conscious balancing of inner and outer. "Pleasures may come about through chance contact and stimulation," he writes, and "such pleasures are not to be despised in a world full of pain." But "a fulfillment that reaches to the depths of our being" requires "an adjustment of our whole being with the conditions of existence."

In discussing this fuller level of living, Dewey emphasizes our connections with our animal cousins. "The live animal is fully present, all there, in all of its actions: in its wary glances, its sharp sniffings, its abrupt cocking of ears. All senses are equally on the *qui vive*." He also emphasized our connections with our 'savage' ancestors. As he writes, "when the savage is most alive, he is most observant of the world about him and most taut with energy."[27] Thus, being fully alive in nature is the basis of, not distinct from, being aesthetically alive, even in the mistaken museumized sense.

In his familiar discussion of "Having an Experience", Dewey begins to elaborate how he sees art fitting into the fabric of satisfactory living. He recognizes that when we study the process of our experience in a statistical sort of way, most of it is pretty dull and an-aesthetic. We spend the bulk of our lives driving, cooking, working, studying, cleaning, and doing who-knows-what in order to get by. Our response to the felt inadequacy of 'normal' experience, however, should not be an attempt to escape from, or to transcend to what is thought to be better than, experience. Nor should it be the time-honored practice of periodic blow-outs to make the dull parts seem more bearable. We need, rather, to study more carefully those occasions when experience is most successful: when we have what he calls "*an* experience" that "runs its course to fulfillment" and that closes in "a consummation and not a cessation." Such full experiences contain no holes or dead spots. Dewey points to such examples as a particular meal in Paris or a storm crossing the Atlantic; we might focus upon an athletic triumph or a successful performance of Eugene O'Neill's *Long Day's Journey Into Night*. "In such experiences, every successive part flows freely, without seam and without unfilled blanks, into what ensues," he writes, "[a]t the same time there is no sacrifice of the self-identity of the parts."[28] In instances where everything hangs together, we have a clearer grasp of the nature of all experience as a process of doing and undergoing.

Dewey expands his discussion of doing and undergoing through a consideration of the interrelated notions of rhythm and symmetry.[29] If we focus on the active sense, or rhythm, we see that he is not emphasizing the usual understandings of the term like regularity or mechanical repetition or literal recurrence.[30] Most of our lives – when not simply shapeless – have this bogus kind of rhythm. Our lives are mechanical cycles of work and play, unappetizing patterns of fast (and often fast-food) eating, and habituated interactions between long-time spouses. None of these even approximate what is necessary for "*an* experience." For him, rhythm is rather an aspect of experience that, while including some repetitive elements for the sake of organization, is moving forward toward consummation. In this way, order is combined with novelty – both of which are aspects of human fulfillment[31] – and the result is aesthetically satisfying. Dewey describes the results of such a full and enriching aesthetic experience in terms of "perception," an active process that goes beyond simple "recognition" to grasp the meaning and implications of the experience.[32] We cannot perceive very much of human significance, for example, during a high-speed tour of a museum, or during our first hearing of a Bruckner symphony; nor can we perceive very much in a mindless factory job or a bureaucratized process of medical treatment.

Dewey assures us that there is no need to go beyond experience – something that is impossible anyway – to attain more fulfilling lives. Our need is to improve experience. For him, the emotional impact of art shows its depth – its rootedness in the human psyche – not its alleged transcendence. This is why he emphasizes that our goal should be to make our 'normal' lives more aesthetic: to enhance life rather than to avoid it. As an example of what he has in mind, he considers a piece of Pueblo pottery as a rooted cultural object.[33] There is clearly a difference between such an object and a mass-produced bowl from one of our discount stores. Sadly, however, many of us overlook this difference as long as our bowl is really cheap. Others of us do notice, but misdirect their hatred of our an-aesthetic discount store lives into individual efforts to move 'up' to where the 'good life' is. A few get it right and emphasize that the perception of the aesthetic power of Pueblo pottery should lead to efforts to create a world that is a more valuable place for everyone.

Aesthetic imagination is crucial here. For Dewey, imagination is misunderstood if it is seen primarily as a means to avoid reality or to engage in open-ended fantasizing. It is, rather, a creative power to make connections. Through art we can imagine how this world might be. Art is neither an escape, nor a salve; art is integrating force. As he writes, "art weds man and nature" and "renders men aware of their union with one another in origin and destiny."[34] Again, we see Dewey emphasizing that aesthetics is not simply an attempt to understand the way that works of art can bring moments of pleasure to our lives. Artistic living is our goal. He writes further that art "enables us to share vividly and deeply in meanings to which we had been dumb," because it can open us up to the experience of others. "Communication is the process of creating participation, of making common what had been isolated and singular," he continues, and "part of the miracle it achieves is that, in being communicated, the conveyance of meaning gives body and definiteness to the experience of the one who utters as well as to that of those who listen." In this way, "[a]rt breaks through barriers that divide human beings," barriers that are "impermeable in ordinary association."[35] The power of art to stir the imagination enables us to recognize good that might have gone unnoticed, Dewey writes; and then "it is by activities that are shared and by language and other means of intercourse that qualities and values become common to the experience of a group of mankind." Since "art is the most effective mode of communication that exists,"[36] the aesthetic approach is essential to understanding the possibilities of living and to enacting them.

4. The Role of Art

Ultimately, Dewey is interested in the role that art has to play in a viable society. "Esthetic experience," he writes, "is a manifestation, a record and celebration of the life of a civilization, a means of promoting its development, and is also the ultimate judgment upon the quality of a civilization." We can separate several strands in his discussion. His first theme is that humans are social beings, whose values are carried in their institutions: their languages, their traditions, and their art. He notes that "individuals are what they are in the content of their experience because of the cultures in which they participate." To be enculturated is a necessary part of being human: without a social place, no one is fully a person. His second theme is that while "[t]he individuals who have minds pass away one by one . . . [t]he works in which meanings have received objective expression endure." These works of art "become part of the environment" that serves to socialize future generations. He also considers how these customary ways are then hallowed by art. "The ordinances of religion and the power of law," he writes, become "clothed with a pomp, a dignity and majesty that are the work of imagination." In human society, "social customs are more than uniform external modes of action," he continues. Through the imaginative power of art "they are saturated with story and transmitted meaning." Fourth, Dewey indicates that the arts used to play a larger social role in Western societies – they were more than just diversions to fill empty hours. The ceremonies of society – rites for the fallen, thanksgivings for the harvest, celebrations of a birth or a new leader – were more than days off from work. These social ceremonies were occasions in which the members of the group experienced, in a heightened sense, their place in the community. "Each of these communal modes of activity united the practical, the social, and the educative in an integrated whole having aesthetic form," he writes. "They connected things that were overtly important and overtly done with the substantial life of the community." These ceremonies were artistic moments, of course; but even though their "esthetic strand was ubiquitous," the social meaning of such experience was primary.[37]

Dewey also emphasized that art has an extra-cultural function. There are many cultures that have existed, and will continue to exist; and we are lucky if we can grasp even a few of them reasonably well. He emphasized that, other than long-term immersion, art is the best avenue for attempting to enter into the life of another culture. Because art, he writes, "is expressive of a deep-seated attitude of adjustment, of an underlying idea and ideal of generic human attitude, the art characteristic of a civilization is the means for entering sympathetically into the deepest elements in the experience of remote and foreign civilizations." He also emphasized the converse: the arts of other cultures can help us to understand ourselves better. An increasing understanding of life beyond our original culture helps us to overcome its blind spots; and the arts are key here as well because they "effect a broadening and deepening of our own experience, rendering it less local and provincial as far as we grasp, by their means, the attitudes basic in other forms of experience."[38]

Just as art offers us assistance in opening up to the possibilities of other cultures, it can also help us by increasing our expectations about the possibilities of life. Art cannot save us by itself, of course; we need to consider the whole social situation "in which works of art are produced and enjoyed." For Dewey, "the moral function of art" cannot be understood "in terms of a strictly personal relation between the selected works and a particular individual." Rather, we need to understand more fully "the *way* in which art exercises its humane function."[39] This brings us to our artists and how they interact with our current customs. We cannot assume "that morals are satisfactory in idea if not in fact, and that the only question is whether and in what ways art should conform to a moral system already developed." On the contrary, we need for our artists to become once again the "moral prophets of humanity," who present us with a "vision of possibilities" of justice and equality and freedom different from our contemporary realities that have "hardened into semi-political institutions." He continues that "[a]rt has been the means of keeping alive the sense of purposes that outrun evidence and of the meanings that transcend indurated habit." The artist's willingness to challenge the customary and the settled with the possibilities of imagination "constitutes the heart of the moral potency of art"; and this aesthetic indifference to the claims of the familiar gives art its "liberating and uniting power."[40]

Because aesthetic experience enables us to better recognize our need for, and our possibilities of, personal fulfillment and social growth, we must better integrate it into our broader philosophical thinking. As Dewey writes, art "provides a unique control for the imaginative ventures of philosophy."[41] By developing a broad aesthetic approach to experience, he believes that it is possible to speak of the moral function of art "to remove prejudice, do away with the scales that keep the eye from seeing, tear away the veils due to wont and custom, perfect the power to perceive."[42]

5. Reconstruction

Our culture at present does not directly control art as the Church and the Monarch and the Party have in the past. Rather, it simply isolates art, and philosophy and the rest of the humanities, from the main stream of social existence. For us, art has become the 'beauty parlor of civilization'; it helps us to relax, or feel better, or otherwise adjust to the negative aspects of 'real life.' Art now plays, however, a minimal role in fostering social advance. Dewey suggests a number of reasons for this unfortunate situation. He indicates that, while our culture is rooted in Greek science and philosophy, Roman law, and Abrahamic religion, it has been most strongly influenced recently by two other social forces that have undermined the possibility of aesthetic living. One of these is science, the impact of which has been mixed. On the positive side, through science the human has been naturalized and presented with a wealth of new materials and experiences; at the same time, nature has been reduced to quantitative mathematical systems and human behavior has been presented as fitting into these mathematical models. The second influence is industrialization, the mechanization of production and living that has separated useful and fine arts. Because "[t]he mechanical stands at the pole opposite to that of the aesthetic, and production of goods is now mechanical," Dewey writes, "[t]he liberty of choice allowed to the craftsman who worked by hand has almost vanished with the general use of the machine."[43] There have been some counter-trends since his day – some perhaps influenced by his thought – but, with the continuing pressures to economize, in production and lifestyles and even education, there would seem to be no solution for us without fundamental change.

It is clear that we are in need of widespread reconstruction – social and political and economic – to redirect our situation toward more aesthetic living. Dewey, writing when he did, focused on the an-aesthetic aspects of industrial society, calling for the participation of the worker "in the production and social disposition of the wares he produces." To achieve this increased participation, he further calls for measures to "reduce the force of external pressure" and to increase "a sense of freedom and personal interest in the operations of production." This reconstruction will require, he continues, the elimination of the "private control of the labor of other men for the sake of private gain" to give rise to forces that liberate "esthetic quality in the experience that accompanies processes of production." In general, Dewey notes that "art itself is not secure under modern conditions until the mass of men and women who do the useful work of the world have the opportunity to be free in conducting the processes of production and are richly endowed in capacity for enjoying the fruits of collective work."[44] Living in post-industrial society, we will need to decide how much of his message we can still follow. On one hand, while we may be less concerned at present with the short-comings of factory production, we are still living amidst economic insecurity. On the other, we need to adopt the methods of ongoing social inquiry to determine what sorts of reconstruction are necessary to deal with the shortcomings of our lives and to help us reach current and future potentials. For this reconstruction to be successful, we will need to adopt a broad philosophical approach.

Philosophy is the love of wisdom; but this wisdom is not the contemplation of a different and better world. Nor is it playing with ideas, however enjoyable we know that practice to be. Philosophy is a tool for making life better. As Dewey writes elsewhere, "philosophy is a form of desire, of effort at action – a love, namely, of wisdom." This wisdom, while too often associated only with the display of "systematic and proved knowledge of fact and truth," is more centrally concerned with "moral values," with a "better kind of life to be led." As a moral term, wisdom "refers not to the constitution of things already in existence," but "to a choice about something to be done, a preference for living this sort of life rather than that." Wisdom ultimately refers, then, "not to accomplished reality but to a desired future which our desires, when translated into articulate conviction, may help bring into existence."[45] As Henry Steele Commager writes: "More fully than any other philosopher of modern times, Dewey put philosophy to the service of society. More, he formed a whole network of alliances – with science, with politics, with education, with aesthetics, all directed toward advancing the happiness of mankind."[46] Art, for its part, can contribute to this desired future by opening our eyes to possibilities for enhancing democratic community among humans that we had not recognized; and aesthetics is thus an essential part of philosophical inquiry. Dewey sees the whole project of democracy to be tentative, of course, describing it as an "experiment in which we are all engaged . . . of living together in ways in which the life of each of us is at once profitable in the deepest sense of the word, profitable to himself and helpful in the building up of the individuality of others."[47] Still, he believes that our trials to advance this social hope through our democratic efforts will be simplified if we keep in mind the moral sense of wisdom and the value of aesthetic living.

While democracy is the core of Dewey's philosophical vision, he repeatedly writes that it had developed serious troubles in his contemporary world. Democracy arose in history because of a mixture of political ideals and accidental social conditions: the development of cities, the invention of the printing press, increased human contact through travel, etc. It advanced greatly in America because of some additional local conditions: the relative equality of wealth, increased opportunities, and the prevalence of isolated, face-to-face communities. These conditions fostered and preserved democracy; and to Americans, it seemed to be 'automatic' and 'self-perpetuating.' During Dewey's lifetime, however, changes in these supporting factors led to fundamental problems with American democracy that called for fundamental changes. It has become necessary, he writes, to rethink and to recreate democracy for the contemporary world. In part, it was necessary to modify the American conception of 'democracy.' No longer was it possible to accept a Jacksonian agnosticism that no elite is more likely to make better choices than the majority. No longer could democracy be a justification for simply doing what we want, or for following whatever others do. He emphasizes instead that "democracy should be a means of stimulating original thought, and of evoking action deliberately adjusted in advance to cope with new forces."[48] Democracy thus requires faith in others, and in our possibilities, as we honestly work together to search out *desideranda*. Dewey believed that we can reach our highest goal of living together if we foster democracy as faith in cooperative life.

Dewey further believes that it is only within the context of a democratic community that human beings can become responsible and fulfilled citizens. For him, the goal of life is neither wealth, nor travel, nor leisure, nor retirement, nor accumulating stuff, nor personal salvation. The goal of life is simply participating in the process of mutual fulfillment; and it is achievable only within a democratic community. In a democratic world, he writes, there are no "fixed order of species, grades or degrees." On the contrary, "every existence deserving the name of existence has something unique and irreplaceable about it, that it does not exist to illustrate a principle, to realize a universal or to embody a kind or class."[49] It is here that we recognize the importance of aesthetics as social philosophy. A focus upon aesthetic living helps us to recognize the unique contribution of each individual in a democratic community. Thus, when art is not just a decoration to living, but permeates life to make living aesthetic, art can perform its moral function to inspire and advance the democratic pursuit of human fulfillment.

NOTES

1. Cf., e.g., Dewey, *Democracy and Education* (1916), *Human Nature and Conduct* (1922), and *The Public and Its Problems* (1926).

2. For some background to the lectures and the volume, see the critical edition of *Art as Experience*, published as volume 10 of the *Later Works*, ed. Jo Ann Boydston, (Carbondale IL: Southern Illinois University Press, 1987), pp. 372-398.

3. Dewey to Sidney Hook, 20 February and 10 March 1930, *Art as Experience*, p. 375.

4. For a bibliography of Dewey's published work on art and aesthetics between 1887 and 1926, see: *Art as Experience*, p. 375 n.12.

5. Edman, "Dewey and Art," in *John Dewey: Philosopher of Science and Freedom*, ed. Sidney Hook, (New York: Dial, 1950), p. 48. Cf. Sidney Hook, *John Dewey: an Intellectual Portrait* (Amherst NY: Prometheus, [1939] 1995), p. 194.

6. Dewey, *Art as Experience*, p. 18.

7. Ibid., p. 278.

8. Pepper, "Some Questions on Dewey's Esthetics," *The Philosophy of John Dewey*, ed. Paul Arthur Schilpp, (LaSalle IL: Open Court, 1939), p. 388.

9. Dewey, *Art as Experience*, p. 346.

10. Ibid., p. 9.

11. Ibid., pp. 16-17.

12. Ibid., p. 9.

13. Ibid., p. 10.

14. Ibid., pp. 10-12.

15. Ibid., p. 16.

16. Ibid., pp. 14-15.

17. Taylor, Charles *Scientific Management* (New York: Harper, 1947).

18. Edman, "Dewey and Art," p. 57.

19. Dewey, *Art as Experience*, p. 26.

20. Ibid., pp. 16-17.

21. Ibid., pp. 52-53.

22. Ibid., p. 34.

23. Ibid., pp. 191-192.

24. Langer, *Problems of Art: Ten Philosophical Lectures* (New York: Scribners, 1957), pp. 110, 27.

25. Dewey, *Art as Experience*, p. 9.

26. Ibid., pp. 19-20.

27. Ibid., pp. 22-25.

28. Ibid., pp. 42-43.

29. Ibid., pp. 187-189

30. Ibid., pp. 168-170.
31. Ibid., pp. 171-174.
32. Ibid., pp. 58-59.
33. Ibid., p. 32.
34. Ibid., p. 275.
35. Ibid., pp. 248-249.
36. Ibid., p. 291.
37. Ibid., pp. 329-331.
38. Ibid., p. 335.
39. Ibid., pp. 348-349.
40. Ibid., pp. 350-352.
41. Ibid., p. 301.
42. Ibid., p. 328.
43. Ibid., p. 344.
44. Ibid., pp. 345-347.
45. Dewey, "Philosophy and Democracy" (1919), in volume 11 of the *Middle Works*, ed. Jo Ann Boydston, (Carbondale IL: Southern Illinois University Press, 1982), pp. 43-44.
46. Commager, *The American Mind: An Interpretation of American Thought and Character since the 1880s* (New Haven: Yale University Press, 1950), p. 99.
47. Dewey, "Democracy and Education in the World of Today" (1938), in volume 13 of the *Later Works*, ed. Jo Ann Boydston, (Carbondale IL: Southern Illinois University Press, 1988), p. 303.
48. Dewey, *Human Nature and Conduct* (1922), volume 14 of *Middle Works*, ed. Jo Ann Boydston, (Carbondale: Southern Illinois University Press, 1983), p. 48.
49. Dewey, "Education and Social Direction" (1918), in volume 11 of the *Middle Works*, p. 52.

Three

THE 'QUALITY' OF PHILOSOPHY: ON THE AESTHETIC MATRIX OF DEWEY'S PRAGMATISM

Robert E. Innis

The category of 'quality' joins the two differently pitched but exemplary pragmatisms of C.S. Peirce and John Dewey at their deepest phenomenological, categorial, and semiotic levels. It can and should play a central role in any exploration of the 'forms of sense' that define the ultimate matrices of world apprehension. It is the uncovering of these matrices that make up one of the main tasks of the types of philosophy so fruitfully and challengingly exemplified in their work.

In what was once a little known article, "Peirce's Theory of Quality," originally published in *The Journal of Philosophy* in 1935 Dewey specified, with brevity and force, the philosophical and methodological nature and implications of Peirce's fundamental discovery of an all-pervading 'firstness' in mental—which for Peirce ultimately meant semiosic—action. In other essential writings, though without making thematic at every point the connection with Peirce, Dewey attempted to develop and apply the notion of quality, especially in two deeply provocative essays, "Qualitative Thought" and "Affective Thought," and in his indispensable *Art as Experience*, in which the category of 'quality' is ever-present.[1] In this focal book quality is systematically connected with the imaginative production of 'expressive' forms and sign-configurations. These are, from a Peircean point of view, properly called 'iconic' and they encompass, within Dewey's framework, the attendant forms of artistic as well as aesthetic experiences. These qualitative forms and the structures within which they are apprehended are by no means restricted to art. Dewey's thinking about quality is in effect not just an 'expansive gloss' on the work of Peirce. It makes explicit and extends, especially, but not exclusively, in a reconfigured aesthetic realm, the complex heuristic fertility of Peirce's fundamental discovery for philosophy and its concerns in general. The philosophical importance of Peirce's theory of quality is indicated by Dewey when he says, near the end of his 1935 article, that "I am quite sure that he [Peirce], above all modern philosophers, has opened the road which permits a truly experiential philosophy to be developed which does not, like traditional empirical philosophies, cut experience off from nature."[2]

In fact, as we shall see, experience, nature, and semiosis are brought into unity by both Peirce and Dewey. The semiotic dimension is foregrounded by Peirce, while the relations between experience and nature are foregrounded by Dewey. But for both, the 'world,' as an internally articulated matrix of sense, is accepted first and foremost as a qualitatively defined web of experienced significance. Dewey forthrightly connects Peirce's phenomenology, which deals, as Dewey puts it, with "the matter of experience as experienced" to "a logical analysis of experience: an analysis based in what he calls Firstness, of sheer totality and pervading unity of quality in *everything* experienced, whether it be odor, the drama of King Lear, or philosophic or scientific systems."[3] Firstness is, in short, "the given permeating total quality of anything experienced."[4] Peirce himself writes, in a passage cited by Dewey, that "a firstness is exemplified in every quality of a total feeling. It is perfectly simple and without parts; and everything has its quality. Thus the tragedy of King Lear has its Firstness, its flavor *sui generis*. That wherein all such qualities agree is universal Firstness, the very being of Firstness."[5] Firstness, then, in Dewey's words, is "total undivided quality."[6] It marks, indeed defines, both poles of experience: the experiencing (self) as well as the experienced (world). It marks also the sign-configurations and artifact-configurations out of which both self and world emerge, and especially those sign-configurations whose specific function is to thematize quality *qua tale* in distinctive ways. There is, on such a position, a certain primacy of the aesthetic in world-building.

This is a crucial issue for Dewey's own philosophy (and conception of philosophy), which culminates, in my opinion, in his reformulation of the role of the 'aesthetic' or 'aisthetic'—not the artistic—dimension in all life, thought, and world-building. There are, first of all, on a Deweyan position, different kinds of experiencings because experiences are differently 'qualified,' by their objects as well as by their mediating signs or roots, even on the perceptual level. I have argued this point extensively elsewhere.[7] This is the central theme of John Herman Randall, Jr.'s rich and provocative Dewey-inspired reflections on "Qualities, Qualification, and the Aesthetic Transaction."[8] Jean-François Lyotard will also make much of this in his *Lessons on the Analytic of the Sublime* in his discussion of 'tautegorical' feelings, feelings that tell us 'how things are with us,' which closely resemble in scope and function Heidegger's *Befindlichkeit* as a fundamental (modal) structure of human existence or being-in-the-world. Understanding, on his position, which is projection of and toward a world, is always both "tuned" and "attuned."

In this sense quality is what makes a form of experience, as a way of taking the world and of being taken *by* it, be just 'this' form and not another. Indeed, just as there are different 'forms' of sign-mediated experiencings, so there are qualitatively different experiencings within each form. "Considered in itself, quality is that which totally and intimately pervades a phenomenon or experience, rendering it just the one experience which it is."[9] But, in this case, Dewey remarks, the quality is "ineffable," in that it cannot be objectively denoted in such a way that it is not embedded in another experience with its own quality, in fact, a "totalizing unifying quality."[10] The reflected upon experience is accessed through, becomes the object of, a reflecting experience, and each has its own qualitative unity. We live forwards, says, but we understand backwards—in, Peirce would add and Dewey would agree, an unlimited spiral of semioses, in which later semioses appropriate prior semioses and so become "funded." Each new understanding—and self-understanding—is a qualitatively distinctive existential structure, defining a unique horizon and 'state' of the individual. To experience the imaginary—say, centaurs—has a different quality from experiencing "horses in the barn,"[11] just as experiencing the difference between the imaginary and the real has a different quality from either of them. Dewey is insistent, through observations such as these, which belong to his 'method of denotation,' that quality as Firstness is not applicable solely or even primarily to such putative primitive non-relational givens such as 'red,' 'hard,' and 'sweet.' These are, in his words, partial phenomena,[12] the results of prescissive abstraction. Rather, the unity and totality of any experience whatsoever, as projection of world, is rooted in the quality of Firstness, "something which characterizes any and every experienced subject matter."[13] So, both the experiencing and the subject-matter experienced manifest quality. Quality is itself 'bi-polar,' corresponding to 'worlding' and 'world.'

It was a cornerstone of Dewey's mature philosophical work, rooted in a deep and pervasive naturalism, that this correlation between experiencing and subject-matter is not based on a fundamentally cognitive or reflective relationship. "Quality belongs to the domain of the occurrences of any single and total experience wholly irrespective of any cognitive or reflective reference."[14] For example, as I have already noted, 'having an experience' and 'describing the having of an experience' are two very different experiences. But, Dewey notes, we cannot infer from this that the describing in some way "alters the quality of what is described,"[15] which dwells in some inaccessible realm of immediacy. In fact, what is described, in *its* distinctive quality, gets embedded in the new quality, the describing quality, or the experienced quality of describing, which for us language animals, narrating creatures that we are, is a protean phenomenon. Dewey, then, subscribes to the Peircean phenomenological description of quality: it is "first, present [immediate, fresh], new, initiative, original, spontaneous, free, vivid, conscious, and evanescent."[16] It is this domain that we try to capture in the web of signs or sign-functions, especially the great self-generating spiral of aesthetic signs which 'iconize' the qualitative determinations of existence, far from the realm of formal art. As Susanne Langer has shown in her deeply argued works, 'feeling' is intrinsically bound to 'forms,' understood as sign-configurations, which 'articulate' the feelings and make them known.[17] Art, as thematic, is derivative from this prior matrix of qualitative world-building.

Dewey acknowledges (a point also foregrounded by Sheriff)[18] that the Peircean thematization of quality is a kind of psycho-logic with cosmological intent, a speculation he, however, perhaps wisely, wants to avoid. This is a consequence of Peirce's consciously held objective idealism. But Dewey wants to stick resolutely to Peirce's "logical analysis of a phenomenon, or any experience."[19] Psychologically, Dewey holds, "it is through feeling (including sensation as such) that qualities present themselves in *experience*,"[20] clearly a pure Peircean position. Eschewing all cosmological speculation, Dewey asserts that, looked at ontologically, the crucial thing is that "existence itself is qualitative, not merely quantitative, is marked by stress and strain, and by continuities,"[21] that is, exemplifies and embodies the Peircean categorial (and semiotic) schema of firstness (immediacy), secondness (duality), thirdness (mediation). But we are not only dealing with a psycho-logic. A key Peircean passage points toward a psycho-semiotic: "The quality of feeling is the true psychical representative of the first category of the immediate as it is, of the present in its direct, positive presentness."[22] Feeling is the psychical representative of the immediate, of the qualitative dimension of experience. It will be the distinctive function of iconic signs, in their aesthetic use, to body forth or flesh out this dimension. The philosophical point in Dewey's analysis is, however, that we do not "define or identify quality in terms of feeling,"[23] which would 'subjectify' quality. The relationship is the reverse: the 'qualifying' of subjectivity. It is the reference to immediate quality that defines a feeling or a 'feeling subject': "anything that is a feeling, whether of red or of a noble character, or of *King Lear*, is of some immediate quality when that is present as *experience*."[24] This is the "sound doctrine" that Dewey thinks is the heart of Peirce's position and to which he himself subscribes. The semiotic point is that "feeling" is a psychic sign of quality and hence the distinctive mode of access to it.

Now, it is clearly the case, Peirce has pointed out, that there is a distinctive set of signs, and sign-types, whose semiotic function is defined by a quality they share with their objects. These are icons (hypoicons), and the Peircean story about them is a familiar, even if not fully developed and at times controversial, one. Images, diagrams, and metaphors—Peirce's primary schematization—are all rooted in a shared quality, or firstness, instantiated as, or exemplified in, resemblance (firstness of a first), schematic isomorphism of relations (firstness of a second), or parallelism (firstness of a third). A classic Peircean formulation runs as follows:

> Hypoicons may be roughly divided according to the mode of Firstness of which they partake. Those which partake of simple qualities, or First Firstnesses, are *images*; those which represent the relations, mainly dyadic, or so regarded, of the parts of one thing by analogous relations in their own parts, are *diagrams*; those which represent the representative character of a representamen by representing a parallelism in something else, are *metaphors*.[25]

Although Dewey does not always draw the explicit semiotic parallel, we can see how his further analysis of quality can illuminate this issue. The pivotal idea of Dewey's superb 1931 essay, 'Qualitative Thought,' and of great philosophical and methodological import for understanding Dewey's conception of philosophy and its role, is that "the world in which we immediately live, that in which we strive, succeed, and are defeated is pre-eminently a qualitative world. What we act for, suffer, and enjoy are things in their qualitative determinations."[26] Indeed, thought as such, in all its forms, is "definitely regulated by qualitative considerations."[27] Dewey notes the oft-remarked strange experienced difference between the way objects are accessed, putatively non-qualitatively, in the propositions of physical science and the way they are accessed and expressed (also in propositions) in our normal, vital, existentially involved dealings with things. That we are dealing with propositions, however, is what makes the issue of qualitative thought pertinent to 'logical' considerations, in Dewey's broad conception of logic as the 'theory of inquiry' quite generally. Dewey argues that classical logic, by which he means Aristotelian logic, "treats qualitative determinations as fixed properties of objects, and thus is committed to either an attributive or a classificatory doctrine of the import of propositions."[28]

The proposition 'The red Indian is stoical'—Dewey's example—which on the traditional interpretation is read as ascribing to the object 'redness' and 'stoicism' as properties or as placing the Indian in the class of stoical objects, must rather be read as expressing the fact that "the indigenous American was permeated throughout by a certain quality, instead of being an object possessing a certain quality along with others. He lived, acted, endured stoically."[29] The chief weakness of the traditional view, in Dewey's opinion, is that it has a false notion of the 'given'—the focal point of the discussion in "Peirce's Theory of Quality"—and, whether in its attributive or classificatory form, it leaves no place for "the integral development and reconstruction of subject-matter effected by the thought expressed in propositions."[30] Indeed, for Dewey, the elaboration of propositions is *out of* a qualitative matrix, the 'originary' subject-matter. The methodological import of this notion is that qualitative thought is present at the beginning of the determination of what the 'subject-matter' at hand actually is. It is that factor in experience toward which thought originally 'turns' and which 'calls' to thought prior to systematic, thematic differentiation and reflection (an echo, here, of Merleau-Ponty's thought).

In general, Dewey thinks that the neglect of qualitative thought is catastrophic not only for epistemology or the theory of knowledge, which he called "that species of continued intellectual lockjaw."[31]but also for philosophy quite generally. It would leave "thought in certain subjects without any logical status and control."[32] Which subjects? First and foremost, aesthetic matters, as well as morals and politics, which in their distinctiveness are by no means reducible to quasi-mathematical treatment. Their logic, Dewey thinks, is a qualitative logic—a qualitative logic opposed to an essentially objectifying world-view.

Dewey's schematization of examples, and the types of lessons he claims are to be learned from them, is extremely illuminating and 'flesh' out, especially in a radical and broadly conceived aesthetic domain, the essentially Peircean implications of the distinctiveness of quality. Pictures (Peircean images), for example, have distinctive qualities just as much as persons and historic events do. "It is something which externally demarcates it from other paintings, and which internally pervades, colors, tones, and weights every detail and every relation of the work of art."[33] This quality "gives meaning" to and "binds" the different aspects or particulars of the painting together. It is this qualitative unity, embodied in and 'de-fining' the image, that "regulates" our intercourse with the painting as a distinctive "subject-matter." It is also the case that the painting, as experienced sign-configuration, regulates our intercourse with *its* subject-matter. Sign and world share a common, indeed unique, quality.

We can formulate the issue by recourse to Dewey's insightful and fundamental distinction between 'situation' and 'object.' The term 'situation' signifies the fact that "the subject-matter ultimately referred to in existential propositions is a complex existence that is held together, in spite of its internal complexity, by the fact that it is dominated and characterized throughout by a single quality."[34] This is the "pervasive and internally integrating quality"[35] that is articulated in the members of a configuration. In the statement "the stone is shaly" both 'stone' and 'shaly' "are determinations or distinctions instituted within the total subject-matter to which thought refers."[36] But they are not purely 'mental' distinctions and the situation out of which they emerge "is not and cannot be stated or made explicit."[37] It constitutes a lived-in, and lived-through, tacit background, taken for granted in all propositional symbolization. This background is the basic way the world appears and imposes itself on us.

Peircean semiotics, for its part, is based on an explosion of the claim of uniqueness and exclusive power of propositional—linguocentric—symbolization.' Indeed, iconization, or iconic 'symbolization,' is a semiotic phenomenon *sui generis*, as Peirce tried to show in many ways. Now it is the progressive 'iconization of sense' that gives the aesthetic dimension quite generally—and art in particular—its distinctive power and role in regulating our interactions with experience in the process of world-building. This is a distinctively Peircean point with a Deweyan twist, for the plain fact is that 'quality' plays a role more thematic and perhaps even more universal and relevant in Dewey than in Peirce. It is not that Dewey did not recognize the point of the Peircean categories or the scope of a Peirce-based semiotics. But in line with his main concern to show philosophy's central task as bringing human beings into satisfactory multiform relations to their environments (qualitatively enriched on both the experiencing and experienced side, Peirce's 'concrete reasonableness') he wanted to establish the ultimate matrices for this lived engagement in which experiences of *every* sort are *had* prior to being cognized or reflected upon. The world 'has' us before we 'have' it. Explicit reflection and thematically methodical 'sense-giving' and 'world-construction' both emerge from a protean and universally pervasive qualitatively defined matrix and are meant to lead us, enhanced, back to such a matrix. 'Being-in-a-situation,' which Randall also foregrounds, is the ultimately defining element of thought, for the distinctions of the situation are *its* distinctions, not those of mere thought that is externally related to it. It is the single pervasive quality that "enables a person to keep track of what he doing, saying, hearing, reading in whatever explicitly appears."[38] Pertinence, relevance, force are *its*. We are captured by it rather than it being captured by us. In this sense we are 'thrown' toward the world.

This qualitative matrix, the affective horizon of world-access is not an isolated object in itself, any more than Heidegger's 'world' is itself a 'thing' or an 'object.' It is, as Dewey points out, "the background, the thread, and the directive clue in what we do expressly think of"[39] This background is felt rather than thought,[40] a constant strand in Dewey's analysis and part of his incessant onslaught against the epistemology industry. "The existence of unifying qualitativeness in the subject-matter defines the meaning of 'feeling.'"[41] Anger, for example, pervades a situation. It is not a separate, isolable element, not an object of "analytic examination."[42] Thought begins, says Dewey, with the apprehension of an unanalyzed whole; indeed, we are grasped by it before it is grasped by us. It is had before it is cognized, and inquiry begins by the embodied inquirer being perplexed in a problematic situation. And, as Randall also noted, different perplexities have different qualities and call forth different methods of rendering them explicit through the construction of appropriate sign-configurations. It is through ramified forms of inquiry that what is problematic about a problematic situation itself even arrives at differentiation and explicitness. This is the persistent theme of Dewey's *Logic: The Theory of Inquiry* and is epitomized in the chapter on 'The Pattern of Inquiry.'

Dewey, moreover, is by no means a fan of mere immediacy and of 'mere' intuition. As he puts it, "Intuition ... signifies the realization of a pervasive quality such that it regulates the determination of relevant distinctions or of whatever, whether in the way of terms or relations, becomes the accepted object of thought."[43] Intuition as 'realization,' then, is an outcome, an eventuation, as Randall emphasized, for whom "'immediate experience' is always 'mediated.'"[44]Intuition here comes close to Polanyi's 'tacit' power.[45] While Dewey admits that "ejaculatory judgments"—'yes,' 'no,' 'fine'—are the simplest examples of pure qualitative thought, their primitive nature does not entail that they are superficial and immature (Wittgenstein made this point, too), for they are able to "sum up and integrate prolonged previous experience and training, and bring to a unified head the results of severe and consecutive reflection."[46] The "failure of language" that Dewey is pointing toward here is due to the richness of the subject-matter. "Language fails not because thought fails, but because no verbal symbols can do justice to the fullness and richness of thought. If we are to continue talking about 'data' in any other sense than as reflective distinctions, the original datum is always such a qualitative whole."[47] I have treated elsewhere the thorny problem of how to frame the failure of language in terms of upper and lower thresholds of sense.[48] It is the function of iconization, as a general process of world-building, and apart from the strictly discursive domain of sense-giving, to open up new forms of sign-defined access to qualitatively defined subject-matters.

The logic of artistic construction and apprehension is for Dewey a singularly perspicuous instance of the logic of qualitative thinking. It exemplifies for Dewey "in accentuated and purified form the control of selection of detail and of mode of relation, or integration, by a qualitative whole,"[49] which it is the job of the artist to reveal and of the aesthetic perceiver to grasp (or be grasped by). But in this, artistic thought is not so much a specification as it is an *intensification* of a characteristic of all thought. Scientific thought is a "specialized form of art, with its own qualitative control. The more formal and mathematical science becomes, the more it is controlled by sensitiveness to a special kind of qualitative considerations,"[50] indeed, those resident in forms of diagrammatic thinking whether of the geometrical or algebraic variety. Peirce explored these matters at great length in his work on existential graphs and on diagrammatic reasoning, another form of opposition to making the 'sense of the world' dependent primarily on linguistic apprehension.

Qualitative thought, Dewey thinks, throws light especially on (a) the nature of predication and (b) the phenomenon that goes under the name of the "association of ideas," which Peirce correlated with the great distinction between icons, indices, and symbols, an issue I have dealt with in a comparative analysis of the work of Peirce, William James, Edmund Husserl, and Karl Bühler.[51] With respect to (a), Dewey's thesis is: "predication—any proposition having subject-predicate form—marks an attempt to make a qualitative whole which is directly and non-reflectively experienced into an object of thought for the sake of its own development"[52] This development is paradoxically both "analytic" and "additive, synthetic, ampliative" at the same time. The quality is *analyzed* while the various differentiated aspects are *related* to one another. In this sense Dewey, as does Randall in his own gloss on Dewey, foregrounds the active, essentially relational, nature of the copula. "That thing is sweet" means it will sweeten some other object. "Men are mortal" means that men die. "The dog is ugly" captures what the dog will *do*, that is, "snarl and bite." Now, it is the work of articulation, exemplified in predication, that raises "dumb qualities" up into a 'symbolic' order where they attain "intellectual and propositional form"[53] *Art as Experience* will definitively show, in agreement with the Peircean project of establishing the *sui generis* nature of iconization, that the 'going out into symbolization' that Dewey refers to can capture quality in forms, sign-configurations, essentially cognate to it. The "given," Dewey insists, the only thing that is "unqualifiedly given," is "the total pervasive quality" in any problematic situation or subject-matter. It is this quality that controls transitions in thought, giving it both limits and direction. The selective character of propositions is not falsifying, for the articulation is always controlled by a surplus of quality. Thought functions through *its* statement, which can never be complete. 'Articulating' the world is laconically shown to be an endless project.

Dewey gives a remarkable account of the so-called association of ideas along the same lines. His main thesis is: "existentially, thinking *is* association as far as the latter is controlled."[54] Association then means "connection of objects or their elements in the total situation having a qualitative unity."[55] So, once again, it is the quality that controls the thinking. Situations, in Dewey's homey example, are 'disassociated' into objects that suggest and objects that are suggested. A bird-in-a-nest, as a "single total object," divides into a bird and nest. They 'belong' together in a way that a nest and branches of a tree do not. This sense of mutual 'belongingness' is the focus of Dewey's analysis. For Dewey it is the total "situational object" that is primary; thereupon follows the dissociation or distinguishing of connected objects. But the connecting is primary and the objects are secondary—rather the 'sense of connecting' or the 'connecting sense,' which is a quality, is primary. The ground of association is the quality of the situation. It is this that produces the functional connection between objects. Dewey foregrounds our sense of relevancy and pertinency, rooted in the unity of quality, that binds objects together. But the objects do not precede the binding. They emerge out of the binding as its articulation. It will be the singular role of the artist to produce artifacts that capture and embody the configurations of qualities, which, to be sure, are not free-floating but intrinsically bound to exemplifying objective structures, which are the art works themselves. Because the qualitative unities are not free but are precisely the unities binding together this configuration of objects or 'objective aspects,' each work of art has its own unique feel and qualitative 'space' that it opens up. The inexhaustibility of art is rooted in this fact. Randall notes that "every aesthetic transaction seems to be ultimately unique."[56] The 'world' of quality and the 'worlds' made present in art have no antecedent 'objective' existence. There is no way of accessing them apart from mediating structures, including feeling structures, all of which are embedded in a qualitative matrix.

Now art is concerned with the production of 'images,' including those types of linguistically defined images called 'metaphors,' although the metaphorical is by no means restricted to the discursive domain. Dewey has an enlightening, though primarily exemplifying rather than extensive, analysis of this that is both dense and rich and, in one sense, almost counterintuitive, and he connects it with so-called association by similarity. "When troublesome thought suggests the sting of an insect, or when change of fortune suggests the ebb and flow of the sea, there is no physical conjunction in the past to which appeal can be made."[57] The first thing to note about Dewey's observation is the direction of suggestion. It seems to reverse what for us would be the natural order from the concrete to the abstract. That the formulation is not a mistake is indicated by Dewey's further (admittedly not original) example of "being reminded of blotting paper by a certain voice."[58] The voice, which is primarily encountered, leads one to the blotting paper, but it does so by the creative drift of the shared quality, which, in Dewey's account, is directly experienced. There is, it is clear, no "external existential identity between two things, but a present immediate quality,"[59] which controls and elicits the apprehension of similarity, identical to Peirce's notion of a 'parallelism' as underlying metaphors. This controlling and eliciting, exemplified here in the linguistic realm, and rather prosaically at that, is "prior to and independent of all reflective analysis"[60] and furthermore is "something of the same nature which controls artistic construction."[61]

Aesthetic judgments, carried out both by the artist and the interpreter, are first and foremost oriented toward the formed or patterned quality that makes up the 'sense' of the 'work.' For Dewey each art work—even the most sophisticated sign-configurations—has its own qualitative uniqueness: aimed at by the constructive activities of the artist and grasped by the interpreter/perceiver. 'Goyaesque' refers to something distinctive, something that 'marks' the artifact as much as its weight or location. This qualitative distinctiveness, say, of a painting, is the "quality of the painting as a whole,"[62] indeed, a *founded* quality, for it does not exist independently of its embodiment, though it is not present in any one of its components but in the "form or pattern," in fact, the qualitative "pattern that connects," identified by Dewey in his essay 'Affective Thought,' in the context of a discussion of Albert Barnes's *The Art in Painting* as the "*integration of all plastic means.*"[63] Dewey remarks that a nose, as a feature of face, differs from 'triangularity' as immediately realized and recognized in a specific triangle. The nose is isolable, though not completely, for "it is characterized by the whole face as well as characterizing that face."[64] The expression of a person's face is "a total effect of all elements in their relation to one another, not a 'single feature among others.'" [65] Dewey thinks that the expression of a human face and the triangularity of a triangle have deep affinities and neither is identified, in our primary apprehension, in the way we identify a person by an examination of finger prints.[66] This is the same distinction between the Gestalt-like method of 'resonance' and the 'indexical' method, that is, the explicit reading of indices, that Karl Bühler drew within the context of his reconstruction of psychology and language theory.[67] Moreover, Dewey was well aware of the phenomenon of "family resemblances," which has so exercised the Wittgensteinian tradition, the point being that while they are often detected one is yet "totally unable to specify the points of resemblance."[68] They, too, belong to the 'tacit dimension,' in Polanyi's sense of that term, and to the Wittgensteinian powers of 'seeing connexions.'

The "significance of dominant qualitativeness" shows "why thinking as an existential processes is all one with controlled association,"[69] and in the case of humans "this operation of quality in effecting results then goes out into symbolization and analysis," that is, becomes shaped by semiotic events of distinct types. Neither existential conjunction nor physical sameness are the defining features of thought as "association." It is the recognition of pervasive quality that controls association. "By its means a voice is assimilated to blotting paper, and in more serious intellectual matters analogy becomes a guiding principle of scientific thought,"[70] a central Peircean thesis, instantiated in the previously remarked heuristic powers of diagrammatic thinking in both its corollarial and theorematic forms.[71] Dewey thinks that this process of assimilation leads to, or can lead to, a further, more explicit recognition of similarity of quality, but that this recognition, leading to judgment, is made possible by symbols, that is, by signs quite generally. The emphasis on judgment here was taken up by Justus Buchler in his notion that art works are exhibitive judgments, that is, exhibitive of their sense, which is located in a qualitative matrix.[72]

Dewey is insistent that the work of primary assimilation of diverse domains to one another on the basis of shared quality is not a thematic action. Thematization of the basis in an explicit judgment of likeness comes later, setting up a formal relation. This process of formalization, coupled with the movement toward explicit construction of sign-constituted artifacts, is one of the functions of art. Seeing *as*, for example, seeing something *as* a promontory, which Dewey identifies with assimilation, is the "net outcome of prior experiences" and it gives present experience its dominant quality. When this seeing as is differentiated or rendered distinct it is the pervasive quality which connects the differentiations. In the case of discursively formed language, that is, the Peircean 'symbolic' order, the result is "an explicit statement or proposition,"[73] which still in itself has its qualitative feel, the feel of discourse, or prose. In the 'iconic' order the result is an image, a diagram, a metaphor, all of which, to be sure, has an articulate content but signify by means of a qualitative affinity between themselves as sign-functions and what they mean. Thinking in icons—of whatever subtype or species—both depends upon the creative apprehension of quality—through assimilation—and at the same time advances the existence of quality in the constructing and apprehending consciousnesses. 'Scientific' as well as 'artistic' construction is dependent upon quality. *All* thinking—in whatever mode—has a qualitative background, point of departure, and regulative principle. But it does not just emerge from and be guided by quality, but it aims toward the exhibition of quality in objective forms. This exhibition is a 'realization,' a 'making real.' Thinking *in* quality is exemplified *par excellence* in artistic construction, which is a case of genuine thought, as well as in "genuine esthetic appreciation,"[74] which in order to be vital must in some way "retrace the course of the creative process."[75] This is our incorporation into, our embodiment, in quality. So: we have the embodiment of quality and the quality of embodiment.

The bearing of this theory of quality upon aesthetic judgment and theory is the express subject matter of Dewey's *Art as Experience*. But it is clear that Dewey's analysis of language, with its implied account of metaphor (Dewey speaks in *Experience and Nature* of meaning being "self-moving to new cases"),[76] and his analysis of inquiry, with its reliance upon schematic formulations and operational symbols, intersect directly with his aesthetic thought and clearly show how 'quality' controls and defines them all.

Dewey's rich and comprehensive aesthetic theory, which I can only gesture at here, further defines and refines the scope and range of quality. Art is the construction of 'sensible totals' marked by distinctive qualities. But it is a central thesis of Dewey's whole aesthetic project that art and the aesthetic are not related to normal experience as veneers. The aesthetic grows out of the fundamental movements of experiencing itself. The very point of art is to increase as well as enrich the qualitative dimension of experience and to give us the proper 'feel' of semiotically defined 'situations' which are expressed in the sign-configurations that are the 'carriers' of aesthetic sense. A consequence of Dewey's approach to quality is an insistence on the utter materiality of aesthetic artifacts, notwithstanding his distinction between the art product and the art work, the latter being the 'work' that the art product accomplishes in the experiencer. But the materiality of each artistic artifact—indeed, of every aesthetic situation—gives it *eo ipso* its own feel. The art work both mediates and embodies, by reason of its distinctive materiality, a structure of feeling. Its semiotic power arises from its capacity to engender configurations of feeling, as the interpretants of its constitutive signs.[77]

Dewey, in effect, in his aesthetics and in his theory of inquiry takes the Peircean 'logic of icons' as a given. Art works function, certainly, as images, they schematize, in systems of signs, logically analogous forms of feeling (the main point of Langer's aesthetics),[78] and they are, in the deepest sense, metaphorical in as much as they assimilate and then thematize the shared qualities joining diverse domains of experience together in a complicated relationship. Any art work is a qualitatively defined image as well as an image of quality. It is clear that the semiotic power of any hypoicon is a function of its qualitative expressiveness. The richness and existential bite of Dewey's analyses here are exhibited in the foregrounding of the body's full participation in all perception and, *a fortiori*, in all properly aesthetic perception:

> It is not just the visual apparatus but the whole organism that interacts with the environment in all but routine action. The eye, ear, or whatever, is only the channel *through* which the total response takes

> place. A color as seen is always qualified by implicit reactions of many organs, those of the sympathetic system as well as of touch. It is a funnel for the total energy put forth, not its well-spring. Colors are sumptuous and rich just because of a total organic response is deeply implicated in them.[79]

Systems of perceptual, and perceptually accessible, signs, have, on this account, their own intersensory 'feels:'

> When we perceive, by means of the eyes as causal aids, the liquidity of water, the coldness of ice, the solidity of rocks, the bareness of trees in winter, it is certain that other qualities than those of the eye are conspicuous and controlling in perception. And it is as certain as anything can be that optical qualities do not stand out by themselves with tactual and emotive qualities clinging to their skirts.[80]

As Dewey put it in "Affective Thought" changes in the surroundings involves correlated changes in the organism, and so the eye and ear gradually become acclimatized. The organism is really made over, is reorganized in effecting an adequate perception of a work of art ... integration in the object permits and secures a corresponding integration in organic activities."[81] This is a central point of connection between Dewey's and Whitehead's approach to "the civilization of experience," where the 'experiential' and the 'semiotic' dimensions stand in complex relationships to one another. [82] The fact is that the formal qualities of an integrated and integrating medium optimally become the "traits which naturally characterize objects when the environment is made over in consonance with basic organic requirements."[83]

Aesthetic perception, then, exemplifies in perspicuous form the centrality and universality of quality in all perception. Indeed, it is there that the preanalytic apprehension lying at the base of our primary encounter with experience is undeniably displayed:

> The total overwhelming impression comes first, perhaps in seizure by a sudden glory of the landscape, or by the effect upon us of entrance into a cathedral when dim light, incense, stained glass and majestic proportions fuse in one indistinguishable whole. We say with truth that a painting strikes us. There is an impact that precedes all definite recognition of what it is about. As the painter Delacroix said about this first and preanalytic phase "before knowing what the picture represents you are seized by its magical accord." This effect is particularly conspicuous for most persons in music. The impression directly made by an harmonious ensemble in any art is often described as the musical quality of that art.[84]

Through indications such as these, which I have only been able to hint at, we can see that Dewey was able to write the general aesthetics of quality that Peirce never got around to and that his work is in full agreement with the central Peircean position: “The *quale*-consciousness is not confined to simple sensations. There is a peculiar *quale* to *purple*, though it be only a mixture of red and blue. There is a distinctive *quale* to every combination of sensations so far as it is really synthesized—a distinctive *quale* to this moment as it is to me—a distinctive *quale* to every day and every week—a peculiar *quale* to my whole consciousness.”[85] A philosophically nuanced semiotics of the qualitative, accordingly, would go beyond art and the aesthetic and the iconic as such and would encompass, as reflective task, all those dimensions in which experience is universally mediated through signs in their qualitative specificities. In this sense both Dewey and Peirce are joined in the common task of sketching, in the words of Arthur Bentley, “the living behavior of sign-using men in a long-time world.”[86]

NOTES

1. John Dewey, 'Peirce's Theory of Quality,' in The Essential Dewey, vol. 2. Edited by Larry Hickman and Thomas M. Alexander (Bloomington and Indianapolis: Indiana University Press, 1998), pp 371-376; 'Qualitative Thought,' in Essential Dewey, vol. 1, pp. 195-205; 'Affective Thought,' in John Dewey, Philosophy and Civilization (New York: G.P. Putnam's Sons, 1931), pp. 117-125; Art as Experience (Carbondale: Southern Illinois University Press, 1987).

2. Dewey, 'Peirce's Theory,' p. 375.

3.Ibid., p. 371.

4. Ibid., p. 371.

5. Charles Sanders Peirce, Collected Papers of Charles Sanders Peirce, vols. 1–6, edited by Charles Hartshorne and Paul Weiss (Cambridge, Mass: Harvard University Press, 1931-1935), 1.531. Cited after volume and paragraph number.

6, Dewey, 'Peirce's Theory,' p. 372.

7. Robert E. Innis, Consciousness and the Play of Signs (Bloomington and Indianapolis: Indiana University Press, 1994) and Pragmatism and the Forms of Sense: Language, Perception, Technics (University Park: Penn State University Press, 2002).

8. John Herman Randall, Jr. Nature and Historical Experience. (New York: Columbia University Press, 1958), chapter 10.

9. Dewey, 'Peirce's Theory,' p. 373.

10. Ibid., p. 373.

11. Ibid., p. 373.

12. Ibid., p. 373.

13. Ibid., p. 373.

14. Ibid., p. 374.

15. Ibid., p. 374.

16. Peirce, Collected Papers, 1.357.

17. Susanne K. Langer, Philosophy in a New Key (Cambridge, Mass: Harvard University Press, 1942; Feeling and Form (New York: Charles Scribner's Sons, 1953); Mind: An Essay on Human Feeling, 3 vols. (Baltimore: Johns Hopkins University Press, 1967/1972/1982).

18. John Sheriff, Charles Peirce's Guess at the Riddle: Grounds for Human Significance (Bloomington and Indianapolis: Indiana University Press, 1994).

19. Dewey, 'Peirce's Theory,' p. 375.

20. Ibid., p. 375.

21. Ibid., p. 375.

22. Peirce, Collected Papers, 5.44.

23. Dewey, 'Peirce's Theory,' p. 375.

24. Ibid., p. 375.
25. Peirce, Collected Papers, 2.277
26. Dewey, 'Qualitative Thought,' p. 195.
27. Ibid., p. 195.
28. Ibid., p. 196.
29. Ibid., p. 196.
30. Ibid., p. 196.
31. Dewey, 'Does Reality Possess Practical Character?,' in Essential Dewey, vol 1, p. 133.
32. Dewey, 'Qualitative Thought,' p. 196.
33. Ibid., p. 196.
34. Ibid., p. 197.
35. Ibid., p. 197.
36. Ibid., p. 197.
37. Ibid., p. 197.
38. Ibid., p. 198.
39. Ibid., p. 198.
40. Ibid., p. 198.
41. Ibid., p. 198.
42. Ibid., p. 198.
43. Ibid., p. 199.
44. Randall, Nature and Historical Experience, p. 279.
45. Michael Polanyi, Personal Knowledge (Chicago: University of Chicago Press, 1958) and The Tacit Dimension, with a foreword by Amartya Sen (Chicago: University of Chicago Press, 2009; originally published 1966).
46. Dewey, 'Qualitative Thought,' p. 199.
47. Ibid., p. 199.
48. Robert E. Innis, 'Language and the Thresholds of Sense: Some Aspects of the Failure of Words,' The Journal of Speculative Philosophy, 22/2: pp. 106-117.
49. Dewey, 'Qualitative Thought,'p. 200.
50. Ibid., p. 200.
51. Robert E. Innis, Die Überwindung der Assoziationspsychologie durch zeichentheoretische Analyse: James, Peirce, Husserl und Bühler. Zeitschrift für Semiotik 10/4 (1988): pp. 149-173.
52. Dewey, 'Qualitative Thought, 'p. 200.
53. Ibid., p. 201.
54. Ibid., p. 202.
55. Ibid., 202.
56. Randall, Nature and Historical Experience, p. 290.
57. Dewey, 'Qualitative Thought,' p. 203.
58. Ibid., p. 204.
59. Ibid., p. 204.

60. Ibid., p. 204.
61. Ibid., p. 204.
62. Ibid., p. 204.
63. Dewey, 'Affective Thought,' p. 122.
64. Dewey, 'Qualitative Thought,' p. 204.
65. Ibid., p. 204.
66. Ibid., p. 204.
67. Karl Bühler, Theory of Language: The Repressentational Function of Language, translated by Donald Fraser Goodwin (Amsterdam: John Benjamins, 1990/original German edition Sprachtheorie (Jena: Gustav Fischer, 1934)), pp. 313-320.
68. Dewey, 'Qualitative Thought,' p. 204.
69. Ibid., p. 204.
70. Ibid., p. 205.
71. Peirce, Collected Papers, 4.233.
72. Justus Buchler, *The Main of Light: On the Concept of Poetry* (New York: Oxford University Press, 1974).
73. Dewey, 'Qualitative Thought,' p. 205
74. Ibid., p. 205.
75. Ibid., p. 205.
76. John Dewey, Experience and Nature, edited by Jo Ann Boydston (Carbondale: Southern Illinois University Press, 1988/1925), p. 148.
77. Robert E. Innis, 'Perception, Interpretation, and the Signs of Art,' The Journal of Speculative Philosophy 15/1 (2001): pp. 20-32 and 'Dimensions of an Aesthetic Encounter,' Semiotic Rotations: Modes of Meaning in Cultural Worlds, edited by SunHee Kim Gertz, Jaan Valsiner, and Jean-Paul Breaux. Charlotte, NC: Information Age Publishing, pp. 113-134.
78. Robert E. Innis, Susanne Langer in Focus: The Symbolic Mind (Bloomington and Indianapolis: Indiana University Press, 2009).
79. John Dewey, Art as Experience, p. 127.
80. Ibid., p. 129.
81. John Dewey, 'Affective Thought,' p. 122.
82. Robert E. Innis, 'Dewey's Aesthetic Theory and the Critique of Technology,' Phänomenologische Forschungen 15 (1983): pp. 7-42, now chapter 5 in Pragmatism and the Forms of Sense, and Aesthetic Rationality as Social Norm,'' Phänomenologische Forschungen 20 (1987): pp. 69-90.
83. Dewey, 'Affective Thought,' p. 121.
84. Dewey, Art as Experience, p. 150.
85. Peirce, Collected Papers, 6.223.
86. John Dewey and Arthur Bentley, A Philosophical Correspondence 1932-1951. Introd. by Sidney Ratner. Selected and edited by Sidney Ratner and Jules Altman with James E. Wheeler as Associate Editor (New Brunswick. NJ: Rutgers University Press, 1964), p. 73.

Four

SOCIAL AND POLITICAL POWERS IN JOHN DEWEY'S AESTHETICS

Krzysztof Piotr Skowroński

The interconnection between thinking and acting is one of the major postulates of John Dewey's philosophy, and of pragmatic philosophy in general. Assuming a close tie between aesthetics, or the philosophy of art (I use these two terms interchangeably in the present text) and social practice, is an inevitable consequence of this attitude. Namely, thinking about any artistic ideas and speculating on any aesthetic concepts can hardly be separated from taking into consideration the preferences of the authors, the expectations of the audience, the metaphysical and religious convictions that prevail in a given society, the moral consciousness of the members of the public, the cultural character of the epoch, and its technological development. Indeed, the bond between social life and aesthetics goes so far or, rather, so deep in Dewey that he applies the term "relation" to both of them, saying that " [a] social relation is an affair of affections and obligations, of intercourse, of generation, influence and mutual modification. It is in this sense that 'relation' is to be understood when used to define form in art."[1]

In the present paper I would like to take a look at aesthetics and the arts in Dewey from the point of view of the role of social and political powers. I will propose a definition of the term "powers" in the context of aesthetic values in order to re-read Dewey in such a way as to see how deeply, or how broadly, he recognizes this interconnection. By doing this I will not, I hope, impose upon his system of thought something completely alien or external to it. More importantly, however, it appears that aesthetics and the arts are not immune to the tensions and influences coming from the areas which are external to them, that is from the social and political centers of power and, as a consequence, I will discuss the aesthetization of political life too. This interdependence does not take place exclusively in the contemporary world; it comes forth from ancient times and, as Dewey observes, "Plato's demand of censorship of poetry and music is a tribute to the social and even political influence exercised by those arts."[2] I will not hide the fact, that in my personal view (which I will just outline here), these external factors are constitutive to the sense, value, and meaning of aesthetic ideas and the status of (great) works of art. I have a feeling that, at least to some degree, my view here aligns with Dewey's thinking; although, I am not sure if he proposes for us any good options as regards what to do when contrasting aesthetics, contrary visions of the arts, and irreconcilable concepts of values, along with their social and political contexts, clash.

1. Why the Topic is Important

There are various levels of "intensity" in the relationship and interdependence between social/political powers, on the one hand, and aesthetics/the arts, on the other. The development of cultural anthropology and social psychology in the 20th century—regarding which some works of Dewey and of George Herbert Mead have been instrumental—makes it possible for us to take into consideration a general and "moderate" character of this relationship. Namely, we can claim that in societies where there exist established, static, and fixed types of social relations, the members of the public will see moral and aesthetic values as fixed, the vision of the universe will be rendered in the "spirit" of eternity, and interpretations of the most important public issues will be ascribed to the competence of the elite of this society. On the other hand, those societies, in which social relations are more dynamic, changing, evolving, and open to debate, the character of values and valuations seems more dynamic and progressive, and the interpretation of what is (aesthetically) valuable, and what is not, is done by various groups within a given community, rather than by one privileged group of people, telling all others what is good or beautiful and what is not.

We can also think about the extremes of this interdependence, that is the aesthetization of political life and the politicization of aesthetics. Thus, if we take a closer look at any political debate on TV we will see how "aesthetic" it really is, I mean, how many aesthetic factors it necessarily employs: a carefully worked out *image* of the politician, the *clarity* of his/her speech, the communicativeness of his/her *gestures*, his/her attempts to *arouse emotions* in the audience, the *theatrical* scenery of the stage on which a politician appears, the *dramaturgy* of the debate, and many others. On the other hand, it was the experience of my generation in my native country to see, how much soc-realism dominated the artistic life (and not only artistic) in former Soviet block states; and, immediately after the collapse of Communism and the subsequent political changes there, how much the introduction of the policy of economic liberalism has converted the arts into a set of commercial products subject to free market economy rules—in both cases, the political circumstances being hugely influential upon the tastes of the audiences, the preferences of the authors, the policies of cultural/artistic institutions (mass media in the first instance), the rearrangement of the hierarchy within the arts, including the status of the artists, and the (diminished) role of what has been previously seen as "high art."

Apart from these experiences, there are some movements and trends in contemporary thought in the West that have contributed to a general knowledge about this interconnection in its various aspects:

feminism (e.g. showing the link between politics and sexism, also within the arts), post-structuralism (e.g. Michel Foucault's idea of *bio-power*, Umberto Eco's idea of *open work*), postmodernism (e.g. Wolfgang Welsch's idea of aesthetization of everyday life), hermeneutics (Hans-Georg Gadamer's destruction of the classical separation of art and reality), the Frankfurt School (Max Horkheimer and Theodor Adorno criticizing the post-Enlightenment cultural industry), and many others.

The aesthetics of American pragmatism—and Dewey is the most significant figure here—joins some of these tendencies, and I would like to focus on how much it recognizes the role of social and political contexts. Let me just state at the beginning, that I neither study Dewey in order to better explain what he wanted to say, nor to reinterpret his aesthetics and his ontology of art, if he produced any. Rather, the present issue has been taken up with the conviction that the relationship between social/political powers and aesthetics is, for various reasons, interesting and important both in the present debate on aesthetics and the arts in general (not just within pragmatism) and, in particular, in the intellectual confrontation with influential movements such as postmodernism, and post-structuralism. Hence, a better recognition of the character of this relationship—without going too deep into the intricacies of Dewey's aesthetics, and without outlining its specificity, something which has been competently presented by other authors in this volume—can tell us more about the origin of the status of the *beautiful* or *aesthetically valuable* works of art, and help us better understand why some arts, some aesthetic ideas, and some works of art are appreciated more than others and why some of them gain importance while others fade away or disappear from public recognition. Below, then, I take a look, selectively, at the relation between social/political powers and aesthetics in Dewey. I examine in Dewey's texts—*Art as Experience* (1934) in the first instance—the perspectives regarding social and political powers, and the aesthetic concepts or artistic ideas in which, or by means of which, these powers are articulated.

Thus, I think, that what is, by some, understood as *aesthetic values*, to a major degree—although it would be difficult to claim exactly what degree—is constituted or emerges outside of aesthetics, in the same way as the status of *the beautiful* or the *aesthetically valuable* is, in a significant way, defined by the non-aesthetic and non-artistic factors, that is social, political as well as cultural, socio-psychological, historic, and economic. Such an approach can be contrary to all or nearly all those philosophies, which adhere to the classical tradition, stemming, directly or indirectly, from the Pythagorean, Platonic, neo-Platonic, Scholastic, and traditionally religious metaphysics. Some of them—like phenomenological aesthetics (e.g. Roman Ingarden's idea of *literary work*), neo-Thomist aesthetics (e. g. Jacques Maritain), Russian aesthetics (e.g. Wladimir Soloviev), aesthetics related to icon painting, not to mention some

Oriental types of aesthetics (e.g. Muslim)—have a prominent status in some quarters, and, this way, constitute a challenge to non-essentionalist aesthetics. I say "non-essentialist" because, if we wanted to identify the most animated controversy within contemporary philosophy of art, it might turn out to be the one between essentialist and non-essentialist aesthetics; the former decreases the role of the external, social context in artworks and looks for the kernel that would constitute their internal or essential meaning and character, while the latter would emphasize the role of the social context and blur the border between the external and internal sources of influence upon works of art.

This debate was initiated at the turn of the 19th and 20th centuries by, among others, Friedrich Nietzsche, Henri Bergson, Sigmund Freud, and representatives of the avant-garde (George Braque, Andre Breton, Tristan Tzara, Emil Nolde, Vasil Kandinsky, Filippo Marinetti, and others). Today the debate is continued in various quarters (for example, within postmodernism). Needless to say, their "rebellion" was directed, among other things, against classical concepts of aesthetics and the traditional understanding of the fine arts, which maintained some of the following attitudes and perspectives: (aesthetic) ideas revered entirely in terms of their intrinsic significance , appeals to absolute values, essentialist visions of works of art, and seeing the beauty of the arts in isolation, that is without a (strong) reference to cultural, social, and political context(s). Dewey periodically joins these "rebellious" movements, and his philosophy of art articulates a very important tendency in the contemporary approach towards the arts, that is its non-essentialism, contextualization, and, most of all, its interdependence with factors commonly conceived as non-aesthetic and non-artistic. Obviously, I do not claim that Dewey shares much with the avant-gardists and postmodernists; for example, he might have abhorred some of them—like those, for whom such terms as "universal", "unity", and "common" are the remnants of ontologically meaningless metaphysical discourse of old times—, by claiming that "rhythm is a universal scheme of existence,"[3] and that "the unity of the arts resides in their common form."[4] I will return to this controversy between Dewey and some of the avant-gardists.

2. Concentration of Powers and Accumulation of Values

As I show in my book, *Values and Powers. Re-reading the Philosophical Tradition of American Pragmatism*[5], the operation of forces, of various kinds (e.g. social, political, psychological, institutional), is constitutive in pragmatic descriptions of the social and non-social world. As I understand it, the meaning of the term "powers" deals with all types of pressure, coercion, domination, control, and influence imposed by some agents and/or centers of powers upon others, which leads to the reorientation of their understanding of what is good or evil, satisfactory or unsatisfactory, worthy or unworthy, important or unimportant, and so forth. More precisely, I have in mind such situation(s) in which an agent or a group of agents, by using pressure, coercion, domination, control, and influence, make(s) it necessary for other person(s) to modify in some way their previously held positions on values, on what is significant, and needed. This imposition can be exercised upon others intentionally, or not, directly or indirectly; nevertheless, without this imposition the previously held position would not have been changed or reconsidered. Such an understanding of the role of powers in pragmatism does not signify at all that pragmatism prefers or justifies using brute force. I claim that a separation of values and powers is hardly possible in the work of most of the representatives of American pragmatism. Any view about the melioration of the social world—so much stressed in American pragmatism—must, implicitly or explicitly, include or assume the way in which this melioration will be efficiently executed. Simply, powers are the factors that make the accumulation of values or any type of melioration possible, and also allow values to "work" in practice. Without powers, values are impotent ideas; on the other hand, powers, without values, become mere brute force. Moreover, if pragmatists understand ideas as programs, or plans of action, this means that powers are necessary to make the action realizable; without the means of the realization of the plan, the ideas become futile and unrealizable. Thus, ideas, to be implemented into the texture of social and political life necessarily need various kinds of power in place in order to be effective, be it "the power of thought and of will" in Ralph Waldo Emerson's proto-pragmatic account in *The Transcendentalist*; human powers or internal energies of men as William James suggested in "Energies of Men"; the loyalty to a worthy cause and provincialism as "saving power" in Josiah Royce's *Philosophy of Loyalty* and "Provincialism"; or the might of the democratic system, as in Dewey's thinking, thanks to which, or by means of which, values can be had and "the social power" strengthened. Dewey's views on the role of powers in social life are not suspended in a vacuum, and his naturalism is a good background for the recognition of operating powers or "energies resisting each other"[6] in the human and non-human world. As we can read in *Art as Experience*:

> Resistance accumulates energy; it institutes conservation until release and expansion ensue. There is, at the moment of reversal, an interval, a pause, a rest, by which the interaction of opposed energies is defined and rendered perceptible. The pause is a balance or symmetry of antagonistic forces.[7]

While talking about the realm of human affairs in the present text, it would be difficult for me to strictly and methodologically differentiate social powers from political powers in Dewey's works, due both to reasons internal to Dewey's system of thought, and to certain external factors. Internally, Dewey does not deliminate strict borders of aesthetic notions and ideas, instead, he tries to show their interdependencies with various factors and their relationship with the environment in which they are saturated and in reference to which they are contextualized. Selection of one of these ideas and making it an isolated object of investigation—isolated from contexts, conditions, and changeable factors—would mean following the classical and metaphysical approach in philosophy. Pragmatists, including Dewey, want to avoid this approach in the conviction that it is false and arbitrary—as are all static classifications, categorical definitions, black and white divisions, and final conclusions—and, moreover, not efficient in understanding life and making it better. Small wonder, that if we assumed Dewey's philosophical stance and implement his philosophy in our investigation, we could hardly provide any clear restriction on various issues. For example, we would be in trouble to clearly see whether the cultivation of a nation's heritage employs social *or* political powers, and whether there are social institutions involved in this cultivation, like schools *or* political institutions, like a Ministry of Education. Both options have to be taken into consideration and, moreover, multilateral interdependencies studied.

Externally, it is difficult to differentiate social powers from political ones because of, among other things, the obscurity of the term "powers" in the context of aesthetic values. Here, I do not intend to reconstruct Dewey's view on this, because I do not believe Dewey operated by means of these terms. I also do not believe that he attempted to construct a theory of the relation between powers and aesthetic values, although, in my view, he recognized this relationship. Instead, I wish to offer my understanding of powers in the context of aesthetic values (and aesthetics in general), hoping that, first, this does not infringe upon Dewey's way of seeing things, and, second, that it will be methodologically more accurate if I warn the reader that I am using my definition of the term "powers" to re-read Dewey rather than suggest how Dewey might have intended to use it in his works. Thus, very briefly, we deal with values and valuations when we hold an attitude towards agents (humans and, perhaps, the higher mammals), living bodies (animals and plants), things (objects), actions (deeds and activities), and states of affairs (facts and situations) that places them in the context of how good, worthy, beautiful, true, useful, important, and satisfactory they are, they can be, and/or they should be. The meaning of the term "powers" which I propose in the context of aestheticsand aesthetic values, deals with all types of pressure, coercion, domination, control, and influence imposed, by some agents and/or centers of powers upon others, which leads to the reorientation of their understanding of the aesthetic, the arts, the beautiful, the aesthetic valuable, the sublime, and the satisfactory. To be more precise, I have in mind situation(s) in which an agent or a group of agents, by using pressure, coercion, domination, control, and influence, make(s) it necessary for other person(s) to modify a previously held position on aesthetics, the arts, the beautiful, the aesthetically valuable, the sublime, the satisfactory, and so on. This imposition can be exercised upon others intentionally or not and directly or indirectly; nevertheless, without this imposition the previously held position by those others on the beautiful, the sublime, and so on, would not have been changed or reconsidered. I do not think, that my understanding of the operation of powers is much different than some of Dewey's claims on energy (or energies), especially when we take into consideration Dewey's following statement:

> Theory can be based *only* upon an understanding of the central role of energy within and without, and of that interaction of energies which institutes opposition in company with accumulation, conservation, suspense and interval, and cooperative movement toward fulfillment in an ordered, or rhythmical experience. Then the inward energy finds release in expression and the outward embodiment of energy in matter takes on form.[8]

I would like to stress at this point that studies on the relation between social/political powers and aesthetics are worth while if we want a better understanding of the meaning of such terms as *progress*, *development*, *success*, *melioration*, *betterment*, and similar terms. Namely, all of these terms refer to a kind of expansion, aggrandizement, and/or amassment of what is seen as valuable, good, worthy, satisfactory, and beautiful, according to the presupposed criteria. If we talk about any kind of moral progress or social development—for example, by promoting the role of the arts and aesthetic education—we must inevitably assume that a portion of the good things will be larger, in one sense or another, than it was before. This amassment can be done in a variety of ways. For example, this can be done by adding valuable things or states of things to the existing one. The world would be bettered in a similar way as a museum is improved by getting successively valuable works of arts in its collection. Another way of amassing the good things is by means of perfection, that is by the betterment of the present thing, or the state of a thing, into the stage which is the closest to the ideal or, if possible, identical with it. This aiming toward full completion was characteristic of the aesthetics (and ethics) of ancient Greece. Still another way, is to make what we have better, without troubling ourselves overtly about whether an ideal perfection is possible and whether we add something to the overall collection of good things in the world. William James's idea of meliorism (as sketched in "The Moral Philosopher and The Moral Life") and the attempt to attain "the greater ideal,"[9] are good examples for making life better in this way.

Irrespective of the ways for the betterment of the world, there must be some general, metaphysical and ontological assumption, without which such an amassment, or "accumulation of values," makes hardly any sense. The precondition for Dewey (as for other pragmatists) is the unity of the structure of the world as a sort of matrix for social development. By this I mean Dewey's conviction that it is possible to find a general framework within which social melioration would be possible. If, ontologically, there is nothing like any unity of the world or, cognitively, if human minds are unable to reach this unity by means of science, and/or by means of any discourse of human language, then the accumulation of values in one place, or at one time, might not be converted into the accumulation of values at another time and place. This will make any accumulation of values local, temporary, and arbitrary, because making, for example, *good* or *beautiful* things, according to the ontological or cognitive criteria of one community would mean making *bad* or *ugly* things, according to the criteria of another community. Moreover, the specification of the works of arts as perfect, complete, and final would be disputable, since, as Dewey writes, the "organization of energies to move cumulatively to a terminal whole in which the values of all means and media are incorporated is the essence of fine art."[10] When, however, we are convinced about this unity, as Dewey seemed to be, then, within this ontological framework, we can measure and objectively or inter-subjectively see that "each step forward is at the same time summing up and fulfillment of what precedes, and every consummation carries expectation tensely forward."[11] A more or less coherent vision of values is necessary if we are to hold a universal way of meliorating the world, and/or cross-cultural ways of accumulating values. As Dewey once noticed, and I agree with him:

> There can be no movement toward a consummating close unless there is a progressive massing of values, a cumulative effect. This result cannot exist without conservation of the import of what has gone before. Moreover, to secure the needed continuity, the accumulated experience must be such as to create suspense and anticipation of resolution. Accumulation is at the same time preparation, as with each phase of the growth of a living embryo.[12]

What I have reservations about, however, is whether this unity is factual or, instead, postulated. As it seems, the problem of unity, or whether unity is postulated rather than factually taking place, is one of the major controversies of contemporary philosophical aesthetics (and outside of it), especially since the time of the avant-garde. Namely, a part of the philosophical message of the cubists (Braque), Dadaists (Tzara), expressionists (Nolde), and others, was whether we should or should not try to construct anew, or to re-construct, the world and its parts. They believed that there are *real* or really existing objects in the world, but had reservations about the truthfulness, or ontological *reality*, of the relations between/amongst these objects; actually, these relations were seen as arbitrarily fixed, conventionally justified, and habitually accepted. Thus, we can interpret a philosophical message within a part of the cubists' arts (e.g. Pablo Picasso's *Girl with a Mandolin*) as disbelief in the structure of the prearranged world and, at the same time, calling for a re-thinking of it afresh and to re-establish some kind of order. Also, we can interpret the Dadaists' message as a protest against the use of the cultural policy of the Institutions to impose their vision of the structured world, along with a clear difference between the *sacrum* and the *profanum*. Thus, installing an urinal in a gallery (Marcel Duchamp's *Fountain*), or placing a bicycle wheel on a stool as if on a pedestal (Duchamp's *Bicycle Wheel*) seemed appropriate manifestations of the rebellion against this artificially constructed and conventionally specified role of objects in the overall structure of the world. Expressionists and surrealists made us realize that thinking about solid objects as the essential parts of the structure of the world—something accepted in classical ontology and metaphysics—rather than emotions and dreams, can be undermined. As a result, the focus on real relations amongst the objects would seem secondary to the relations taking place within the world of imagination, dream, and emotions. Polish avant-gardist Witold Gombrowicz, to provide one more example, in his philosophical novel *Cosmos*, shows a mechanism, according to which a collection of arbitrarily linked objects create new meanings. These new meanings are not convertible into the meanings these same objects have for other individuals; consequently, the world and the objects within it, constitute an area for constructing innumerable interpretations and new vistas, without the possibility of translating these interpretations, and these vistas, into one common language/discourse and into one coherent system of values. In Gombrowicz, the unity, or the structural whole of the world, is an assumption or a postulate deprived of ontological grounding.

3. Social/Political Powers in the Context of Aesthetic Ideas

To avoid any misunderstanding, I am not claiming that Dewey suggests that social and political powers are alone, or primarily, responsible for the structure of aesthetics and the arts, and it would be methodologically unjustified to put forward and defend such a type of reductionism. Nor is it to be claimed that Dewey's aesthetics is political aesthetics; this would make political thought the aim and the focus of his aesthetics, which is not true. Instead, there is a clear reference to political life (to democracy in the first instance) and to the social (communal character of human actions) that can be found in Dewey's aesthetics or philosophy of art, which makes it hardly possible for us to avoid involvement in some kind of social and political, not to mention ideological stance, or attitude, even if we wanted to stay preoccupied exclusively with aesthetic ideas and works of art.

If we turn to Dewey's texts to study how much non-aesthetic and non-artistic factors embody the power of the fine arts and the might of aesthetic doctrines, we will have fertile ground for intellectual exploitation. For example, if democracy—understood as a form of associated life, not just, narrowly, as a political system—for Dewey is something important, his aesthetic concept(s) will reflect it. Thus, he will stress, while talking about the arts and aesthetics, the importance of such democratic values as pluralism, freedom, openness, tolerance, activism, discussion, and others. His aesthetics will emphasize the role of subjective or individual activity, the aiming for free expression of thoughts and of feelings, plurality in the form of this expression, a potental to full development of aesthetic experience in most everyone, not just in those dedicated to the fine arts, open acceptance and offering of criticism to help create better solutions in the arts, and so on. Also, he will show how significant it is to educate people (possibly all people) in appreciating and approaching these ideals in their communal life. This way, he will have to—implicitly or explicitly, directly or indirectly—oppose the absolutistic concepts of values, the objective canons in aesthetics, the "closed in museum" idea of the arts, the reference to past masters who seek the exclusive truth about what is beautiful, the contemplative aesthetics, which sanctifies the (social) reality instead of trying to change it into something better, and, most importantly, he will criticize the institutions which petrify and promote such traditional ideas. In other words, Dewey, by constructing his aesthetics would try to reform the institutions of public life, and, in this way, encroach upon the area of social and political life or, which is the same, by reforming the social life he will encroach upon aesthetics. Then, it appears that promoting openness and pluralism in the arts means that we have to, willy-nilly, encourage others to exclude, diminish, and marginalize the social and political role of the types of aesthetics that are not pluralistic, not democratic, and not open to discussion. This embraces: all or nearly all elitist and aristocratic concepts of the arts; all or nearly all absolutistic concepts of values; and all traditionally religious types of aesthetics that assume a transcendental approach. Needless to say, all these aesthetic concepts claim to possess true access to the realm of beauty, and all these concepts have been vindicating their claims philosophically throughout the centuries.

Educationally, Dewey shows us that much depends upon the type of schooling one gets in the education system in a democratic or non-democratic political system. In non-democratic states knowledge and skill are to a degree established and transmitted from generation to generation with hardly any modification, as if aesthetics and the arts were to reflect the eternal truths. In a democracy, however, the tremendous access of youth to artistic education causes them to become sensitive to things impossible for them to know in social and political systems with restricted access to a free (artistic and aesthetic) education. Likewise, the opening of the art world to everyone who would like to participate elevates the role of popular culture (and popular aesthetics) and extends the role of aesthetic experience resulting in the participation of many more people in cultural (and artistic and aesthetic) life as well as political life. This does not take place in the non-democratic systems. If one of the main messages of an educational system in a democracy is that access to the arts is not restricted, aesthetic experience is not exclusive, and everyone can try to participate. This, inevitably, will depreciate all those philosophies of art which have claimed the opposite throughout ages. If, in the process of education, popular arts will be democratically appreciated side by side with the classical arts, and the recipients of high art will be addressed side by side with the receivers of popular or low art, inevitably the traditional canons in aesthetics will be undermined, because they strictly delineate(d) the high from the low, and the sophisticated and the common, in the arts and aesthetics.

I suspect, Dewey would agree that, morally, the arts can serve as a litmus test for the scope of liberty within a given political system. Free execution of one's artistic choices and an open articulation of aesthetic ideas signify that a given liberal democracy functions. The understanding of what liberty or liberal democracy is crucial here; namely, if we lived in a hierarchically ordered society and understood liberty Platonically, we would treat the arts as a way of manifesting the pre-arranged structure of powers; then, we would have to accept the present state of things, along with the elevated role of the social and political elite—the contents of the aesthetic/artistic elite and the way of its arrangement is a separate topic, one that includes the role of social and political powers—, whose members know best how the world should be interpreted and teach others accordingly. However, if we understand liberty pragmatically (and democratically), we will treat the arts as a free co-creation of the structure of power, along with making better and more just the arrangement of power within society. In this way, one of the aims of the arts will be an expression of the injustice being done by some people to others, and the articulation of the social demands for the betterment of life. Additionally, the artist will, most probably, be future oriented rather than past oriented in the cultivation of his or her tradition by selecting such elements in his or her heritage that will best suit the rearrangement of social and political life to make it better than it was previously. Dewey has drawn our attention to how much the promotion of a given model of freedom, responsibility, and education can influence our perception of the arts and the role of the arts in our social life. The promotion of pluralism in the area of politics and the promotion of democracy in the area of the social life, convert into the promotion of pluralism and democracy in the arts and aesthetics. Simply, if you promote pluralism in the artistic, aesthetic, and cultural institutions you, at the same time, influence the arts by showing that pluralistic aesthetics has a meaning and can be seen as worthy and inspirational.

This way, Dewey stimulates us to ask whether it is possible to have an art that is not—indirectly, if not directly—ideologically and/or politically involved, and we can use these suggestions to interpret various trends present in contemporary aesthetics. Pop-art, for example, expresses, as I re-read it, a sort of ideology which says that it is the popular, not elitist, themes that should be taken into consideration and that everyday reality, accessible to everybody, irrespectively of their social status, should be given priority while dealing with the arts and artistic perception. Recently and even more convincingly than in Dewey, it has been articulated by Andy Warhol in his art and in his aesthetics, of which the following text is a famous manifestation:

> What's great about this country is that America started the tradition where the richest consumers buy essentially the same things as the poorest. You can be watching TV and see Coca Cola, and you know that the President drinks Coca Cola, Liz Taylor drinks Coca Cola, and just think, you can drink Coca Cola, too. A coke is a coke and no amount of money can get you a better coke than the one the bum on the corner is drinking. All the cokes are the same and all the cokes are good. Liz Taylor knows it, the President knows it, the bum knows it, and you know it.[13]

Warhol shows us how, on the one hand, democratic ideals are applied to the arts and, on the other, how blurring the social contrasts makes the arts a social and political factor of great significance. Consuming coca cola—which, by the way, became the subject for his own paintings—"consuming" the arts by the receivers takes places irrespectively of their social status, and both the President and a poor immigrant will be deeply moved, say, by the films starring Marilyn Monroe and the music of Elvis Presley.

4. Social/Political Powers and the Role of Artistic Institutions

Dewey, perhaps the most significant educator in America, fully recognized the power of educational institutions in the context of aesthetic perception and artistic sensitivity. He must have seen the link between the arts and social/political powers and its role in aiming at something which he calls, in *The School and Society* ("The School and Social Progress"), "the development of social power."[14] At the same time, he sees educational development as "the direction of powers into special channels,"[15] in order to prevent a dissipation of energy and to promote self-development better realizable in democracy than elsewhere. Indeed, the educational system of any country, and of any national tradition, must not avoid the affirmative reference to the heroic poems of the tradition, the noble ceremonies cementing the spirit of the given community, and promoting the sensitivity as to what is noble and great. It, also, must not be indifferent to those pieces of art, which were socially disruptive and non-patriotic. Apart from this, schooling is a strong, if not the strongest, factor in the development of aesthetic perception because it gives the mass of learners and students the sense as to what areas in human experience are responsible for good taste (e.g. cultural institutions), what kind of activity deserves artistic attention (e.g. self-fulfillment), and where we should find the experts that will tell us what is beautiful, and also where we should not search for such experts.

I think that Dewey would agree with the claim that, if we want to refer to any aesthetic ideas, as well as aesthetic canons, standards, norms, and fashions, we immediately get involved, directly or indirectly, in the problem of operating institutions of various sorts, which support or persecute, vindicate or disqualify, promote or nip in the bud, and tolerate or are disrespectful to these ideas. This institutional support, or vindication, is usually done by means of some kind of a cultural policy of the prevailing system of education, the mass media, by public opinion, of the authority figures' approval and disapproval, of the legal systems, which will "control" whether artistic publications will be in accordance, or will not be in accordance, with it. Without such support or vindication, these ideas are not seen as serious in the public reception of them, and, unless the conditions change, they will not be introduced into the texture of social life or, they will stay impotent and vacuous. For example, it is hard to imagine a traditionally religious type of art—along with the moral obligations it proffers to show and transcendental commitments it calls to—to reach any level of significance in a secular society, unless this art, like some old icon pictures, gains a commercial or mercantile significance. Or, it is hard to imagine a piece of a secular art—with its declared freedom to one's own expression of earthly issues and indifference to the heavenly—to be (institutionally) promoted, fully accepted, and widely recognized in traditional religious cultures. Likewise, it is difficult for a contemplative type of art to be recognized by the public submerged in a daily rush for a multitude of things, unless this type of art is used, by some, for relaxation purposes. The same can be said about *political* factors in an aesthetic and artistic world. A totalitarian vision of the arts—I mean the arts supported, vindicated, and legalized by a totalitarian political system, such as soc-realism under Communism—can (could) hardly be appreciated in a democratic society, in the same way as the arts articulating democratic themes were anathema for the cultural policy makers in the former Soviet states, and as a result the prisons there were full of artists who dared to raise their voices in the name of freedom and liberty.

As was said above, Dewey recognizes the role of political powers in the arts and aesthetics, and this is visible when, for example, he pays attention to the fact that "Most European museums are, among other things, memorials of the rise of nationalism and imperialism."[16] Indeed, it would be difficult for us to indicate which role of national museums (and other artistic institutions of this type) in many countries is more important: to collect unique works of art or to show the greatness of the given national tradition. Here the artistic function of aesthetics—more precisely: an institution promoting the arts and the aesthetic ideas—and the ideological and/or political one intertwine. This is even more visible, when we deal with the art collections that have been taken from distant lands and are shown to the public as if to prove the international, if not global, aspirations of the given country.

Also, while stressing the importance of the collective character of the arts, Dewey recognizes various kinds of tensions and influences within the given community or society, and one of them is ascribing the high or low status of the arts and "shaping" their characters accordingly. Indeed, a strong separation of the high arts from the low arts in 19th century Europe reflected a strong class divide of the day, its aristocratic structure, with the elite at the very top, and lack of movement of the masses of people up or down the social ladder. On the other hand, the non-aristocratic structure of American society has been articulated by the democratic tendencies and popular character of American art. This does not mean that we have to talk about a direct imposition of a sense and a mood of the arts by a given center, or by given centers, of social/political power. It is enough to say, that the arts and aesthetic concepts refer to "the emotions and ideas that are associated with the chief institutions of social life."[17] These chief institutions are, among others, museums and galleries, schools and universities, church and state, the courts and the legal system along with the seriousness and respect they enjoy. Also, I would add to this list the metaphysical assumptions prevailing in the group and the moral convictions of the given society fixed in customs and habits, and seen as proper in various institutions, such as the mass media, schools, and the law, which regulates the scope of the freedom of artistic expression. The arrangement of social and political forces, irrespective of the system, make such institutions fixed in a top-down hierarchy. Thus, in a country with a more or less stable social system, school is an institution respected enough to be trusted in the transmission of knowledge and the ways of receiving it. The reference to these institutions or, more precisely, to the values, knowledge, and sensitivity transmitted there, is a strong factor in making aesthetic ideas, artistic concepts, and works of art serious or not serious, important or trifle, and capable of being recognized as true or fake.

This creates some difficulties, and let me just give, very briefly, three aspects of the discussion about aesthetics and the arts, where the reference to the political or politically controlled institutions must be taken into consideration. Hence, if we have The Church (of whatever denomination) as the highest, or as one of the highest, institutions in a given culture, for example as it was in the Middle Ages of Christian Europe or the Orthodox Church in Eastern Europe now, or in some Muslim countries in the Middle East, the religious art and the religious point of reference in aesthetic theories and concepts will have a more elevated role than in any secular society. Then, many people will listen to theologians ideas about the meaning of the arts, the sense of aesthetic ideas, and the meaning of artistic/aesthetic education. In such a case, our stance—by "us" I mean liberal democrats living in the West—would be bemused; on the one hand, since this artistic, aesthetic, and cultural status quo should be seen as a genuine manifestation of what people cherish over there, we would have to respect this; on the other hand, this needs rearrangement since it is, according to the views of some of us, a result of an unfair power structure within a given society.

The division of high vs. low arts does not only refer to the preference of the people. This also—if not predominantly—refers to the problem of *who* is going to decide about the criteria used to distinguish that this or that work of art belongs to high or low art, and deserves or does not deserve any special attention, be it in the process of the promotion of the arts or in the process of teaching the arts. Likewise, when Dewey writes so much about the significance of aesthetic experience, he assumes that it is *our* experience, not exclusively the experts' experience, the critics' experience, the professors' experience, the theologians' experience, and the artists' experience; the *democratic* assumption that *everyone* can participate in the world of the arts is taken for granted. It is hardly possible to avoid some further perplexities. For example, within the existing educational system in the West, we accept that it is the free market, rather than selected aestheticians, which will ultimately decide on the worth of the works of art, we automatically reduce the meaning of those who would see such criteria as barbaric. If we assume that books are instruments, or tools, serving various cognitive and hedonistic aims, we, willy-nilly, degrade all those traditions, in which (some) books are sacred and tell the truth. In this instance, the authors' messages *should* be cultivated, and the norms they provide obeyed with respect and trust by all the members of the public. If, in the process of education, we appreciate the individual efforts of the students in shaping their sensitivities and likes or dislikes, we will depreciate the great masters, with their sense of exclusive sensitivities, their exclusive standards, and their exclusive views on what is beautiful.

This takes us to another point of interconnection between aesthetics/the arts and the social and political powers, namely *authority figure(s)*. By this term I understand a "personalized" center of power recognizing aesthetic value. Be it noted, that an authority figures cannot exist outside institutions which give him or her support. Authority figures in traditional societies have been respected and highly appreciated due to the restricted knowledge they possessed, which made them mightier than the rest, and nowadays, in the industrial and democratic societies, the authority of institutions is supported by the political system, with the effective execution of law. Now, if we, the liberal democrats living in the West, want to follow Dewey in ascribing the role of aesthetic and artistic judgment to the public and the critics, we, willy-nilly, decrease the role of those authority figures, which, in some traditions, enjoy the status of undisputed opinion-makers as regards aesthetics, including the scope of freedom associated with aesthetic expression. A tragic illustration of this type of clash between "the West" and "the world of Islam" was the reaction of some of the Muslim communities after the publication of Muhammad's caricatures in a Danish newspaper.

5. Social/Political Powers in the Context of the Status of Great Work of Art

In the present text I use the phrase "external" for a work of art. I am not sure if Dewey would appreciate such a constant use of this term; this inevitably suggests that there are also some "internal" relations within a work of art, or some kind of a kernel, or essence, having its own autonomous status and its own specificity, independent from the outside. I will not hide the fact that I have some sympathy with such a way of seeing works or arts and, perhaps, I will differ from Dewey as regards the meaning, or significance, of the internal relations. I partly follow those (e.g. Ingarden, George Santayana) who claim that a work of art has its internal structure and that some part of this internal structure is hardly dependent—although it would be hardly possible to talk about a complete independence—upon the outside (social) world, and this is *very* important while talking about any work of art. For example, one can find in a work of art its own space (I mean, specific arrangement of spatial relations, different than those in physical space) and its own time, different than those in astronomical time, conventional time measured by our clocks, and different than subjective time "produced" in each of us, as a part of our reaction to various events.

Without going further into the intricacies of the problems of the ontology of art, I place the status of a work of art—as "excellent" or "great"—in the context of the influence or impact of social and political powers, and I think Dewey would have been supportive of this idea. As I show elsewhere,[18] we can understand the greatness of a work of art in terms of its capability to encompass an expressive and condensed substance given to it by, or from, an author or authors together with the elements of the historical and cultural background transmitted while making the work of art. It also means its capability to exercise an influence upon its readers/viewers due to its profound axiological depth and strength of impact upon the receiver/viewer in one or more areas. In other words, a great work of art, or a masterpiece, contains a unique internal world, which radiates its message amongst those able to perceive it (if not influencing them in the process). I will absolve myself in this text from discussing the "power" of great works of art which is manifested by its factual or possible influence upon its receivers, for example, by shocking them, making them think about various things in a different way than they did previously (before perceiving the given work of art), and stimulating them to committing or absolving from certain actions. Instead, I think that a part of the "greatness" of the work of art results, partly, from the external factors (I mean, external to the work itself), most importantly, from the educational, cultural, and artistic institutions, not to mention the media, promotion, and business aspect of the whole thing.

This can be well illustrated by what takes place currently in the 21st century West with roots that go back into the 20th century. Nowadays, in the age of the "crisis" of traditional art, we can witness the situation in which excellent works of arts, I mean, excellent according to classical or traditional criteria, are seen by many as hardly anything more than touristy relics of past times. These works do not produce any emotions amongst the viewers, nor do they stimulate the receivers to reflect upon deeper levels of understanding. Yet, we have a multitude of great works of art within pop culture, rock'n'roll, and film, which are seen as great, despite the fact that they—in comparison to the great works, either from the past or from the present—seem less elaborate, less deep, and less complex, although some may claim that this is not the case. For example, it is an important message of Richard Shusterman's aesthetics (*Pragmatist Aesthetics*), that the popular arts can meet the most crucial standards of aesthetic tradition.[19]

I think that the "greatness" of a work of art can be explained, party, by the influence it has upon the receivers, and I think Dewey proposes a good way of understanding this, by employing the term "energy" or "energies" into the realm of aesthetics. Namely, that a work of art—which Dewey defines as "an organization of energies"[20]—is not an independent or autonomous entity. Instead, it necessarily involves interaction with the outside, and is for Dewey a matter of course:

> When the structure of the object is such that its force interacts happily (but not easily) with the energies that issue from the experience itself; when their mutual affinities and antagonisms work together to bring about a substance that develops cumulatively and surely (but not too steadily) toward a fulfilling of impulsions and tensions, then indeed there is a work of art.[21]

As we can interpret Dewey on this, without any "support" from the external factors, a work of art would not be able to speak to its audience in the same way a great work of literature written in Chinese is not able to speak to readers who cannot read Chinese, or as Chinese theater would leave cold all of those who do not know its history, symbols, and context. A receiver without any preparation as regards the history of European painting, is hardly able to grasp the meaning and the message of Leonardo da Vinci's *Mona Lisa*—although s/he most probably can recognize the picture and identify its author—and the main reason why s/he cares to see it in the Louvre is its fame, and the fact that it has become a cultural icon and the most recognizable picture in Western painting. I do not believe *Mona Lisa* speaks to most of the visitors who come to see it; rather, I think, it is its splendor and its omnipresence (I mean, the omnipresence of *Mona Lisa*'s image) in cultural institutions that convinces people of its possible worth and its artistic greatness. The very same picture, presented on a city street, without its promotional glamour would not, I suspect, arouse anybody's attention. It is the special place (the Louvre), the special promotion (Western media), a touristic attraction ("while in Paris you have to see it"), the education (in which teachers tell the students that *Mona Lisa* is a great work of art), and the common conviction in various quarters (stimulated by some centers of education), that a cultivated person should be delighted to see *Mona Lisa* and other representatives of the high arts. To provide another argument that, as Dewey writes, "The 'center' of the picture is not spatial but is the focus of interacting forces,"[22] is that, every so often, we learn that a work of art has been elevated to the level of a masterpiece due to the discovery of its authorship by a famous painter. In the same way, the esteemed value of a recognized work of art decreases dramatically and its greatness collapses due to the discovery that it is a forgery instead of the original, although no part of the work was destroyed, spoiled, or injured, and it is the same work as before this fatal discovery. In such cases, the outer world of a work of art has a decisive influence upon the estimations of its value and its charm. Scandal, shock, provocation, and similar events can combine a message about values, articulated in an artistic and/or aesthetic form, and put forward by means of power or energy accumulated in a shocking scene or a provocative statement. It is possible to speculate that a great work of art is recognized as "great", as "work", and as "art" mainly because of the influence of social contexts—one that includes teachers, parents, authority figures, educators, critics, professors, the mass media, publicity, promotion, and so on.

Let me conclude by referring, shortly, to the role of education, this time in the context of the greatness of works of art. Very frequently, one of the major social and political roles of institutions of education at all levels, is to explain to the students what books (and other pieces of great art) should be interpreted; how books should be read by providing interpretations; which authors are eminent and noble, and why, by providing the values, according to which nobleness can be measured. Such a contribution by education into the recognition of which works of arts are "great" and which are not, cannot be ignored while discussing the value of works of art; and what is more, I think, it is one of the most essential factors in deciding the status of any work of art.

NOTES

1. John Dewey, *Art as Experience* (New York: Minton, Balch, and Co., 1934), p. 134.
2. Ibid., p. 328.
3. Ibid., p. 150.
4. Ibid., p. 191.
5. Cf. Krzysztof Piotr Skowroński *Values and Powers. Re-reading the Philosophical Tradition of American Pragmatism* (Amsterdam-New York: Rodopi, 2009), pp. 13-19.
6. Dewey, *Art as Experience*, p. 155.
7. Ibid., p. 155.
8. Ibid., pp. 159-160.
9. William James, "The Moral Philosopher and the Moral Life," in John J. McDermott (ed.), *The Writings of William James. A Comprehensive Edition* (The University of Chicago Press, 1997), p. 627.
10. Dewey, *Art as Experience*, p. 172.
11. Ibid.
12. Ibid., p. 137.
13. Andy Warhol, *The Philosophy of Andy Warhol (From A to B & Back Again)*, San Diego: Jovanovich, 1975), p. 100.
14. John Dewey, *The School and Society* (Chicago: Chicago University Press, 1899/1907), p. 32.
15. John Dewey, *Democracy and Education* (1916), *World Wide School*, (http://www.worldwideschool.org/library/books/socl/education/DemocracyandEducation/toc.html), 3, Pt.1.
16. Dewey, *Art as Experience*, p. 8.
17. Ibid., 1934, p. 7.
18. Cf. Skowroński, *Values and Powers*, p. 131.
19. Richard Shusterman, *Estetyka pragmatyczna. Żywe piękno i refleksja nad sztuką* (Wrocław: Wydawnictwo Uniwersytetu Wrocławskiego), p. 219.
20. Dewey, *Art as Experience*, p. 191.
21. Ibid., p. 162.
22. Ibid., p. 180.

Five

THE DEWEYAN AESTHETIC OF CHARLES IVES

Richard Anthony Spurgeon Hall

Now I will do nothing but listen,

To accrue what I hear into this song, to let sounds contribute toward it.

I hear bravuras of birds, bustle of growing wheat, gossip of flames, clack of sticks cooking my meals, . . .

I hear all sounds running together, combined, fused or following,

Sounds of the city and sounds out of the city, sounds of the day and night, . . .

—Walt Whitman

Philosophers can influence art, as in the cases of Aquinas' influence on Dante, Locke's on Wordsworth, Plato's on Joshua Reynolds, and Schopenhauer's on Wagner. But what of John Dewey's influence on art in the twentieth century? His philosophy of art as found in the monumental *Art as Experience* does not lack for philosophical disciples such as, most recently, John J. McDermott, Richard Shusterman, and Thomas Alexander who have developed a pragmatic theory of aesthetics. But what of the artists who have taken their cue from it? I know of none offhand, though there may have been those who were inspired by Dewey's aesthetics, but only indirectly or without acknowledgment. However, there are several American artists who, though apparently uninfluenced by Dewey, may be said to work out of a Deweyan aesthetic. Among them are the architects Louis Sullivan and Frank Lloyd Wright, the painters Jackson Pollock and Andy Warhol, the contemporary installation artists Christoand Jeanne-Claude, and the composers Charles Ives and John Cage (Indeed, looking at the art of any one of them as refracted through the prism of Dewey's philosophy of art would, I think, make an interesting study.) Here, though, I shall focus exclusively on Charles Ives: "There is a great Man living in this Country—a composer. / He has solved the problem how to preserve one's self and to / learn / He responds to negligence by contempt. / He is not forced to accept praise or blame. / His name is Ives"[1]

—thus did Arnold Schoenberg aptly epitomize the soul of the man. Charles Edward Ives (1874-1954) was the son of George Ives, a bandmaster in the Union Army during the Civil War. He was his son's first music teacher and, Charles reckoned, his best. In his youth he earned money as a church organist. Ives matriculated at Yale where he studied music theory and composition with Horatio Parker, the then German-trained doyen of American composers, but chafed under the restrictions of the conventional forms of musical composition. As a musical innovator far ahead of his time, and not wishing to compromise his standards and aims, he knew that he could not earn a living from composition. Consequently, he became a senior partner in the insurance firm of Ives & Myrick, where he had a very successful career, and wrote music in the evenings and on weekends. It was not until after his death that his music began to receive the recognition and the performances it deserved. Much of Ives' music is program music inspired as it was by extra-musical ideas. He saw himself as a spiritual heir of Emerson, whose literary art he emulated in his *Essays Before A Sonata* and whom he commemorated musically in the first movement of his Second ("The Concord") Piano Sonata. He was a Transcendentalist at heart with music as his métier instead of poetry and prose. Emerson's injunction, "Whoso would be a man, must be a nonconformist,"[2] might serve as an apt motto for Ives' art of musical composition.

Why Ives, though, instead of, say, Sullivan or Cage, as an exemplar of a distinctively Deweyan aesthetic? For one thing, Ives and Dewey are comparable in stature and reputation: Dewey is one of America's greatest philosophers, and Ives is arguably America's greatest composer, and certainly her most original and influential. For another, they were contemporaries: Dewey was born in 1859, Ives in 1874, and both lived into the fifth decade of the twentieth century; consequently, they lived through, and responded to, a formative period of American history. Further, (and this, I think is particularly significant) both were profoundly influenced by Emerson. And finally, Ives and Dewey were endowed with a strong social conscience and actively sought to bring about a more democratic polity; both were Progressives. What Jan Swafford writes of Ives, that "in his music and his life he embodied a genuine pluralism, a wholeness beneath diversity, that in itself is a beacon for democracy and its art,"[3] could as well be written of Dewey.

My thesis is that Ives' philosophy of music is fundamentally Deweyan in character, and his music may be fruitfully understood as a practical enactment, as it were, of Dewey's aesthetics. What makes Ives' philosophy of music "Deweyan" is its radical and pervasive anti-dualism. Ives sought to collapse (1) the dualism of art and life as represented by the sequestration of pictures and sculpture in museums and of music in concert halls—he believed that music should be integral to life as lived; (2) the dualism of so-called aesthetic experience and ordinary experience—Ives insisted on their continuity; (3) the dualism of popular (vernacular) music and classical (cultivated)—he integrated them. In what follows, I shall juxtapose the anti-dualistic elements in Ives' musical thought with similar elements in Dewey's philosophy of art.

A prescient soothsayer would perhaps have found a harbinger in John Dewey's birth in 1859, the year of publication of Darwin's *On the Origin of Species.* Dewey was destined to become a major pragmatist, and evolutionary theory is to pragmatism much as Newtonian mechanics is to early British empiricism—indeed, it would perhaps not be far Darwin fetched to characterize pragmatism as so many footnotes to Darwin, and this is specifically true of Dewey's pragmatic philosophy of art. Dewey's anti-dualistic philosophy of art is rooted in his endeavor to explain art—its nature, genesis, appreciation, and value—in terms of Darwin's theory of evolution. A Darwinian idea informing Dewey's aesthetics is that all living things form a seamless web of life insofar as the evolutionary process goes on at every level of organic nature and so ensures the continuities within it. This idea of the continuity of life informs Dewey's affirmation of the continuities between aesthetic and ordinary experience and between art and life. Another evolutionary idea governing Dewey's philosophy of art is the importance of the environment in determining an organism's state, and the dependence of its survival on its flexible adaptation to ever-changing environmental conditions. I shall begin with the idea of the continuity among living forms and its impact on Dewey's aesthetics.

The fine arts (to be vital) must be fully embosomed in the lives of individuals and their communities. To put them at arm's length from our daily lives as actually lived by safely consigning them to museums or confining them to concert halls and theaters is to enervate, kill and embalm them, depriving them of their living force. Museums are little more than mausoleums for artworks. Swafford describes as follows the situation of the fine arts in Dewey's America:

> The American sense of the arts, tied to the Gilded Age lifestyle, remained conservative. The "high" arts were preserved as in a museum, behind glass, supported and attended mainly by women and the rich. High art was a matter of European paintings, European statues, European singers, European music, all imported to this country with great expense and ballyhoo.[4]

Works of art should not be as lights hidden away in baskets and thus made objects for occasional disinterested contemplation, but should be as leaven that enables our personal and social lives to rise to a flourishing state; to do this they must be allowed to operate in the total human environment—"life goes on in an environment; not merely *in* it but because of it, through interaction with it,"[5] and this applies no less to the life of the arts than to that of organisms.

Dewey attributes this lack of organic connection between fine art and culture to the precipitous rise in modern times of science and technology. This has led to overspecialization and the segmentation of human life, so that art has now become a specialized activity in its own right cultivated and appreciated by the few and sequestered from the rest of life. Industrial mass production has widened the gap between the so-called fine arts and the useful arts that first opened in ancient Greece. Artifacts that were once hand-crafted and integrated into the daily life of craftsmen bore the stamp of their individual aesthetic sensibilities. But now they are mass produced by machines and severed from the daily lives of their manufacturers and their aesthetic impulses. Furthermore, science has disclosed an alien cosmos that is deeply impersonal and apparently indifferent to human aspirations, hopes and values, a view contrary to that held by traditional religion and metaphysics—"the things of the physical world and those of the moral realm have fallen apart, while the Greek tradition and that of the medieval age held them in intimate union."[6] In this diremption of the physical and the moral worlds Dewey finds the ultimate source of all the dualisms bedeviling philosophy since the early modern period.

Art's remoteness from daily social living—as much a fact regrettably in our day as in Dewey's—Dewey interprets as a symptom of imperfection. "In an imperfect society," he observes, "fine art will be to some extent an escape from, or an adventitious decoration of, the main activities of living." By contrast, art's centrality to daily living is not only the sign of a flourishing society but also a *means* thereto: "Works of art that are not remote from common life, that are widely enjoyed in a community, are signs of a unified collective life. . . . In the degree to which art exercises its office, it is also a remaking of the experience of the community in the direction of greater order and unity."[7]

However, Dewey thinks that the reintegration of art into the larger culture requires radical changes in the socio-economic arrangements of society thereby creating an environment hospitable to aesthetic experience: "The values that lead to production and intelligent enjoyment of art have to be incorporated into the system of social relationships. . . . art itself is not secure under modern conditions until the mass of men and women who do the useful work of the world have the opportunity to be free in conducting the processes of production and are richly endowed in capacity for enjoying the fruits of collective work." We might begin, according to Dewey, with the practical task of eliminating the profit-motive involved in land transactions in order to regenerate the art of architecture and, more indirectly, all the other arts. Why is the architecture of our large cities so unworthy of a fine civilization?" Dewey asks. He answers, "Their character is determined by an economic system in which land is used—and kept out of use—for the sake of gain, because of profit derived from rental and sale. Until land is freed from this economic burden, . . . there is little hope for the rise of general architectural construction worthy of a noble civilization . . . . the social forces that affect the buildings in which we subsist and wherein we do our work operate upon all the arts." Moreover, the modern scientific world view need not be put art at arm's length from common experience. The disclosure by biological science of humanity's intimate relation to nature may benefit art in the long run: "For the closer man is brought to the physical world, the clearer it becomes that his impulsions and ideas are enacted by nature within him. Humanity in its vital operations has always acted upon this principle. Science gives this action intellectual support."[8] Dewey's mission in *Art as Experience*, then, is "recovering an organic place for art in civilization" as it once enjoyed, for example, in the ancient world.

No less strenuously than Dewey, Ives stipulated that art be organically united with life, particularly the collective life of a community; but if not, then life would be diminished and art itself would atrophy and die: "An interest in any art-activity from poetry to baseball is better, broadly speaking, if held as a part of life, or *of* a life, than if it sets itself up as a whole—a condition verging, perhaps, toward a monopoly or, possibly, a kind of atrophy of the other important values, and hence reacting unfavorably upon itself."[9] Moreover, Ives practiced what he preached. By day he was an insurance executive, composing music in evenings and on weekends and holidays. (In this respect, interestingly, he resembles T. S. Eliot, William Carlos Williams and Wallace Stevens, three American poets for whom poetry was an avocation.) Music, then, for Ives, was not an ivory tower activity pursued for its own sake and apart from and unrelated to the other activities of life. "The fabric of existence weaves itself whole," affirms Ives echoing Emerson. "You can not set art off in the corner and hope for it to have vitality, reality and substance. There can be nothing '*exclusive*' about a substantial art. It comes directly out of the heart of experience of life and thinking about life and living life. My work in music helped my business and my work in business helped my music."[10] In this context he comments on the music of a fellow composer and contemporary: "Debussy's content would have been worthier his manner if he had hoed corn or sold newspapers for a living, for in this way he might have gained a deeper vitality and a truer theme to sing at night and of a Sunday."[11]

Like Dewey, Ives believed that society might be regenerated through art. He foresaw the time when art would revivify the common experience of everyman, when it "shall beat around and through his ordinary work—shall strengthen, widen, and deepen all his senses, aspirations, or whatever the innate power and impulses may be called." Art is the birthright of all people. As Emerson exhorted all men and women to be their own Jesus, Ives exhorted them to become their own composers: "The instinctive and progressive interest of every man in art, . . . , will go on and on, ever fulfilling hopes, ever building new ones, ever opening new horizons, until the day will come when every man while digging his potatoes will breathe his own . . . symphonies."[12] In an afflatus of democratic zeal, he went so far as to speculate that even the art of Beethoven "is as nothing in comparison with the future product of some coal-miner's soul in the forty-first century."[13] "Every normal man," explains Ives, "has, in some degree, creative insights . . . and an interest, desire and ability to express it," and these are the "source of strength and supreme joy," and "a value that will supplement the other values and help round out the substance of the soul."[14] Moreover, the "true art" of the future will not be the self-expression of some paltry individual, but "the direct expression of the mind and soul of the majority"[15] of people.

Ives' mission to integrate his art with life and his urge to express in music real-life experiences indicate that he, as Wiley H. Hitchcock remarks, "believed in music as a re-creation in sound of life itself."[16] This judgment is seconded by Henry Cowell who writes, "His music is marked by an unprecedented drive and breadth and inclusiveness, an overwhelming sense of the fullness of life." And on his own admission, Ives said of his music, "It comes directly out of the heart of experience of life and thinking about life and living life."[17] As such, Ives music is programmatic; that is, it is about something extra-musical—albeit a physical scene, an event, an idea, an emotion, or a state of mind—to which it seeks to give musical expression. However, it is not programmatic in the way that much program music is, say, for example, the symphonies of Berlioz or the tone-poems of Franz Liszt or Richard Strauss. What Ives was trying to capture is the ebb and flow, the dynamic pulse, the very life of experience, that is, of experience in the making—"not something that happens, but the way something happens,"[18] as Ives puts it. Here is Cowell's description of a typical Ivesean program:

> So he pondered the relations of things, testing out music by life and life by music, and building abstract musical structures like concrete events. This makes his particular kind of program music, in which the flow of musical relationships derives from the pattern of activity he saw around him. The music therefore records not a *thing* that happens but the *way* things happen. Because events don't move by singly, but carry memories and forecasts with them, colliding and conflicting with other events too, Ives' music moves in many directions at once and is built on many levels, in the way that experience comes to the mind.[19]

Underlying Dewey's insistence that art be integrated with life and made central to our social lives is his understanding that all experience is implicitly aesthetic. For him, the experiences we discriminate as "aesthetic" are not something distinct or separate from the rest of our experience, but represent aspects of our ordinary experience abstracted, intensified, and consummated. Dewey's express aim in *Art as Experience* is, in his words, "to restore continuity between the refined and enhanced forms of experience that are works of art and the every day events, doings, and sufferings that are universally recognized to constitute experience." Art, as he variously describes it, "develops," "accentuates," "idealizes" qualities found in common experience—it is "the complete culmination of nature."[20]

The continuity between aesthetic and ordinary experience can be shown by considering what Dewey identifies as "*an* experience." Now for Dewey any experience is germinal aesthetic if it involves fulfillment through struggle—"Because experience is the fulfillment of an organism in its struggles and achievements in a world of things, it is art in germ." However, *an* experience stands out from the stream of common experience when its specifically aesthetic qualities stand out and arrest our attention. *An* experience, in Dewey's description, is aesthetic insofar as "its varied parts are linked to one another, and do not merely succeed one another. And the parts through their experienced linkage move toward a consummation and close, not merely to cessation in time." Dewey further explains, "An experience has a unity that gives it its name, *that* meal, that storm, that rupture of friendship. The existence of this unity is constituted by a single *quality* that pervades the entire experience in spite of the variation of its constituent parts."[21]

For striking examples of what Dewey meant by "*an* experience" we need look no further than Ives. Ives was particularly prone to having such experiences which he would describe and subsequently express in music. Here is his description of a boy's awakening to the early sounds of a holiday morning, a reminiscence undoubtedly of his own childhood experience:

> In the early morning of a Memorial Day, a boy is awakened by martial music—a village band is marching down the street . . . he seems of a sudden translated—a moment of vivid power comes, a consciousness of material nobility—an exultant something gleaming with the possibilities of this life—an assurance that nothing is impossible, and that the whole world lies at his feet. But, as the band turns the corner, at the soldier's monument, the boy's vision slowly vanishes—his "world" becomes less and less probable—but the experience ever lies within him in its reality.[22]

This experience has all the hallmarks of *an* experience—it is demarcated, unified, complete and consummatory—thereby qualifying it as eminently aesthetic. Ives evokes that boyhood experience in the second movement of his *Holidays Symphony* of 1913, subtitled, "Recollections of a boy's holidays in a Connecticut country town." Like Dewey, Ives clearly discerned the aesthetic potential of common experience, seeing within it a musical program. Indeed, Ives had a gift for feeling immediately and vividly the continuity between common experience and art (and the intimacy between a program and its musical expression) and memorializing it in music.

A result of Dewey's breaking down the arbitrary barriers erected between art and other forms of life is the enfranchisement of all materials as potential art objects and of any human experience as potentially aesthetic. Dewey insists that there is no object too lowly or unsuitable for transmutation into art, and no bit of experience lacking latent aesthetic qualities. "Impulsion beyond all limits that are externally set inheres in the very nature of the artist's work. It belongs to the very character of the creative mind to reach out and seize any material that stirs is so that ths [*sic*] value of that material may be pressed out and become the matter of a new experience." The only limitation as to use of materials is the artist's vital interest; he must be governed by the "necessity of sincerity" which stipulates that "he shall not fake and compromise." Dewey attributes this modern trend of broadening the domain of materials deemed suitable for artistic expression to those mental conditions—namely, sharpened curiosity and close observation—which gave rise to the scientific revolution of the seventeen century: "the extension of painting and the other arts to include matter that was once regarded as either too common or too out of the way to deserve artistic recognition is a permanent gain. This extension is not directly the effect of the rise of science. But it is a product of the same conditions that led to the revolution in scientific procedure."[23]

Ives too breaks down artificial barriers, in his case the barrier between what is variously called "classical" or "art" music of the cultivated tradition and so-called popular music of the vernacular tradition, thereby widening the domain of what is considered appropriate for expression in music of the cultivated tradition. Cowell writes, "He was the first composer in the United States to commit himself unreservedly to the vernacular for the grammar of a new symphonic speech."[24] A salient and distinctive feature of Ives' music is his weaving of folksongs, hymn tunes, and marches into the musical fabric of his classically conceived compositions. "If a man finds that the cadences of an Apache war-dance come nearest to his soul," advises Ives, "let him assimilate whatever he finds highest of the Indian ideal so that he can use it . . . , in his symphonies." And in the Gospel hymns that he fondly remembered hearing at revival meetings as a child in Danbury, he found "a vigor, a depth of feeling, a natural-soil rhythm, a sincerity." From this he generalizes to "whatever excellence an artist sees in life, a community, in a people, or in any valuable object or experience, if sincerely and intuitively reflected in his work—his work, . . . is, in a way, a reflected part of that excellence."[25] Hitchcock detects an Emersonian note in Ives' inveterate borrowing from vernacular music in composing his own. "Such borrowings," he remarks, "were a transcendentalist's acceptance of the validity, even divinity, of all the things under God that he had known, felt strongly, and believed to have vitality and 'substance.'"[26]

Ives' aesthetic inclusiveness went beyond simply marrying vernacular music with the classical forms of the cultivated tradition. Ives enfranchised all sounds, man-made and natural, as potential material for music; Ives made no distinction between "musical" and nonmusical sounds which might interact so as to form an aural ecosystem of sorts. An example of this is his own description of a walk he would take through New York's Central Park at night and the music it inspired, *Central Park in the Dark* (1906):

> This piece purports to be a picture in sounds of the sounds of nature and of happenings that men would hear some thirty or so years ago . . . when sitting on a bench in Central Park on a hot summer night. The strings represent the night sounds and silent darkness-interrupted by sounds from the Casino over the pond—of street singers coming up from the Circle singing . . . the tunes of those days—of some "night owls" from Healy's whistling the latest or the Freshman March—the "occasional elevated," a street parade or a "breakdown" in the distance—of newsboys crying "uxtries"—of pianolas having a ragtime war in the apartment house "over the garden wall," a street band and a street car join in the chorus—a fire engine, a cab horse runs away, lands, "over the fence and out," the wayfarers shout—again the darkness is heard—an echo over the pond—and we walk home.[27]

Needless to say Ives' nocturnal experience is unmistakably *an* experience in Dewey's sense—he heard the aesthetic quality in the cacophony of that night and made of it music. Dewey also did not fail to miss the aesthetic qualities of such mundane things as "the sights that hold the crowd—the fire-engine rushing by; the machines excavating enormous holes in the earth; the human-fly climbing the steeple-side; the men perched high in air on girders, throwing and catching red-hot bolts."[28] For Ives, the sounds of nature in and of themselves may be experienced as music as well as being incorporated in a musical composition. Ives cites the example of Thoreau who records in *Walden* his experience one evening of hearing the distant peals of a bell mingling with the sound of wind among trees. In the spoken introduction to his song, *Thoreau* (1915), Ives quotes Thoreau's own description of that occasion:

> I heard the bells, . . . ,[a] natural melody, worth importing into the wilderness. At a sufficient distance over the woods this sound acquires a certain vibratory hum, as if the pine needles in the horizon were the strings of a harp which it swept . . . . , a vibration of the universal lyre, just as the intervening atmosphere makes a distant ridge of earth interesting to our eyes by the azure tint it imparts to it.[29]

Thoreau's aural experience is yet another paradigmatic example of *an* experience as described by Dewey. Such was Thoreau's attunement to nature's own music that Ives thought him a natural musician who did not have to go to Boston to hear music. Anyone, remarks Ives, might develop this gift of hearing so that "he will look up over the mountains and see his visions in their reality, will hear the transcendental strains of the day's symphony resounding in their many choirs, and in all their perfection, through the west wind and the tree tops!"[30] Ives would have agreed with Dewey's remark that "the [a]esthetic is no intruder in experience from without, whether by way of idle luxury or transcendent ideality, but that it is the clarified and intensified development of traits that belong to every normally complete experience."[31]

A corollary of Dewey's conception of art as the clarification and intensification of common experience is that in order to accomplish that end, together with enfranchising more and more experience for expression, an artist needs to be a fearless experimenter:

> Yet one of the essential traits of the artist is that he is born an experimenter. Without this trait he becomes a poor or a good academician. The artist is compelled to be an experimenter because he has to express an intensely individualized experience through means and materials that belong to the common and public world. . . . Only because the artist operates experimentally does he open new fields of experience and disclose new aspects and qualities in familiar scenes and objects.

"Experimentalism in the arts with respect to using novel materials, forms and techniques is not just an incidental matter of style; it is a manifestation, no less, of a valuable evolutionary trait insofar as it represents a new mode of "the interaction of the live creature with his surroundings"[32] which improves its chances of adapting and flourishing.

Ives was a born experimenter if ever there were one. Like Berlioz, he was one of music history's inspired innovators. He experimented with polytonality, microtonality, atonality, dissonance, and polyrhythms years before others like Schoenberg and Stravinsky would do so thus making them staple devices of twentieth-century music. One area where Ives experimented is musical form. Anticipating the charge that his music was formless and confused, Ives had a ready answer. "An apparent confusion if lived with long enough, may become orderly," he advises, and "what is unified form to the author, or composer, may of necessity be formless to his audience." He observed that traditional musical forms were based upon the principle of repetition, which helped the listener follow them. Repetition in music helps give it unity, a traditionally prized constituent of aesthetic form, but which over time may become stale and lull the listener into passivity. "Initial coherence to-day may be dullness to-morrow," admonishes Ives, "probably because formal or outward unity depends so much on repetition."[33] Moreover, the traditional canons of aesthetic unity may be merely conventional, not natural, and should not be identified with form itself. "And unity is, too generally conceived of, or too easily accepted as analogous to form; and form is analogous to custom; and custom to habit," explains Ives. "Perhaps all unity in art, at its inception, is half-natural and half-artificial, but time insists, or at least make us, or inclines to make us feel that it is all natural."[34] Ives sought to break our customary habits of listening.

Rather than to the frozen compositional forms of the past, Ives looked to nature for models of musical form: "Ives concerns himself first of all with the forms of nature and their mysterious and complex behavior, rather than with any mathematically symmetrical and balanced repetition and permutation; these he believes may lead to distortion."[35] More specifically, he proposed that musical forms might instead be constructed on the principle of analogy which, he believes, is nature's *modus operandi.* Echoing Emerson, Ives affirms, "Nature loves analogy and hates repetition."[36] More particularly, says Cowell, Ives "relates his music by analogy to the individual's experience of life. The sense of unity is not brought about through exact repetition, either of motifs or of sections, but is established though relationships." Thus, when Ives "wants to convey an emotion about something, he reproduces the behavior of the sounds that are associated with it, their approach and departure, their pace and drive, interweaving and crossing—all this by analogy, which seems to be the way he approaches reality, rather than by description or literal imitation." This is well illustrated in the second movement of his *Holidays Symphony* where he seeks to evoke a boy's feeling of "an exultant something gleaming with the possibilities of this life." The boy's experience is elicited by the sounds of a village band passing outside. To evoke this experience in the listener Ives mimics in the symphony the sounds of a marching band approaching and then receding in the distance. However, "Ives prefers that these relationships should not be too obvious. There must always be something for the mind and feeling to work on, some new aspect of relationship to be found. If everything is self-evident, that spiritual inactivity that Ives so abhors might be induced."[37] Ives justifies the resultant calculated vagueness of his music by again appealing to Emerson: "She [nature] dislikes to explain as much as to repeat"— because "vagueness, is at times, an indication of nearness to a perfect truth."[38] Ives "feels that music, like other truths, should never be immediately understood; there must always remain some further element yet to be disclosed. A complete musical statement, in all its clarity and simplicity, like any absolute truth, is an ultimate, not a beginning. Ives reserves it, therefore, for the culmination of a work."[39]

Yet sometimes an Ives piece does not culminate in a complete musical statement, but remains vague and open-ended;

> His open-endedness suggests that the music cannot yet be completed, that more is to be said. In his setting of the revival hymn "Shall We Gather at the River?" Ives ends with those words

> on an open harmony, leaving the question to resonate in our minds: *Shall* we gather at the river? He invites us to complete the question, to seek further, to listen for the yet-unheard.

"(One is put in mind of Keats' lines, "Heard melodies are sweet, but those unheard / Are sweeter.") Swafford neatly sums up the intentional vagueness, open-endedness and incompleteness of Ives' musical forms as follows:

> In his greatest music Ives created, for the first time in Western history, a masterful art of suggestion, of ambiguity, of incompleteness. He created forms open, in process, incomplete and incompletable in the world he lived in. Rather than a circle, a closed system, he built arcs that imply the possibility of completion.[40]

Another aspect of Ives' experimentation with musical form is to synthesize into a higher unity contrasting themes, as in a sonata. This reflects Ives' conception of reality in terms of an ongoing Hegelian dialectic where the dualistic opposition of ideas—actually different facets of the same thing—is overcome in approximation to the truth. Cowell puts it this way:

> He finds one principle of unity in the sonata, because it deals with the resolution of two contrasting themes. The two-sidedness of reality, all the paradoxes of existence, are present to all Ives's thinking. . . . he believes that full expression of the opposing aspects of any idea whatever is a necessary step on the way to perfect truth, since only in this way can the common basis for the integration of these opposites be found. Their contrasting facets will ultimately, he believes, be seen as different aspects of the same thing. Ives's interest is in this *process toward* integration; with Emerson he abhors the spiritual inactivity that comes from the conviction that one possesses the truth in its final form. He envisages a series of integrations of dualities, each of which as it is achieved is seen as a sort of partial or temporary truth, a truth which then becomes only one aspect of another set of opposites which sooner or later must be resolved in its turn. This struggle toward truth and integration is the nearest man can come to absolute truth, in Ives's view; but he feels the very effort required imparts a certain unity and coherence of its own.[41]

Interestingly, this sense of incompleteness or open-endedness, of possibilities yet to be fulfilled, hovering over Ives' music comports well with the theory of truth espoused by Dewey and his fellow pragmatists. According to them, the truth about things can never be complete, final and absolute. The totality of truth can be approached only asymptotically. Truth is not fixed but fluid, a process ever evolving in response to fresh experience.

As Dewey did in his philosophy, Ives assimilates in his musical thought the theory of evolution by natural selection—"Darwin told but another of God's parables,"[42] he remarks. In justifying the necessity of new forms and procedures in music, Ives used as precedent the ongoing evolutionary changes in nature. "Botany," he writes, "reveals evolution not permanence." Like Emerson, Ives believed that the closer art was to nature, the more the creative processes of the artist resemble those of nature, the better. "Natural selection is but one of Nature's tunes," he wrote. Paraphrasing Emerson he continues, "All melodious poets shall be as hoarse as street ballads, when once the penetrating keynote of nature and spirit is sounded—the earth-beat, sea-beat, heart-beat, which make the tune to which the sun rolls, and the globule of blood and the sap of trees."[43]

Furthermore, Ives mimics, as it were, the evolutionary process itself in developing his musical themes. Rather than boldly announcing at the start the themes of a piece and then developing them, as was done in the traditional sonata form, he leaves the listener in suspense at the beginning as to what they are. Scraps of themes are introduced and gradually take a recognizable and definitive shape; it is only at the end of the movement that they emerge fully-formed. His music thus exhibits what J. Peter Burkholder calls "cumulative form" which Swafford describes as follows:

> A movement or an entire piece grows from the motivic (and sometimes phrasing and harmonic) elements of a single melody. This melody gradually makes its way from background to foreground until it is proclaimed straightforwardly at the end. There is a symbolic dimension to cumulative form too, in the idea of an underlying generative force slowly making its way to the surface.[44]

Ives' use of cumulative form is particularly evident in his violin sonatas, as well as in the finales of his third and fourth symphonies. Thus, in the third movement of the fourth sonata for violin, "The theme is not *given*; it is *sought* and eventually makes its full appearance."[45] And here is Ives' own comment on this formal procedure as found in a program note for his third violin sonata): "The last movement is an experiment: The Free Fantasia is first. The working-out develops into the themes, rather than from them. The Coda consists of the themes for the first time in their entirety."[46] As described above, Ives' cumulative form exemplifies a movement towards that consummatory end Dewey identifies as constitutive of aesthetic experience. Moreover, there is yet another echo of Dewey in Ives' idea of cumulative form. What we have here, it seems to me, is the makings of an evolutionary philosophy of music parallel to Dewey's evolutionary philosophy of art in general.

But beyond being the mode of his musical thought and a model for his musical forms, the evolutionary idea played a role in Ives' creative process and his stipulation of how his compositions should be performed. For example, he never quite finished composing his Concord Sonata for piano, tinkering with it well after it was scored. Moreover, he encouraged pianists to improvise while playing it, and envisioned their successive performances as contributing to a continually evolving tradition of performance over time: "In fact, these notes, marks and near pictures of sounds etc. are in a kind of way a platform for the player to make his own speeches on," instructs Ives, and to "change and grow, rise to this mountain, then to that, as the years go on through time to the Eternities."[47] More particularly, in the "Emerson" movement of that sonata, "Ives created a work conceived as evolving through time and feeling and performers and performances."[48] Ives even thought of the history of music in evolutionary terms where progress was inevitable. "In the history of this youthful world the best product that human-beings can boast of is probably, Beethoven—but, maybe," Ives foresees, "even his art is as nothing in comparison with the future product of some coal-miner's soul in the forty-first century."[49]

As mentioned earlier, another key evolutionary idea informing Dewey's aesthetics is that an organism's total environment largely determines what it is, and its adaptation to that environment is necessary to its health, survival and flourishing. "Life goes on in an environment," says Dewey, "not merely *in* it but because of it, through interaction with it." Since an organism and its environment undergo constant change, to live and flourish it must continually adapt and readapt itself to its environment. The life of an organism, then, consists of periods of its being ill adapted to its environment rhythmically alternating with periods of its being well adapted. A period of ill adaptation, moreover, gives the organism an opportunity of achieving a re-adaptation superior to the one it attained previously where it now further expands into and flourishes in its environment and so attains "a more extensive balance of the energies of the organism with those of the conditions under which it lives." This is growth. Now these vital processes form the basis of aesthetic experience. From our optimal adaptation to our environment comes our sense and appreciation of balance and harmony achieved through rhythm; from the stable but volatile balance or equilibrium between our energies and between them and our environmental conditions emerges our sense of order or form. For an organism to thrive in an orderly environment, particularly after a period of disruption and conflict, is to invite its having a consummatory experience bordering on the aesthetic. The organic conditions for aesthetic experience is the alternation of change and rest—a period of disruption, conflict, crisis and suspense followed by one of resolution and rest. In interacting with our environment, our "energy gathers, is released, dammed up, frustrated and victorious." The experience of passing "from disturbance into harmony is that of intensest life." Art is implicit in organic processes: "A bird builds its nest and a beaver its dam when internal organic pressures cooperate with external materials so that the former are fulfilled and the latter are transformed in a satisfying culmination."[50] Art proves that we use the stuff of nature to expand and enrich our life, in conformity with our organic structure.

Ives too seems to assimilate something like Dewey's "environmentalism" into his art. He thought that music was inseparable from the space in which it is performed—music, like life itself, was umbilical to the place of its performance. Ives experimented with making music a function of space as well as time; that is, he sought to orient music in space, with its spatial orientation and direction becoming as much a part of its overall structure as its temporal sequence. Thus, in his *The Unanswered Question* Ives places three groups of instrumentalists in different parts of the hall. The spatial orientation of the performers affects the perspective—based upon direction and distance—from which the audience hears them. The music is heard omni-directionally as coming from one direction or another and as sounding near or from afar, with the directionality and distance being integral to the experience of the music. For the performance of what would have been his "Universe Symphony," had it come to fruition, he abandoned the concert hall entirely by envisioning different orchestras and choirs playing and singing outside from hillsides, mountain tops and valleys. Furthermore, the positioning of musicians in various parts of the hall had an effect on the sound of the music:

> In the Fourth Symphony, . . . , the matter of relative intensities requires, to Ives's mind, that members of the various instrumental sections be placed on different parts of the stage, and even in the auditorium, to control the distance the sound travels from the sounding body to the listening ear. This is for the sake of variety in intensity, since the effect of tones played loudly at a distance is very different from the same tones played softly close at hand.[51]

Complementing his experimentation on the composer's side was his experimentation on the listener's side, with the way music is heard. It occurred to Ives, while vacationing in the Adirondacks, that he might compose a piece of music that one would listen to analogously to the way one would look at a landscape. For want of a better phrase it might be called "bifocal" listening. Here is Ives' own description of a personal experiment he performed in this new way of listening:

> When we were in the Keene Valley, . . .I started something that I had had in mind for some time: trying out a parallel way of listening to music suggested by looking at a view. First, with the eyes toward the sky or tops of the trees, taking the earth or foreground subjectively (that is, not focussing the eye on it), and then. . .Second, looking at

> the earth and land and seeing the sky and the top of the foreground subjectively. In other words, giving a musical piece in two parts, but both played at the same time . . . the whole played through twice, first when the listener focusses his ears on the lower or Earth music, and the next time on the upper, or Heavens music.

Ives then goes on to describe in some detail how his experience of the landscape might be transmuted into music:

> The earth part is represented by lines, starting at different points and different intervals. A kind of uneven and interlapping counterpoint, sometimes reaching nine or ten different lines representing the ledges, rocks, woods and land formations, lines of trees and forest, meadows, roads, rivers, and undulating lines of mountains in the distance, that you catch in a wide landscape. . . The part of the orchestra representing the Heavens has its own chord system, but its counterpoint is chordal.[52]

I think this passage is particularly interesting insofar as it describes how *an* experience Ives had in Keene Valley, having its own aesthetic integrity, determined the structure of a musical composition. It well illustrates the continuity between the aesthetic quality of an ordinary experience and that of a musical work of art. It also illustrates, significantly, the continuities (or analogies) between the different modes of aesthetic experience—in this case, between the visual experience of contemplating a landscape (or a picture) and the aural experience of listening to music. Ives saw no end to what experimentalism and innovation in music might produce. Music, he envisioned, "will develop possibilities unconceivable now,—a language, so transcendent, that its heights and depths will be common to all mankind."[53]

What accounts for Ives' Deweyan philosophy of music? Certainly Dewey did not influence Ives because all of Ives' major works were written between 1898 and 1918, well before the publication of *Art as Experience*. And it is unlikely that Dewey had even heard of Ives since most of Ives' music did not receive public performances until after their deaths. Moreover, I have found no references to Dewey in Ives' published writings or to Ives in Dewey's. Ives' highly original art predated Dewey's philosophy of art.

As Cowell remarks, "a system of esthetics is born after the event, derived by students from the new worlds made by artists." It is as if Dewey had derived his own aesthetic ideas from contemplating the new musical world opened up by Ives. Perhaps there was something in the cultural air both breathed (the Zeitgeist) that explains their aesthetic kinship—after all, there were others (mentioned above) with similar conceptions of art. My own hunch (and it is only that) is that it may be explained in part by their common devotion to Emerson. Whatever the answer, though, Dewey's philosophy and

Ives' music are equally expressions of the American spirit of experimentalism and exploration—they are both native to the American experience. I shall leave the last word with Cowell:

> To experiment and to explore has never been revolutionary for an American; he is unaffectedly at home in the unregulated and the untried. In a vast new country experience is direct, intense and various, and so grass-roots creative activity in the United States has been marked by an exuberance and a diversity that are shocking to sensibilities developed in older cultures whose essence is refinement and selectivity. In all the arts Americans quite naturally bring together elements that elsewhere appear as irreconcilable canons of radically opposed schools of thought. Inherited traditions, with all their subtleties, are necessarily pushed aside when the time comes to reinvigorate art with a transfusion from more immediate experience."[54]

NOTES

1. Quoted by Henry and Sidney Cowell, *Charles Ives and His Music*, (New York: Oxford University Press, 1955), p.114 n.

2. Ralph Waldo Emerson, “Self-Reliance,” in *Ralph Waldo Emerson: Selected Essays, Lectures, and Poems*, ed. by Robert D. Richardson, Jr. (New York: Bantam Books, 1990), p. 151.

3. Jan Swafford, *Charles Ives: A Life with Music* (New York: W. W. Norton & Company, 1996), p. 434.

4. *Ibid.*, p. 147.

5. John Dewey, *Art as Experience* (New York: Minton, Balch & Company, 1934), p. 13.

6. *Ibid.*, p. 337.

7. *Ibid.*, pp. 80-81.

8. *Ibid.*, pp. 344, 339.

9. Charles E. Ives, *Essays Before a Sonata and Other Writings*, ed. by Howard Boatwright (London: Calder and Boyars, 1961), p. 124.

10. Henry Bellamann, “Charles Ives; the Man and his Music,” *Music Quarterly*, XIX (1933), 45-58, quoted by H. Wiley Hitchcock, *Music in the United States: A Historical Introduction*, Prentice-Hall History of Music Series (2nd ed.: Englewood Cliffs, New Jersey: Prentice-Hall, Inc., 1974), pp. 150-51.

11. Ives, *Essays*, ed. by Boatwright, p. 82.

12. *Ibid.*, pp. 128-29.

13. *Ibid.*, pp. 88-89.

14. *Ibid.*, p. 126.

15. *Ibid.*, p. 91.

16. H. Wiley Hitchcock, *Music in the United States: A Historical Introduction*, Prentice-Hall History of Music Series (2nd ed.: Englewood Cliffs, New Jersey: Prentice-Hall, Inc., 1974), p. 155.

17. Cowell, *Charles Ives*, p. 9.

18. Quoted in Swafford, *Charles Ives*, p. 299.

19. Cowell, *Charles Ives*, pp. 6-7.

20. Dewey, *Art as Experience* pp. 3, 11, 26.

21. *Ibid.*, pp. 19, 55, 37.

22. Ives, *Essays*, ed. by Boatwright, 1961), pp. 30-31.

23. Dewey, *Art as Experience,* pp. 189, 340.

24. Cowell, *Charles Ives*, p. 5.

25. Ives, *Essays*, ed. by Boatwright, pp. 79-80, 81.

26. Hitchcock, *Music in the United States*, p. 153.

27. From Charles E. Ives, “Postface” to *Central Park in the Dark*, quoted by Stuart Feder, *Charles Ives “My Father’s Song,” A Psychoanalytic Biography* (New Haven: Yale University Press, 1992) p. 195.

28. Dewey, *Art as Experience*, p. 5.

29. Henry David Thoreau, *Walden*, in *Works of Henry David*

Thoreau, ed. by Lily Owens (New York: Avenel Books, 1981), p. 136.

30. Ives, *Essays*, ed. by Boatwright, pp. 128-29.

31. Dewey, *Art as Experience*, p. 46.

32. *Ibid.*, pp. 144, 15.

33. Charles E. Ives, *Essays before a Sonata*, in *Three Classics in the Aesthetic of Music* (New York: Dover Publications, Inc., 1962), pp. 120, 121.

34. Ives, *Essays*, ed. by Boatwright, p. 98.

35. Cowell, *Charles Ives*, p. 142.

36. Ives, *Essays*, in *Aesthetic of Music*, p. 120.

37. Cowell, *Charles Ives*, pp. 173, 144, 173-74.

38. Ives, *Essays*, in *Aesthetic of Music*, p. 120.

39. Cowell, *Charles Ives*, p 142.

40. Swafford, *Charles Ives*, pp. 232, 365.

41. Cowell, *Charles Ives*, pp. 142-43.

42. Charles Ives, Memos, p. 215, quoted by Swafford, *Charles Ives*, p. 210.

43. Ives, *Essays*, in *Aesthetic of Music*, pp. 120, 124.

44. Swafford, *Charles Ives,* p. 157.

45. Stuart Feder, *Charles Ives "My Father's Song," A Psychoanalytic Biography* (New Haven: Yale University Press, 1992), p. 185.

46. Ives Papers 45/7, quoted by Swafford, *Charles Ives,* p 456, n. 59.

47. Ives, *Memos*, pp. 191-92, quoted by Swafford, *Charles Ives*, p. 300.

48. Swafford, *Charles Ives*, p. 300.

49. Ives, *Essays*, in *Aesthetic of Music*, p. 172.

50. Dewey, *Art as Experience*, pp, 13, 14, 16, 17, 24.

51. Cowell, *Charles Ives*, p. 21.

52. *Ibid.*, pp. 201-02.

53. Ives, *Essays*, p. 109.

54. Cowell, *Charles Ives*, pp. 9, 5-6.

Six

QUALITATIVE THOUGHT, THINKING THROUGH THE BODY, AND EMBODIED THINKING: DEWEY AND HIS SUCCESSORS

Sebastian Stankiewicz

At first, it seems that the philosophical problem of thinking – as an activity of the mind that embraces every act and process of consciousness, and consisting of the ability to create and employ concepts, and pass judgments – is not an aesthetic issue. Rather it is associated with epistemology, logic, or the theory of knowledge. Dewey's philosophy calls into question this traditional description of thinking focused only on conceptual and propositional structures. To clarify the problem of thinking and its connections with art and aesthetics, I would like to start with an instructive example from the art world.

"Think with the senses – feel with the mind." In this way American curator Robert Storr formulated the deeply provocative idea of the 52nd Venice Biennale in 2007. There was nothing new for artists and practitioners of the arts in the sentiment. Those who saw the exhibition could also hardly find anything that especially forced them to "think" with their sensory receptors and to "feel" in terms of conceptual and propositional structures. As a matter of fact, the art presented in the exhibition was not exceptionally different from the art presented in other contemporary art exhibitions shown at the same time in Europe and the USA. Rather the difference of this exhibit from others was in the idea, that, as Storr said:

> From Plato onwards philosophers have divided and compartmentalized human consciousness more or less explicitly pitting one faculty against another; mind versus body, thought versus feeling, intellect versus the senses, the conceptual versus the perceptual. At best such dichotomies have served to sharpen our understanding of the different capacities at our disposal of comprehending the world and making our place in it. At worst they have deprived us of some of those abilities by setting up false hierarchies that causes us to mistrust or disparage one for the sake of another."[1]

Storr elaborates that art and artists have never been guilty of establishing such dichotomies, because for artists it is quite obvious that they are unnecessary and do not contribute to the development or understanding of art. The problem addressed by Storr was not directed at artists and art practitioners, but rather against philosophies and philosophers, who still remain in the Kantian paradigm of how human faculties are conceived, which according to Storr, deprives us of some of our abilities.

The aim of this paper is to show that Dewey's pragmatist aesthetics gives us a plausible reconciliation of Storr's concern, by means of a significantly wider conception of thinking which appreciates all human abilities without "disparaging one for the sake of another." Dewey's reconciliation can also be a solution for the other aspect of Storr's query, the compartmentalization of human faculties. Using Dewey's philosophy we can bridge the gap between the practice of art and philosophy, where the latter is based prominently on the "strategy of false hierarchies" that in its traditional formulation obscures human potential. The tool used in this reconciliation is Dewey's conception of "qualitative thinking," which posits that on a primordial, naturalistic level there is a unity of the qualitative, affective, volitional and discursive. Thus, Dewey's theory of "qualitative thinking" alters the traditional philosophical discourse on the problem of thinking, and not only in the domain of aesthetics.

This examination is divided into three sections. The first will be devoted to the reconstruction of Dewey's conception of qualitative thinking; in the second section – proceeding from Dewey's theory – I will elucidate the idea of "thinking through the body," a concept developed in the work of Richard Shusterman, and the idea of "embodied thinking", expressed by Mark Johnson. The last section will be dedicated to showing some of the consequences of changing the "thinking concept" in the philosophy of art. More specifically, I argue that Dewey's notion of qualitative thinking may help solve the central riddle of his aesthetics: how is it possible that aesthetic experience, being an immediate unreflective experience, can be at the same time a self-contained and self-legitimizing experience—"*an* experience"— that gives us growth of meaning?

1. Modes of Thinking in Aesthetic Experience

For Dewey the problem of thinking is not limited to pure discursiveness, but also involves qualities, emotions, and needs, and every thought or inference is a product of their reciprocal cooperation that manifests itself as a complex and multi-layered unity. Dewey's thesis is not obvious if we consider our everyday or cognitive experiences, which are usually focused on reflective inquiries. The thesis becomes more credible if we consider aesthetic experiences, because in the domain of aesthetics the role of qualities, emotions, and needs are recognized as crucial.

In comparison with other fields of philosophy, aesthetics has always dealt with qualities and emotions. It is a natural consequence of the fact that our contact with art products seems to be pervaded by qualities and emotions. In other fields of philosophy and even in everyday human interaction, we are prone to diminish the qualitative and emotional aspects of experience, and value them less than the "rationality" of expressing opinions with the exclusive help of conceptual-propositional structures. That situation is quite natural when we realize that qualities and qualitatively felt emotions, feelings, and needs present themselves to us in their own existential idiom that embraces much more than is possible to differentiate by means of concepts. If we cannot grasp those qualities and make an isomorphic translation of them into verbal signs or words, we rather resign from them and push them aside to the underlying dimension. This is quite common with aesthetic experiences. Qualia, or felt qualities, are self-evident, hence it is easier to describe an aesthetic experience in terms of its structure than to give felt qualities adequate expression. What is Dewey's approach to the problem of qualia and the role they play in thinking?

In brief, Dewey's approach is a functional one, because he is quite aware of the fact that quality is "just that which it is and not anything else."[2] In *Art as Experience* and two very informative essays, "Qualitative Thought" and "Affective Thought," Dewey deals with the problem of qualities, emotions, needs, and their integral unity with the discursive mode of thinking. In the first essay, Dewey gives a crucial statement of how he understands thinking:

> Because perception of the relationship between what is done and what is undergone constitutes the work of intelligence, and because the artist is controlled in the process of his work by his grasp of the connection between what he has already done, and what he is to do next, the idea that the artist does not think as intently and penetratingly as scientific inquirer is absurd. A painter must consciously undergo the effect of his every brush stroke or he will not be aware of what he is doing and where his work is going. Moreover, he has to see each particular connection of doing and undergoing in relation to the whole that he desires to produce. To apprehend such relations is to think, and is one of the most exacting modes of thought. (...) Any idea that ignores the necessary role of intelligence in the production of works of art [it means: aesthetic experiences – note S.S.] is based upon the identification of thinking with the use of one special kind of material, verbal signs and words. To think effectively in terms of relations of qualities is as severe a demand upon thought as to think in terms of symbols, verbal and mathematical. Indeed, since words are easily manipulated in

> mechanical ways, the production of a work of genuine art probably demands more intelligence than does most of the so-called thinking that goes on among those who pride themselves on being 'intellectuals.'[3]

The most general way to specify the concept of thinking is, according to Dewey, "to apprehend consciously the particular connection between doing and undergoing in relation to the desired whole." As we can see, thinking is not confined to using only specific information, like verbal signs and words. In the case of an artist, the individual can think effectively through "relations of qualities," and according to Dewey, this person's logic" is the logic of what I have called qualitative thinking."[4] Through analysis we can name the properties of an art product as symmetry, harmony, rhythm, measure, or proportion. Sometimes they may be formulated mathematically. But as Dewey says, "the apprehension of these formal relationships is not primary for either the artist or the appreciative spectator. The subject-matter formulated by these terms is primarily qualitative, and is apprehended qualitatively."[5]

Three conclusions can be drawn from Dewey's statements. Firstly, Dewey casts doubt on the idea that thought can be intrinsically conceptual and propositional; secondly, he questions the idea that processes of thinking are based on such conceptual-propositional structures. Finally, he asserts that there is a felt qualitative modality of thinking involving qualities, emotions, and needs. Being aware of the fact that it is impossible to account for qualia directly, Dewey gives us a functional description of them. To get a deeper insight into this issue we need to consider the functional interdependence of qualities, emotions and needs.

Function of quality. To consider the function of quality in human thinking we need to focus on an aesthetic experience. The primary thesis of "Qualitative Thought" is that quality lies in the heart of human experience, because, according to Dewey, "the world in which we immediately live is preeminently a qualitative world."[6] Automatically we are prone to associate this quality with the aesthetic quality of the completed aesthetic experience, but as Dewey says, there is also "a unifying pervasive quality of the primary phase of experience." This is not another quality but it is also not exactly the same quality. As Dewey explains in his other essay "Aesthetic Experience as Primary Phase and as Artistic Development,"[7] this is the same quality but in another stage of development.

According to Dewey, the primary locus for every human experience is not atomistic sensations, which are subsequently put together by the synthesizing activity of mind, but one global (also: total or massive) unifying quality, which constitutes and regulates a situation as a whole without being a property of things. Trying to describe this quality in its functions, Dewey contrasts a "situation" with the concept of an "object." In his account, a situation is a complex part of an existential, qualitative world separated by a single global quality. A situation characterized by a single total quality is immediately "had" and it is a global subject-matter, towards which we direct our attention during the development of a transactional experience. An "object" is something that we differentiate from that existential whole. In doing this we receive some hints from the global quality, which are the only clues to make distinctions. This means that objects or events are not ready-made existences found in the extra-experiential domain (that is, the objective world consisting of things and events) as "givens" to which we can ascribe similarly "given" properties. Based on Dewey's philosophy, the ability to differentiate objects and events is an ability acquired by means of inquiries, which take place within prior experiences. Going further, we can say that our ability to see a table, a dog, a painting or an eclipse of the sun, which are the objects or events differentiated from the existential complexity of a situation, is learned by us during our growth from infancy.

Regarding aesthetic experience, there are several processes involved in its consummation. Dewey enumerates these processes as: selective emphasis, controlling, and ordering immediate materials in an organized, meaningful flow of experience. The basis for these processes is a situation specified by the unifying quality. We can reject some elements and choose others as central and important only by means of that quality. We do not differentiate a self-contained object, because there is no such thing in Dewey's experience. There is only an object in its multitudinous relations with the other elements of a situation, or, an object functionally interconnected within a situation. Such processes like selective emphasis, controlling, and ordering, are well expressed by the phenomenological terms of "adequacy" or "inadequacy." Dewey stresses that, "things, objects, are only focal points of a here and now in a whole that stretches out indefinitely. That is the qualitative 'background' which is defined and made definitely conscious in particular objects and specified properties and qualities."[8]

Let us enumerate some of Dewey's statements concerning the function of the global quality in "Qualitative Thought." (i) Unifying quality is a quality of a situation, within which runs an organic-environmental transaction, therefore we can ascribe this quality neither to an organism nor to its surroundings exclusively, rather – as Dewey says – it is a quality of the transaction; hence such a quality is not subjective and private. (ii) Experience comes to us as a permeating qualitative unity, not as atomistic sensations described as properties of objects; we grasp and are seized rather by the overall quality. (iii) Every thought starts with such a quality, and selections and distinctions are regulated by it and because of it; we can differentiate objects and events in that they are secondary to the totalizing quality. (iv) Therefore such a quality is the only "given" in the existential complexity of a situation and objects emerge only because of the function they play in the process of experience. Global quality constitutes a situation but quality and situation "cannot be stated or made explicit, [they are] taken for granted, 'understood', or implicit," therefore the quality is conscious not by itself but is conscious as a background and directional clue in everything we do.[9] Constituting the beginning of every thought, the global quality is a departure point for the processes of selecting, controlling and ordering; as Dewey says: "this quality enables us to keep thinking about one problem without our having constantly to stop and to ask ourselves what it is after all that we are thinking about."[10]

In considering the function of the global quality, it is necessary to pose a very important question: Where does such a quality come from? Dewey gives an explicit answer at the end of his essay: "By some physiological process, not exactly understood at present but to which the name 'habit' is given, the net outcome of prior experiences gives a dominant quality."[11] Our past experiences accumulate themselves and produce the net outcome which starts our present experience. In another essay, "Body and Mind," Dewey makes the same point in another way: "Any particular thing at any particular time is a stimulus, evoking adaptive response and use, only in virtue of the enacted biography of the organism."[12] Briefly speaking, the enacting of organic biography is deeply connected with a situation and a global quality at a particular time. Thus Dewey's statements shed light on the problem: How are those results of prior experiences, which create a net of interconnections, stored? We find some hints to the answer in the quotations mentioned earlier, but in another essay, "Affective Thought," Dewey directly addresses how we accumulate the outcomes of past experiences and subsequently how we use those results. The ways of retaining and using previous outcomes are not confined to the pure mental processes of "memory," but rather our mechanisms of remembering are deeply rooted in our bodily systems and depend on the conditions of our surroundings. This means that the appearance of a dominant quality of the primary phase of experience is deeply connected and embedded in physiological processes operating within our immersed-in-the-world organisms at all times.

For example, when we experience Francis Bacon's paintings, we feel something specific, which does not permit us make a mistake as to the author, and which, simultaneously, integrates all activity. What we feel is a product of past reflective experiences, the results of which have begun to function in our present ongoing experience, while at the same time the burden of past processes of reflective inquiry, completed sometimes a long time ago, is omitted. Past processes of inquiry are remembered by means of physiological processes and in this way are embodied and in some respects fixed within structures of our sensorimotor equipment. In "Affective Thought" Dewey asserts, "Past experiences are retained so that they may be evoked and arranged when there is need to use them in attaining the new end set by the needs of our affective nature. But the retention is not intellectual. It is a matter of organic modifications, of change of disposition, attitude and habit."[13]

The last remark refers to the fact that each global quality has its own developmental tendency, and that the primary phase of such a quality is not self-sufficient. As Dewey describes it, although the quality is dumb, it is part of a complex whole that has movement or transition in some direction.[14] Such a dominant quality, though it specifies and individualizes a situation, does not possess full definitiveness; it still requires a development with distinct terms and relations to transform itself into full consummatory quality, namely: the self-contained aesthetic quality of completed experience.

Function of emotions. The issue of emotions is related to the issue of quality, because of the fact that a global quality, independently of what stage it is in, always has an emotional character. In Dewey's words: "If we designate this permeating qualitative unity in psychological language, we say it is felt rather than thought. Then when we hypostatize it, we call it a feeling. But to term it a feeling is to reverse the actual state of affairs. The existence of unifying qualitativeness in the subject-matter defines the meaning of 'feeling'."[15] The noun, "a feeling" denotes something that is named or differentiated; meaning something that is already a stable product of reflection. Hence, no feeling in this sense is applicable on the level of naturalistic transaction, where everything is fluid in its own way to completion, to consummation. On the level of an ongoing naturalistic transaction, there only functions "feeling" (without article) and "felt," which "are the names for a *relation* of quality."[16]

Significant statements concerning the roles played by emotions are provided by Dewey in *Art as Experience*, where we find important traits ascribed to emotions: (i) emotions, if they have sufficient significance, are qualities of a complex experience; (ii) emotions are a part of our broadly conceived sensuality, they are modes of our sense organs; (ii) an emotion absorbs, subdues, and digests all kind of materials, even that which is merely intellectual; (iii) an emotion is always "at," it needs something concrete beyond itself.

This means that emotions are not self-contained and separate parts of experience, apprehended as separate psychological entities; rather, emotions always join events and objects which are in movement. Even if we consider emotions, at first sight, objectless, nevertheless they are always "emotions-at." (Dewey admits the existence of self-enclosed emotions, but he characterizes them as entirely pathological); (iv) emotions are crucial for the development of experience only when they have become part of a persistent situation with an implied interrelation to things and affairs; emotions which are not part of an inclusive and enduring situation are strictly automatic reflexes (e.g. when we are scared or ashamed, we suddenly jump or blush); (v) a persistent emotion selects appropriate elements and rejects inappropriate elements, and the chosen ones give a qualitative unity to materials which are at first glance incommensurable with each other; (vi) in aesthetic experience, sight, smell, and touch operate as tools of an emotionally moved, living organism.[17]

From such various descriptions it is important to highlight two issues. The first is that according to Dewey, emotions are not subjective or private. Emotions are traits of an ongoing organic-environmental transaction, hence they are neither in an organism nor in its surroundings, but are rather part of a transaction. Such objectiveness of emotions is expressed by Dewey in the idea of "emotion-at," which conveys the idea that emotions demand something beyond themselves; in this way, emotions are both in an organism and at the same time in its surroundings. The affective, joined with the global quality of aesthetic experience defines the whole of a transaction, not its parts. Dewey appealed to scientific evidence from his own day to support his views concerning emotions. See for example "Affective Thought," which to some degree is a short report from the field of biological sciences concerning the physiological functions of an organism.

The second point that should be stressed is that emotions are connected with meanings. According to Dewey, emotions are responses that *make sense*, they are a meaningful and are in significant interaction with surroundings. Emotions mean something, but not in the same sense as concepts "mean" something in philosophical theories of meaning. In transactional cooperation, some elements of appreciation and evaluation are hidden in the background (they can be expressed, e.g., in terms of "adequacy" or "inadequacy"). Dewey describes these elements as primary processes of aesthetic appreciation. These primary, or hidden, processes of aesthetic appreciation and evaluation, whose existence has received additional scientific support in contemporary neuroscience, are crucial elements involved in human understanding and apprehension of a given situation. While these affective processes have a deep influence on "what" and "how" we perceive, they are almost never consciously recognized in our experiences. The problem of hidden affective appreciation raises the issue of the function of needs.

Function of needs. In "Affective Thought" the notion of affectiveness is developed to indicate the fact that an organism has some basic needs which can not be satisfied without activity changing its surroundings. Any obstacle disrupting an organism's path to equilibrium with its surroundings (meaning, harmonious experience), can cause anxiety and desire.[18] According to Dewey, some needs are involved in every mode of experience, hence Dewey is inclined to say that every experience is volitional and voluntary, even in the case of an aesthetic experience. We experience aesthetically by means of our volitional and voluntary activity; not because we intended to do it deliberately or self-consciously: aesthetic experience happens naturally and constantly, and looking at *Mona Lisa* by Leonardo does not have to be the point of departure of our aesthetic experience.

In this context, the ordering of the flow of experience goes from need to fullfilment, where the point of arrival is achieving a state of equilibrium between an organism and its surroundings. In aesthetic experience emotions are conditioned by a conscious perception of relations, in which certain means are taken for chosen ends. But this does not necessarily mean that ends are chosen deliberately or self-consciously, because aesthetic experiences teach us that "not to be deliberately chosen" does not exclude an activity which is volitional and voluntary. In the case of aesthetic experience, being aware of perceived consequences in terms of "adequacy" and "inadequacy" does not mean the conscious identification of needs and emotions either. As Dewey argues, in aesthetic experience needs and emotions are not explicitly stated but are incorporated within the materials of immediate interaction; they are present on the level of the corporeal background of a transaction. Such embodiment makes it difficult to consider needs, desires, and purposiveness in the immediate aesthetic experience. On the level of having experience (the naturalistic level of ongoing transaction) needs are satisfied by means of underlying bodily participation and because there is constant organic-environmental feedback and constant linkage.

After pointing out the functions of qualities, emotions, and needs, let us summarize some general conclusions. Dewey, emphasizing that there is a continuous development from the lower functions to higher, expresses this tendency as "the principle of continuity." According to this principle, qualities, emotions, and needs, are just natural facts, as is discursiveness. Dewey defines "thinking" in general as "conscious apprehension of particular connection between doing and undergoing in relation to the desired whole," and describes it in detail as the functions of qualities, emotions, and needs, meaning, using the overarching term: "qualitative thinking." We can distinguish and give details to each aspect of qualitative thinking, namely the strictly qualitative mode, affective mode, or volitional mode. But being strongly intertwined, they could also be called submodes of an overall qualitative mode of thinking as well. All modes or submodes are deeply connected to each other, including the discursive mode of thinking, which is – according to Dewey's beliefs – impossible without the qualitative one. To summarize, according to the principle of continuity we can discuss jointly the qualitative and discursive modes, and also the submodes of qualitative mode – strictly qualitative (in terms of the global quality), affective (in terms of emotions), and volitional (in terms of needs). But these are only aspects of something that is inseparable. A quality, an emotion, a need, and discursiveness, all these are equally natural phenomena, which go together in every process of experience; they are constituted in relation to each other, and are directed towards reaching new meaningful senses or meanings.

If we compare aesthetic and strictly cognitive experience, it can be said that in both are present the discursive mode of thinking (in terms of concepts and propositions) and also the qualitative mode of thinking (in terms of qualities, emotions, and needs). The only difference between these experiences is that in strictly cognitive experiences the former mode is *more evident* and in aesthetic experiences, *more evident* is the latter. But using the qualitative mode of thinking in an aesthetic experience –Dewey ascribes this mode especially to artists and recipients of the arts – does not exclude the usage of elements which we are prone to associate with the discursive mode of thinking. To make it clearer, we need to note that "using concepts and propositions" is a half-truth about discursiveness; in describing reflective experience (it means, inquiry), Dewey asserts that the main role of symbols or words is to aid us in discriminating during the ongoing flow of transaction. By reaching such distinctions through a process of difficult inquiry (as it often is), we can become sufficiently accustomed to use this result as a part of our immediate experience.. If we consider a concept as a result which enables us to distinguish, this concept is – according to Dewey's words – "comprehended," which means, (a) it is intellectually understood, and, (b) it is felt immediately, without reconstructing all the steps of previous inquiry.[19] Hence, every discrimination we use during an immediate aesthetic experience is a product of past reflections. We thus reach a paradoxical conclusion: in an immediate aesthetic experience knowledge is actively involved, even in its honorific sense as "warranted assertions." To address this issue we need to redefine the concept of "immediacy." We usually conceive of immediacy as something what is not mediated, but on the other hand we know that every kind of knowledge is a mediation. What does it mean that there is knowledge involved in an immediate aesthetic experience? If we realize that we are all the time on a bodily level of naturalistic transaction, it means simply that prior experiential outcomes are embodied and incorporated without the burden of a strenuous reconstruction of past inquiries. As Dewey argues omitting the reflective reconstruction of inquiry is, from the standpoint of an organism, quite economical and a natural phenomenon, because making constant assertions to the familiarity of things would be an unnecessary loss of energy and probably be unproductive to completing any experience. But within this is a very specific understanding of the immediacy of experience, because prevalent in Dewey's view is that immediate experience is always mediated by previously acquired knowledge, and therefore knowledge is always involved. But the paradox of such an experience is illusive, because as John R. Shook writes in his book *Dewey's Empirical Theory of Knowledge and Reality*: "'immediate' means, there is no awareness of mediating factors."[20] It means that in an immediate experience we refer to previous experiences but are not conscious of that reference. This kind of reference would be impossible if we take the standpoint of pure reason. But it is possible because these outcomes are embodied; they are recorded in our bodily attitudes, tendencies, or habits. As Dewey argues, outcomes have been written into our bodily habits on how to interact in particular environmental conditions.

Dewey's theory of qualitative thinking is a re-definition of the traditional philosophical view of conceiving mind as separate from the body, which is the same as the separation of reason from the senses, feelings, emotions, and needs. The main traits of qualitative thinking in such a theory are both: its ubiquity and its bodily grounding. As Dewey implies in "Qualitative Thought," "Affective Thought," and *Art as Experience*, the only source of such a theory may be found in our own embodiment. The theory of qualitative thinking broadens traditional conceptions of thinking confined to conceptual-propositional forms. Dewey points out that we should not separate thinking in conceptual-propositional terms from thinking in terms of qualities, emotions, and needs, because the qualitative mode, though usually tacit, nevertheless is ever-present even in strictly intellectual experiences. Dewey stresses that to use concepts intelligently we need to feel them. Consequently, embodied in "comprehension" connected with, e.g., a warranted assertion, there is a necessary element of immediacy, but only when the assertion is a result of an experience. Dewey's standpoint is further explored in a passage from *Art as Experience*: "But only a twisted and aborted logic can hold that because something is mediated, it cannot, therefore, be immediately experienced. (...) Different ideas have their different 'feels,' their immediate qualitative aspects, just as much as anything else. One who is thinking his way through a complicated problem finds direction on his way by means of this property of ideas. Their qualities stop him when he enters the wrong path and send him ahead when he hits the right one. They are signs of an intellectual 'Stop and Go.' If a thinker had to work out the meaning of each idea discursively, he would be lost in a labyrinth that had no end and no centre."[21]

2. The Successors: Shusterman and Johnson

Though Dewey's conception of qualitative thinking found its fuller expression in the thirties of the past century, hints of it can already be found in his earlier writings, e.g., in the essay "The Reflex Arc Concept in Psychology" (1896). Only recently, though, have successive theories appeared that treated the problem of qualities and thinking together (not to mention the philosophy of Mereau-Ponty). Elaborating his own version of a pragmatist aesthetics, Richard Shusterman presented the conception of "thinking through the body" and Mark Johnson proposed the theory of "embodied thinking." Shusterman's theory appears to be an indirect version of Dewey's conception. Johnson's seems to be the same as Dewey's, but utilizes proofs taken from contemporary neurosciences that strengthen the theory of qualitative thinking by endowing it with the authority of cognitive sciences. Let us briefly describe these conceptions by underlining similarities and dissimilarities.

Shusterman's idea of "thinking through the body" was formulated as a title of an essay published in *The Journal of Aesthetic Education* but elements of the conception can be found in many of his writings, especially in the last book: *Body Consciousness*.[22] Generally his standpoint is the same as Dewey's; his idea of *somatic turn* suggests to "go down" beneath the level of language and to present the notion of rationality in a new and different manner. Shusterman argues: "Stressing the phenomenological immediacy and affect does not exclude the semantic dimension of experience. Meaning is not incompatible with *qualia* and affect."[23] In his conception he also allows for discussion about quality, emotions, needs, habits, and ends, and about problems of how bodily processes influence our thinking and feeling. He accepts Dewey's notion of an aesthetic experience which he conceives as a fully integrated, unified process where sensual, intellectual, emotional and volitional human faculties are present and cooperating together. His main thesis is "we think and feel our bodies" and that even though our mental life is based on bodily processes we can neither reduce it to them nor can we separate it from them.

There is, of course, a controversy between Dewey and Shusterman – in my opinion quite unnecessary— deeply rooted in the Rortian heritage of Shusterman. Namely, Shusterman does not want to treat an aesthetic experience on the naturalistic functional level as Dewey does. He argues that to do this would be a step towards foundational philosophy because it tries to answer the question: What is the true state of affairs in a *mute* aesthetic experience? Shusterman accuses Dewey of subtle epistemological foundationalism – described by him as "a residual myth of the given" – in his description of an immediate experience. He questions Dewey's statement that the experiencing of immediate, non-discursive quality lies at the root of *every* discursive thought and directs it.[24] Actually, Shusterman denies the strength of the underlying pervasive quality, that existing in every stage of discursive thought without being a clear object of consciousness, "determines necessarily" a coherence of thought, a structure of discourse, and an appropriateness of concepts in a proposition. The problem with Shusterman's extended argument lies in the two logical terms of "determination" and "necessity." In Dewey's writings, concepts of causation may function in two different orders: in the logical order of causation and in the existential order of causation. Dewey warns us about mixing these incomparable orders, and he stresses that first one is not about time relations, while the second is.[25] On the naturalistic functional level of transaction, from the standpoint of an ongoing process, what is in the future is not determined and necessary, although from the perspective of completed experience all elements we perceive seem logical consequences in the whole structure of an experience. All elements are seen as determined and necessary, though during the ongoing process *determination* is rather *potential* and necessity is rather a *possibility*. Hence, there is no logical determination and necessity in the existential order of a naturalistically conceived transaction, as Shusterman claims. He imposes a logical order upon an existential one, and despite Dewey's warnings, he constructs practically all his arguments on this claim. Moreover, another part of his objection is built on the thesis that the origin of a global quality is unknown, which means for him metaphysical. But Shusterman is not coherent in this case, because on one hand he denies that global quality can be explained other than in metaphysical terms, but on the other hand, he stresses that Dewey gives an adequate, not-fundamentalist, description of global quality in terms of habits, ends, and needs. But if we consider global quality as "a net outcome of prior experiences," what we get is just such an explanation in terms of habits, ends, and needs. Thus Shusterman accepts the same antifundamentalist solution as Dewey proposes in "Qualitative Thought".

The main reason Shusterman raises such criticism against Dewey is because of the differing standpoints of both philosophers. Dewey treats an experience on the naturalistic functional, primordial level, where qualities, emotions, and needs can be revealed in their mutually connected functions, while Shusterman is prone to talk about qualitativeness, affectiveness, needs, knowledge, cognition, and ends to be reached in experience; nevertheless he does it in an empirical language giving a commonsensical view of interrelations of those bodily embedded elements. To talk empirically means in Dewey's philosophy to talk generally about an aesthetic experience (as cognitive, affective, qualitative, and volitional), without a deeper metaphysical insight into the naturalistic transaction. By resigning from a naturalistic explanation, Shusterman accepts the standpoint that philosophy is sometimes helpless in explaining certain philosophical problems, e.g., an aesthetic experience. But it does not mean that philosophy should not point out undeniably crucial notions. Though aesthetic experiences are philosophically unclear, we should share a common belief that they are important and therefore should treat them philosophically by constantly exercising them practically (literally) and reflecting upon them theoretically. Choosing his own philosophical path, Shusterman's procedure of inquiry is based on two notions: *bodily introspection* with the use of proprioceptive sense, and *self-use* which can be conceived of as self-learning or self-reshaping.

Despite criticisms against Dewey, Shusterman's idea of "thinking through the body" is in its conclusions almost the same as Dewey's theory of qualitative thinking (even as to the reconstruction of the "immediacy" concept [26]). In both views qualitativeness, affectivenes, needs, and discursiveness, are deeply interconnected. In both, the qualitative and discursive mode of thinking depends upon bodily grounding, and the modes of thinking in both go together. But Shusterman is reluctant to admit that the global quality is adequate to control and regulate *every* discursive distinction of thought. As a matter of fact, he ascribes to the weaker version of Dewey's thesis, that qualitativeness only *sometimes* controls and regulates discursive distinctions of thought. Dewey's version of the thesis is for him too categorical, hence foundational.

It is also worth mentioning that besides abandoning the naturalistic functional dimension of description, other criticisms can be formulated on the basis of Dewey's own standpoint. Shusterman postulates somaesthetics as an interdisciplinary field of shared research, for example, shared with the cognitive sciences, but actually rarely engages in it. Such common interdisciplinary research is found more easily in Dewey's writings.

Shortly after Shusterman appeared open to re-reading Dewey from a somatic and naturalistic perspective, he paradoxically stopped doing this on the naturalistic and functional level. Mark Johnson, coming from another philosophical tradition, did what Shusterman had an opportunity to do. After Johnson's book *The Meaning of the Body*[27] the notions often-used by Dewey especially in *Experience and Nature* or *Art as Experience*, such as: embodiment, incarnation, or incorporation, should not be read metaphorically as we sometimes used to do, but rather literally.

Johnson develops Dewey's ideas in the direction of, explicitly stated: embodied meanings and embodied forms of thinking. Appreciating Dewey's notions of the underlying pervasive quality as global apprehension of a situation and the non-subjective view on emotions and needs, he supplies some, taken from the domain of contemporary neurosciences, evidence in support of Dewey's claims. Doing this, Johnson notes, that Dewey (and also James) knew quite well the best scientific knowledge taken from biology, physiology of the human body, brain, and nervous system. The good example of Dewey's use of such kind of knowledge might be "Affective Thought" which can be conceived, in many respects, as a report from the state of scientific inquiry in these fields.

Johnson reveals the real potential hidden in Dewey's ideas, and develops naturalistic functional description of qualities, emotions, and needs in processes of thinking, supplementing them with scientific research. He concludes that processes of thinking, like acts of understanding, of making inferences, and of constituting of meanings, are rudimentary embodied. Even the most subtle achievements of abstract thought, as logic or mathematics, are coming from our embodiment (embodied minds intertwined by transactions with our surroundings). Taking the path of contemporary neuroscience, Johnson accepts Dewey's view that there is no ontological difference between body and mind, organism and surroundings, lower and higher functions of an organism. However, he stresses that the traditional view contradicts Dewey's and neuroscientists' conclusions. It is because it seems that we experience things dualistically, and we acquire a kind of commonsensical, "objective metaphysics." According to such an "objectivist" model of the world: "The world is 'out there' waiting to be experienced, consisting of middle-sized objects that possess objective properties and that stand in definite relations to other objects, independent of beings for whom those objects *are* objects."[28] Only the deep insight given by neuroscience reveals that the main trait of the body is that the body hides itself. Thus, Johnson elucidates Dewey's conception of the non-subjective nature of qualities, emotions, and needs by presenting the outcomes of Antonio Damasio's and Joseph LeDoux's researches. He tries to work out the possible neuronal mapping of the global quality, and to show the primacy of the global quality in thinking processes, using – respectively – Don Tucker's studies and the conception of "felt situation" by Eugene Gendlin. In presenting the scientific studies and their results, Johnson is quite aware as to the status of such scientific conclusions. He shares his opinion with Dewey, and considers them as "working hypotheses," which are open for further reconstruction or verification.

Dewey elaborated his qualitative thinking conception with deep knowledge taken from the "neurosciences" of his own time. Shusterman has read Dewey as a somatic philosopher, recognizing that bodily rooted habits, ends, needs, and our discursiveness, go together in an aesthetic experience. However, he presented the indirect and weaker version of Dewey's theory. Nevertheless, explicitly developed theory has appeared in the book of Johnson, who tried to extend the ideas of embodied thinking and embodied meanings beyond the domain of aesthetics.

3. Aesthetic Experience as Self-Contained

If we return to the long quotation from *Art as Experience* given in section 2, we can find words like: "To think effectively in terms of relations of qualities is as severe a demand upon thought as to think in terms of symbols, verbal and mathematical; (...) the production of a work of genuine art probably demands more intelligence than does most of the so-called thinking that goes on among those who pride themselves on being 'intellectuals.'" As we already know in an aesthetic experience the results of past inquiries may function as immediate with no awareness of mediating factors. But if we consider such result as a meaning – the point of arrival of specific inquiry – there are also some inferential procedures required to complete an aesthetic experience. Dewey calls these inferences in an aesthetic experience: selecting, controlling, and ordering. They are also written in our bodily patterns of interacting with our surroundings and therefore they could take place in the immediate stage of our transaction. An aesthetic experience is qualitative but it does not mean there are not immediate meanings and immediate inferences involved in it. Only by means of embodied meanings and embodied inferences, a new *meaningful sense* can emerge. Therefore, it can be said that aesthetic experience is self-contained or self-explanatory.

According to Dewey, aesthetic experience is the exemplar of other modes of experiencing: "Had not the term 'pure' been so often abused in philosophic literature (...) we might say that esthetic experience is pure experience. (...) To esthetic experience, then, the philosopher must go to understand what experience is."[29] In the arts we appreciate the fullness of quality, and in science preciseness of concepts and propositions. But it does not mean that thinking is either qualitative or discursive, for it is always both, qualitative and discursive, because, as Dewey asserts, "in a distinctively esthetic experience, characteristics that are subdued in other experiences are dominant."[30]

NOTES

1. Robert Storr, *Think with the Senses – Feel with the Mind. Art in the Present Tense*, in *La Biennale di Venezia. 52. Esposizione Internazionale d'Arte. Short Guide* (Venezia: Marsilio 2007), p. 8.

2. John Dewey, *Art as Experience*, in *The Later Works of John Dewey*, vol. 10, ed. Jo Ann Boydston (Carbondale and Edwardsville: Southern Illinois University Press, 1987), p. 196.

3. Ibid., p. 52.

4. John Dewey, "Qualitative Thought," in *The Later Works of John Dewey*, vol. 5, ed. Jo Ann Boydston (Carbondale and Edwardsville: Southern Illinois University Press, 1985), p. 251.

5. Ibid., p. 251.

6. Ibid., p. 243.

7. John Dewey, "Aesthetic Experience as Primary Phase and as Artistic Development," in *The Later Works of John Dewey*, vol. 16, ed. Jo Ann Boydston (Carbondale and Edwardsville: Southern Illinois University Press, 1989), p. 396.

8. Dewey, *Art as Experience*, p. 197.

9. Dewey, "Qualitative Thought," pp. 247-8.

10. Ibid., p. 248.

11. Ibid., p. 261.

12. John Dewey, "Body and Mind", in *The Later Works of John Dewey*, vol. 3, ed. Jo Ann Boydston (Carbondale and Edwardsville: Southern Illinois University Press, 1984), p. 34.

13. John Dewey, "Affective Thought," in *The Later Works of John Dewey*, vol. 2, ed. Jo Ann Boydston (Carbondale and Edwardsville: Southern Illinois University Press, 1984), p. 106.

14. Dewey, "Qualitative Thought," p. 254.

15. Ibid., p. 248.

16. Ibid.

17. Dewey, *Art as Experience*, pp. 36, 37,48, 49, 56.

18. Dewey, "Affective Thought," p. 105.

19. Cf. John Dewey, *Logic: The Theory of Inquiry*, in *The Later Works of John Dewey*, vol. 12, ed. Jo Ann Boydston (Carbondale and Edwardsville: Southern Illinois University Press, 1988), p. 156.

20. John R. Shook, *Dewey's Empirical Theory of Knowledge and Reality*, (Nashville: Vanderbilt University Press, 2000), p. 172.

21. Dewey, *Art as Experience*, p. 125.

22. Richard Shusterman, "Thinking Through the Body, Educating for the Humanities: A Plea for Somaesthetics," Vol. 40, No. 1, Spring 2006, pp. 1-21; Richard Shusterman, *Body Consciousness. A Philosophy of Mindfulness*

and Somaestethics (New York: Cambridge University Press, 2008).

23. Richard Shusterman, *Performing Life* (Ithaca: Cornel University Press, 2000), pp. 23-24.

24. Richard Shusterman, *Practicing Philosophy. Pragmatism and Philosophical Life* (Routledge: New York, 1997), pp. 164-165; cf. all Chapter 6.

25. Dewey, *Logic*, p. 444; Cf. also: Robert Reuter, “The Radical Agent: A Deweyan Theory of Causation,” *Transactions of Charles S. Peirce Society*, Vol. 29, No. 2, 1993, pp. 248-249.

26. Cf. Richard Shusterman, “Complexities of Aesthetic Experience. Response to Johnston,” *Journal of Aesthetic Education*, Vol. 38, No. 4, Winter 2004, pp. 99-100.

27. Mark Johnson, *The Meaning of the Body: Aesthetics of Human Understanding* (Chicago-London: University Chicago Press, 2007).

28. Johnson, *The Meaning of the Body*, p. 46.

29. Dewey, *Art as Experience*, p. 278.

30. Ibid., pp. 61-62.

ETHICS

Seven

JOHN DEWEY AND THE ETHICS OF RECOGNITION

Ramón del Castillo

Comparisons between Dewey and contemporary political philosophers are well-known. While some interpreters have perceived similarities with deliberative models of democracy others, in turn, have related Dewey's ideas to communitarian and republican perspectives. However, most of them would agree that Dewey's own position is not easily reducible to such polarities. Dewey's political philosophy is neither a mere anticipation of Jürgen Habermas's theory of communicative action, nor an early example of the sort of communitarism that proliferated in the United States in the last twenty-five years of the 20th century. Among such interpreters, Richard Bernstein and Axel Honneth occupy a special place, since they have tried to see Dewey as an alternative to both models. Even though Habermas has acknowledged some of Dewey's ideas on the formation of public opinion, in my view Bernstein's and Honneth's interpretations of Dewey's legacy have been much more relevant.

In this paper I will only focus on Honneth's interpretation.[1] More specifically, I will argue that Honneth's reading is particularly useful for reconnecting Dewey to contemporary politics, in spite of the fact that Dewey never used "recognition" as a key term. Among other advantages, for example, Honneth's reading invites one to see the old *motto*, according to which Dewey was the American representative of the Hegelian Left, in a new light.

1. Spheres of Culture

My first suggestion is that Honneth should have put his insightful readings[2] in the context of the debate that Dewey posed in works like *Freedom and Culture:*

> Is there any one factor or phase of culture which is dominant, or which tends to produce and regulate others, or are economics, morals, art, and science, and so on only so many aspects of the interaction of a number of factors, each of which act upon and are acted upon

> by the others? In the professional language of philosophy: shall our point of view be monistic or pluralistic?[3]

Dewey's own answer to this last question is remarkable and it is clear that he embraced an anti-monist attitude. There is, he said, no single force that determines the development of a historical formation:

> The state of culture is a state of interaction of many factors, the chief of which are law and politics, industry and commerce, science and technology, the arts of expression and communication, and of morals, or the values men prize and the ways in which they evaluate them; and finally, though indirectly, the system of general ideas used by men to justify and to criticize the fundamental conditions under which they live, their social philosophy.[4]

It is important to remember that Dewey expressed such a pluralistic view in many other places over the course of his life. Modern society, he said in *Democracy and Education*, is not a single system but rather several diverse spheres of conjoint action, loosely interconnected by different causes and functioning at different levels of proximity: families, circles of friends, playmates, cliques, clubs, gangs, business crowds, companies and corporations, trade-unions, political affiliations, ethnic groups, religious communities, villages, towns, blocks, districts, big cities, states, nations… "Even the guild of artists, the republic of letters, or the professional learned class scattered over the face of the earth" —Dewey observed— constitute forms of community. Commerce, transportation, intercommunication, and emigration, he added, explain the essential heterogeneous character of modern societies.[5] All of these modes of association, Dewey said, are different in the way in which they inspire in their members a sense of sharing ends and aspirations. Each mode exercises a particular influence over the dispositions of its members, a specific cultivation of habits, ends, and modes of evaluation. Furthermore, in works such as *The Public and Its Problems,* Dewey described in detail the dynamic of industrial democracy in terms of the interconnection between different realms of social action, particularly between the State, the public sphere and what he designated as "publics": a variable set of associations and groups better articulated than communities.[6] In *Freedom and Culture*, he emphatically attacked the monistic view of some social philosophies, (particularly the economic determinism defended by some versions of Marxism), and one more time emphasized the social influence of a variety of spheres: science and technology, arts and symbols, and religious practices and rituals even in the context of a more secularized society.[7]

According to this narrative, modern society is not a homogenous system, and there is no single factor that explains social structure and dynamics. But how then, should we interpret Dewey's emphasis on the encompassing and pervasive influence of the *moral factor* in all spheres of social action?

> The moral factor is an intrinsic part of the complex of social forces called culture [...] human beings hold some things dearer than they do others, and that they struggle for the things they prize, spending time and energy in their behalf: doing so indeed to such an extent that the best measure we have of what is valued is the effort spent in its behalf. Not only so, but for a number of persons to form anything that can be called a community in its pregnant sense there must be values prized in common. Without them, any so-called social group, class, people, nation, tends to fall apart into molecules having but mechanically enforced connections with one another.[8]

Ideals of self-realization along with images of a good life, Dewey suggests, shape social formation as strongly as any other factor. Or maybe he is implying something else? Beliefs about what is good and valuable can be diverse within a society and variable through time, but they always play a special function in social life. Over other factors, the moral one has a deeper influence, since all spheres of interaction are *dyed* by it: family and the State, industry, technology and labour, art and religion —Dewey observes— rely on something apparently less objective than the structures and institution associated with them. Any form of association, and Society as a whole, depends on basic modes of valuation and tacit conceptions about what could be a good human life. In other words, even when, in his narrative, modern industrial society is represented as a heterogeneous system, Dewey attributes the moral factor with an influence over the *totality* of the social field. In my view, this would explain the distance between Habermas and Dewey which has been marked on many occasions and from different perspectives.[9] But, might it also justify Honneth's tendency to paint Dewey as a precursor to his own theory on the moral grammar of social conflicts?

2. Democracy and Morals

Since I cannot reconstruct Dewey's narrative in its full scope here, I will concentrate on a particular aspect which concerns the theory of democracy: the relation between deliberation and recognition.

According to Dewey, political formations are always grounded on perceptive and evaluative dispositions shared by individuals. Democracy, as Dewey already observed in early works such as "The Ethics of Democracy" (1888) must be understood not as a form of government, but as "an ethical conception, and upon its ethical significance is based its significance as governmental. Democracy is a form of government only because it is a form a moral and spiritual association."[10] As a form of government it depends on majority rule, but as a form of human association it depends (1) on the process by which the majority is obtained and (2) on the "vast mass of sentiments, many vague, some defined, of impulses, aspirations, ideas, hopes and fears, and purposes" from which it springs. "It is their reflex and incorporation, their projection and outgrowth."[11] Democracy, then, is essentially linked to a manner of developing and adjusting habits, customs, attitudes, sentiments, feelings, modes of perception and action, beliefs on well-being and expectations, i.e., about a good life. Democracy requires a moral background, but not as something fixed and complete in itself but rather as a repertoire of practical mediations to handle new situations and conflicts.

As it is also well known, the middle Dewey transformed his early view (inspired by Hegelian motifs) into a psychological theory of socialization, and reformulated his view of democracy. Maybe this new view of social interaction is less "organic," but the idea that there is a moral background on which many forms of social organization are grounded resurfaces one more time. As he said in "The Problem of Truth" (1911), "Unless people understand each other, they cannot co-adjust their respective act and offices; all assignment of task, all distribution of labour and cooperation is impossible."[12] Later, in *Democracy and Education* (1916) Dewey insisted on the idea that "democracy is more than a form of government; it is primarily a mode of associated living, of conjoint communicated experiences".[13] Communication, he also remarked, is an essential element of such a mode of association, not for the decision-making procedure, but as a process which contributes to common understanding:

> The communication which insures participation in a common understanding is one which secures similar emotional and intellectual dispositions— like ways of responding to expectations and requirements […] To be a recipient of a communication is to have an enlarged and changed experience. One shares in what another has thought and felt and in so far, meagerly or amply, has his own attitude modified. Nor is the one who communicates left unaffected. Try the experiment of communicating, with fullness and accuracy, some experience to another, especially if it be somewhat complicated, and you will find your own attitude toward your experience changing; otherwise you resort to expletives and ejaculations. The experience has to be formulated in order to be communicated. To formulate requires getting outside of it, seeing it as another would see it, considering what points of contact it has with the life of another so that it may be got into such form that he can appreciate its meaning. Except in dealing with commonplaces and catch phrases one has to assimilate, imaginatively, something of another's experience in order to tell him intelligently of one's own experience.[14]

Therefore, communication does not consist in the expression of interests or claims, but rather in the creation of a broader context of experience and interplay in which social actors overcome barriers and become transformed and enriched. In *Reconstruction in Philosophy* (1920), Dewey also affirmed: "Democracy has many meanings, but if it has a moral meaning, it is found in resolving that the supreme test of all political institutions and industrial arrangements shall be the contribution they make to the all-around growth of every member of the society."[15] However, what measures such contribution to the common growth is not a method, but an interaction between individuals which allows them to recognize each other and to perceive the range and extension of their respective ideas of the good.[16]

Some years later, in *The Public and Its Problems* (1927) Dewey offered a complex description of the development of modern heterogeneous society which also contained some normative conclusions. Among other things, he reiterated that political procedures presuppose pre-political bonds and interactions. As in his early writings, Dewey said: "Majority rule, just as majority rule, is as foolish as its critics charge it with being. But it never is *merely* majority rule [...] The means by which a majority comes to be a majority is the more important thing: antecedent debates, modification of views to meet the opinions of minorities...The essential need, in other words, is the improvement of the methods and conditions of debate, discussion and persuasion. That is the problem of the public."[17]

It is arguable that Jürgen Habermas could have been inspired by passages like this for his theory[18], but only by not taking into account all that Dewey said about the *conditions* of deliberation and the need of a public sphere that retrieves *moral* convictions. Human growth, he also said, required much more than social planning and the control of experts. It also called for more than deliberative methods to solve social conflicts. The challenge of modern democracy was not just a matter of how to develop more rational methods of debate, but it was also a question of how to cultivate moral virtues, and how to establish new commitments between individuals in the context of mass society, depersonalized labour, lobbies of experts, and over-professionalized politics.

3. Democracy and Differences

Up to this point, some comments on Honneth's reading might shed light on the significance of Dewey's moral approach to democracy. As I have remarked, Dewey's idea of democracy as a mode of social organization that relies on human growth distinguishes his position from models that take deliberation as the keystone of a participatory democracy.

One issue that exposes the distance between the two models is the weight of discursive exchange in each. Even when Dewey affirmed that the capacity to speak a language is a precondition of social grouping,[19] he always conceived linguistic actions as sustained on broader modes of intercourse and interaction. Language helps individuals to reach agreement, Dewey would say, primarily because it enables them to reach and transform their tacit acquaintance with a situation or context.[20] Can we really associate the heart of democratic life with pure communicative actions, ideally detached from a background of ideals, desires, needs, affective attitudes and habits of perception, by which individuals actually rule their lives? The deliberative stance is an essential part of the game of democracy, and Dewey never denied this. But could we really consider the claims made by others if we do not understand the situation in advance, namely, the idea of a good life from which such claims follow?

Of course, the activities that Dewey associated with such understanding require language, reflection, and intelligence. However, they are not are guided by the ideal end of achieving an agreement between individuals who pay attention to the opinion of each other, weigh up available alternatives, and make decisions. In such a scenario the only moral prerequisite would be this: the recognition of others as equally rational beings able to judge the value of our claims and show us the rationality of theirs. But is not some other mode of recognition between us necessary, namely, a way of seeing others not as equal rational agents, but as particular individuals with convictions that are equally valuable to them as ours are to us?

In a deliberative model it could be accepted that the discursive procedures presuppose certain capabilities, habits and virtues like role-playing, reciprocity, sensitiveness, openness, and empathy. But this would not be placed at the core of democracy, and it would only be regarded as some of its *functional* requirements. The normative force of those habits and virtues would be reduced to their contribution to the grasp of articulated claims of those affected in a situation.[21] Some attitudes which concern the existential dimension of others could be accepted as conditions of social integration, but only to the extent to which they affect the consciousness of an individual's membership in an ideal community. But, what if those virtues are precisely what define democracy as a mode in which conflicts and differences between individuals are part of the political foreground? What if full recognition could not be conceived without an element of *particularity*?

Even if Dewey sometimes projected an image of social unity and harmony, and even if he saw education as a factor of integration beyond differences, he did *not* ignore conflict, or at least not all forms of conflict. And maybe this is the reason why he conceived democracy not only as a sort of social inquiry guided by rational discussion, but also as a mode of acknowledging the existence of differences. It is surprising, then, that after being so sensitive to Dewey's political legacy, Honneth did not included Dewey in such an interesting piece as his "The Other of Justice".[22] In this work Honneth criticises Jürgen Habermas's idea of welfare of the other as a presupposition of communicative action. Such a principle of solidarity —Honneth says— would not really cover the recognition of the others in their difference and the particularity of the goals that move individuals to act. The critique is right, and Honneth invokes numerous opinions to sustain it. But maybe he fails to think of Dewey, one could suggest, because he tends to think that Dewey ignored conflict and antagonism as essential elements of democracy.

In fact, he could have included some of Dewey's opinions which are helpful in understanding his model of democracy. Democracy, Dewey said in "Democracy and Education in the World of Today" (1938) is the moral cause of the "dignity and the worth of the individual. Through mutual respect, mutual toleration, give and take, the sharing of experiences, it is ultimately the only method by which human beings can succeed in carrying on," the experiment of "living together in ways in which the life of each of us is at once profitable in the deepest sense of the word, profitable to himself and helpful in building up the individuality of others."[23] In "Creative Democracy: The Task before Us" (1939) he would add:

> Democracy is the belief that even when needs and ends or consequences are different for each individual, the habit of amicable cooperation —which may include, as in sport, rivalry and competition— is itself a priceless addition to life. To take as far as possible every conflict which arises —and they are bound to arise— out of the atmosphere and medium of force, of violence as a means of settlement into that of discussion and of intelligence is to treat those who disagree —even profoundly— with us as those from whom we may learn, and in so far, as friends. A genuinely democratic faith in peace is faith in the possibility of conducting disputes, controversies and conflicts as cooperative undertakings in which both parties learn by giving the other a chance to express itself, instead of having one party conquer by forceful suppression of the other —a suppression which is none the less one of violence when it takes place by psychological means of ridicule, abuse, intimidation, instead of by overt imprisonment or in concentration camps. To cooperate by giving differences a chance to show themselves because of the belief that the expression of difference is not only a right of the other persons but is a means of enriching one's own life-experience, is inherent in the democratic personal way of life.[24]

According to these passages, democracy has to do not only with the expression of demands in the public sphere, or the political representation of groups with different interests, or even the semi-articulated vindications of publics. It is basically related to the conditions for a previous *visibility* of individuals and groups in their differences. Democracy would require the mutual acknowledgment of something heterogeneous and even incommensurable: the recognition of each existence "with its own account, not a something capable of equation with and transformation into something else."[25] Recognition of differences, then, would be not a question of adopting a symmetrical or reciprocal consideration (in fact, equality could be achieved at some level, but recognition would still be denied in a deep sense). Recognition would demand the virtue of understanding others in relation to a particular mode of existence. It would neither be captured in the formula: "I accept others as they are, since their claims, until it is known, don't clash with my own ones", nor in the phrase: "I admit that they have equal right to do this or that"; not even in the expression: "I recognize them as equal fellows". Instead, it would be articulated in the question: "What mode of life do they embody if they aspire to this or that? What sort of life do they experience if they do this or that?

If human growth is the final end of democracy, then democracy needs a process of cultivation by which individuals can develop a mutual understanding of others as something more than participants in a decision-making process or in a solving-problem procedure. To participate in a democratic way of life is to conceive of deliberation as part of a broader variety of actions, since people achieve things together by more than just being aware of others as equal rational agents.

Thus, a democratic community should include not only the sphere in which individuals judge their respective claims in terms of public consequences, but also the sphere in which they recognize each other in their differences. However, detached benevolence, or tolerance, or respect, are not the same as recognition. Dewey is clear: when we are asked to see others in their difference, our perception could lead to some consequences, like the transformation of our self-understanding. This point is extremely important. In Dewey's model, selfhood is always developed in an intersubjetive way and, consequently, mutual recognition between individuals does not mean recognition of self-sustained identities. For Dewey, of course, self-reliance is essential to human growth, but not because it reinforces the identity of individuals, but rather because it increases their capacity to interact with other individuals and to face situations together. What we recognize when we see others in their difference is not an idenity fixed and defined in advance. Rather, we recognize a way of life or a type of existence which would need particular conditions to be actually developed.

4. Recognition and Growth

According to Dewey, the supreme test of democracy consists in its contribution to the growth of every member of the society. This raises some questions. Did Dewey regard social integration as something overly dependent on the habits and structures that nurture and protect human growth? Would this assumption not be another form of monism? If so, would Dewey not be in need of a definition of human growth as the final end democracy?

Dewey's notion of growth is extremely complex, as many interpreters have shown. Here I only would like to make clear that recognition of differences is intrinsically connected with growth. Why should recognition of individuals be a basic presupposition for growth? To begin with, to suppress or to repress differences is not a simple lack of social courtesy. It implies imprisoning individuals in reduced modes of being and limited spheres of action; a harm and injustice that —as Dewey observes— can be instigated by means of institutionalized discrimination or by ordinary humiliation, by material privation or psychological depreciation. To suppress differences means to deprive individuals of freedom and therefore destroy essential conditions of growth since, according to Dewey, real growth always requires free intercourse between individuals. An individual grows —as James Gouinlock said— not like a balloon swelling with air.[26] Rather, the individual achieves the condition to develop more inclusive and complex modes of behaviour, and more flexible and sustainable habits of interaction with others. Recognition enhances the range and extent of the experience that individuals obtain with each other only by free interplay, and not an expansion of the self as something isolated and self-sufficient.

To summarize: growth prospers best when a democratic way of life takes the form of a process of mutual recognition in which individuals are asked to discern and to appreciate the worth of diverse individuals. It does not happen —at least not at the same level— when democracy only takes the form of a process of deliberation in which individuals see each other merely as ideally equal rational beings. Democracy depends more on a variety of virtues which help to perceive the lives of others and to alter the self-understanding of one's own life, than it does on the capacity to accept justified claims by others and to modify the preferences. The cultivation of the abilities necessary to understand different modes of life, life histories, and alternative ways of conceiving the good —Dewey could say— contributes to that process of "maturing, perfecting and refining", to that type of growth which democracy aspires to cultivate.[27]

To some extent, then, Dewey was *not* defending any sort of monism; he said that the final end of democracy is the free growth of all individuals. Even he seemed to suggest that to take care of this primary good or basic need (growing, flourishing) is the moral presupposition of all forms of justice (including economic ones). However, he reasonably refused to provide a substantive definition of growth. Rather, he described *conditions* of growth. And the virtue of acknowledging differences seemed to be one of them, maybe more important than rational deliberation. To say that free interplay between different and even antagonistic individuals and groups is a basic condition of democracy sounds more like a reconstruction than like a principle. It is both a generalization from past experience (an experience which includes conflicts), and a tentative orientation with which it could be possible to face new situations in the future. But it is not a supreme norm.

Even when Dewey suggested that growth is a characteristic trait of human nature and that democracy is the way of life more suitable to developing such human nature, he was not claiming a definition of some unalterable essence through history. It is true that he considered it necessary to describe the "large and constant features of human sufferings, enjoyments, trials, failures and successes" that "together with the institutions of art, science, technology, politics and religion which mark them, communicate genuine features of the world within which man lives."[28] But he did not take such a description as the revelation of a fixed or eternal human essence. What needs to be recognized is not something ahistorically given.[29] Experience, in its retrospective sense, seems to show (1) the persistence with which some conditions for growth have been misrecognized, unseen, repressed, forbidden, suppressed or devaluated, and (2) how such a lack of recognition has been always *interwoven* with diverse factors (economic, ethnic, religious, cultural). Experience seems to show, in brief, that suppression or misrecognition of others is endemic to human habits and to political institutions, and some traits seems to be constant. However, to define in advance *what* must be recognized in others could preclude the very experience and, as Dewey would say, the experiment.

We are confronted by basic experiences that have different meanings in different moments and places. And Dewey would admit that a mere abstract description of generic traits of the human condition is not helpful in confronting the problematic situation to which those experiences are linked, since it is the particular combination and interaction between those traits of experience that brings to such a situation its particular quality and effects.[30] To acknowledge what is being taken for granted in each circumstance, or to recognize the tacit patterns of experience, then, would be the task of social inquiry, which requires a retrospective understanding of the particular features of different modes of life.

5. Recognition and Emotion

As I said above, what Honneth calls "recognition" is part of what Dewey labelled as "experience". It would be impossible to explain now all that Dewey associated with the idea of "experience", but I think that at least it is important to point out what he did *not* mean by social experience. As Honneth himself was, Dewey was frequently misunderstood for his insistence on the role that the affective and the emotional have in social interaction. But Dewey never said that experience excludes reflectivity, or that it is indefectibly devalued when it turns into an object of thinking. What he actually said is that when deliberation is related to a broader background of experience, it can function as an agent of social integration. Emotions, sentiments, and passions, of course, are an essential element of such a background, but they are neither intractable data, nor are they free from interpretation, separable from habits, customs, and social structures. For Dewey the necessity of affective attachments in no way reveals a gap between solidarity and citizenship, between the private and the public, or even less, between the spontaneous and the elaborated, the authentic and the inauthentic.

If the idea of experience was excluded from the ideal of a rational and deliberative politics, it was probably because mass society and market reduced it to its more uncontrollable forms. Dewey said in *Art as Experience:* "zeal for doing, lust for action, leaves any persons, especially in this hurried and impatient human environment in which we live, with experience of an almost incredible paucity, all on the surface. No one experience has a change to complete itself because something else is entered upon so speedily. What is called experience is so dispersed and miscellaneous as hardly deserve the name".[31] Of course, this remark had a deep political meaning, as some interpreters point out,[32] but not only as a critique of the reification of aesthetic experience into a commodity or a spectacle. Dewey was also saying that a *social* experience is always an experience of communal life, an experience of attachments, but "attachment is a very different function of life from affection [...] The very things which stimulate and intensify affections may undermine attachments. For these are bred in tranquil stability; they are nourished in constant relationships. Acceleration of mobility disturbs them at their root. And without abiding attachments associations are too shifting and shaken to permit a public readily to locate and identify itself".[33] Inflation of affections, then, could amount to being a sign of a *lack* of the social structures of recognition that helps a public to become a great community.

In fact, Dewey observed ironically, many American virtues would be symptoms of certain discontentment in culture. "Generosity, good will and altruism —they are probably more marked features of American life than of that of any other civilization at any time. Our charity and philanthropy are partly the manifestation of an *uneasy conscience.*"[34] Even the 'joining' habit of the average American, Dewey added, reveals a social void. "We should not be so averse to solitude if we had, when we were alone, the companionship of communal thought built into our mental habits. In the absence of this communion, there is the need for reinforcement by external contact. Our sociability is largely an effort to find substitutes for that normal consciousness of connection and union that proceeds from being a sustained and sustaining member of a social whole."[35] "The American habit of "joining', is a tribute to the reality of the problem but has not gone far in solving it".[35] To some extent, then, some extreme forms of communal life and some versions of recognition could be seen as defensive reactions, or, as Dewey says, as substitutes of the social integration that a society is not able not create.

In short, a democratic community, in which individuals see each other, requires much more than deliberative procedures, but also much more than affections. It requires a good deal of receptiveness, and some enduring practices of imaginative observation and reflection. It does not require, however, the type of compulsive bonds which characterize the most parochial forms of communal life.[36] The process of sharing common values or a mode of life —Dewey would say— never can be grounded in a mere re-enactment of something given: it needs to be the result of a prospective experience.[37]

6. Dewey after Honneth

What I have said should not be read as if I wish to subsume Dewey's idea of social integration under Honneth's model of justice (with his pivotal notion of recognition). If there are good reasons for not calling Dewey a precursor to deliberative models such as Habermasian ones, there are also good reasons for not converting him into an imitation of Honneth's approach to politics and culture. Certainly though, Dewey's model (even in later works as *Freedom and Culture*) is closer to Honneth's single perspective than to Habermasian dualism of moral sphere and deliberative sphere. However, Dewey never formulated a monistic definition of the good, and probably Honneth himself has never endorsed the monistic view of justice that some of his opponents sometimes attribute to him. Both perspectives have in common the insistence on the moral dimension of social action, but not the will to design a moral philosophy.[38]

Honneth's reading, however, is quite debatable, but not because it occasionally could inspire a reading of Dewey as the enemy of procedural democracy that he was *not*. The divergence concerns other issue: the connection Honneth establishes in *Reification* between Dewey's philosophy and some European thinkers. Dewey was right not only in providing a moral approach to politics —Honneth comes to say— but also in describing in existential terms the process by which relations of individuals with nature and society turn reified. In fact, he goes so far as to suggest that Dewey's notion of experiencecould be seen in continuity with Heidegger's early notions, such as *Sorge*. There is, in my view, a highly relevant discrepancy: Heidegger's philosophy does not provide the basis for a moral understanding of democracy as Dewey's does.[39]

Even in his most "metaphysic" expressions (*Experience and Nature*), Dewey's idea of experience is related to something less sublime but much more civilized that Heidegger's vision of *Dasein*. Heidegger, —Honneth should have remained in mind— detested Max Weber and other sociologists and anthropologist of his time. By contrast, Dewey's generic view of existential engagements, no matter what degree of generality it requires, was *not* the sort of rebellion against the social empirical sciences that many European philosophers like Heidegger enthusiastically set out since the early years of the past century. In other words, the difference between Heidegger and Dewey is that even such theoretically burdened works like *Experience and Nature* can be read as an imaginative combination of philosophy, cultural anthropology, social psychology and cultural history, or as a set of descriptions of human concerns with some normative recommendations. But it is not difficult to imagine Heidegger scorning such a mixture and eventually Dewey's own moral perspective of political affairs.

It is also not difficult to imagine why Dewey appeared in Honneth's own philosophical anthropology, developed in early books as *Soziales Handlen und menschliche Natur* (1980, in collaboration with Hans Joas), and how such anthropological views were connected with a political horizon in his *Kritik der Macht* (1988), *Die Zerrissene Welt des Socialen* (1990) or his famous *Kampf um Anerkennung* (1992). Among other reasons, Dewey tried to do something similar in his own terms: to merge social psychology and politics into an overall view. The merit of Honneth consists in contributing positively to Dewey's rehabilitation in the context of the debates between liberalism and multiculturalism, or between redistribution and recognition, while other Hegelians who were also interested in the relation between human agency and politics (Charles Taylor, for example) ignored Dewey.

Obviously, Dewey's idea of democracy is far from that of more radical contemporary thinkers who have insisted on antagonism and difference. However, as I hope to have shown here, Honneth's reading of Dewey opened a route which goes beyond regarding him as an edifying moralist who did not trust in veils of ignorance and put democracy in hands of good sentiments and habits of heart. Honneth could have made use of Dewey's view of the moral structure of democracy, as well as recognition as experience of others in their individuality.

Interestingly, in his *Reification* Honneth also refers briefly to William James's ideas about certain *blindness* in human beings. The fact is that, keeping in mind what Dewey really said about the moral dimension of democracy, one must say that his gigantic work constituted a magnificent attempt to think seriously about the social conditions which help to avoid such types of blindness.[40] The truth is that William James didn't like Hegel too much (although he demonstrated he knew to read him more creatively than many other readers) but Dewey was philosophically educated in a Hegelian atmosphere, and even when he distanced himself from Hegel, he transformed James's trust in the capacity of sensitive *individuals* to see others, into the *social* art of experiencing the lives of others, an art without which a democratic way of life could develop better and more just institutions.

Honneth's interpretation of Dewey, complementary as I have proposed here, offers a unique guide to understanding this legacy. As I said above, there is only one other thinker (who has dialogued permanently between American and European traditions) capable of reinstating Dewey in the context of contemporary political philosophy with a comprehensive perspective, and his name is Richard Bernstein. But I must leave the discussion of his Dewey for some other occasion.[41]

NOTES

1. On Richard Bernstein's reading see my introduction to the compilation of his essays on Dewey (Herder, Barcelona, 2010, forthcoming). Of course, Hans Joas, *Pragmatism and Social Theory* (Chicago, Chicago University Press, 1993) is also indispensable to understanding the reinsertion of Dewey in the European contemporary debate. James Kloppenberg's *Uncertain Victory: Social Democracy in European and American Thougth, 1870-1920* (Oxford University Press, NY, (1986) offered a widely held view of the transatlantic links between Dewey and the European political *milieu.* John P. Diggins's *The Promise of Pragmatism* (The University of Chicago Press, Chicago, IL, 1994), did also tender a variety of connections between American pragmatists and European political philosophers. More recently, Robert Westbrook's *Democratic Hope: Pragmatism and the Politics of Truth* (Cornell University Press, Ithaca, NY., 2005), relocates Dewey in the context of contemporary debates on democratic deliberation and education.

2. See Axel Honneth, "Democracy as Reflexive Cooperation: John Dewey and the Theory of Democracy Today" (*Political Theory*, vol. 26, 6, 1998) pp. 763-783, and "Between Proceduralism and Teleology: An Unresolved Conflict in Dewey's Moral Theory" (*Transactions of the Charles S. Peirce Society*, vol. XXXIV, 3, 1998), pp. 689-711. See also Honneth's references to Dewey in *Reification. A Recogntion-Theorethical View*, The Tanner Lectures on Human Values, University of California, Berkeley, March 14-16, 2005, reprinted as "Reification and Recognition. A New look at an Old Idea", with comments by Judith Butler, Raymond Geuss and Jonathan Lear, in Martin Jay, ed., *Reification. A New Look at an Old idea,* (New York, Oxford University Press, 2008).

3. *Freedom and Culture* (1939), *The Later Works of John Dewey, 1925-1953* (Carbondale: Southern Illinois University Press, 1981-1991), Vol. 13, p. 74.

4. Ibid., p. 80.

5. *Democracy and Education* (1916), in *The Middle Works of John Dewey 1899-1924* (Carbondale: Southern Illinois University Press, 1976-1988), Vol . 9, p. 25.

6. *The Public and Its Problems* (1927), in *The Later Works of John Dewey*, Vol. 2. See my introduction to the Spanish version, *La opinión pública y sus problemas* (Madrid, Morata, 2004)

7. *Freedom and Culture,* pp. 69-70. See also pp. 67-68 and p. 186, on political institutions as effect of diverse cultural forces and not as the ultimate factor of social life.

8. Ibid., p. 71.

9. On Habermas and Dewey see: Larry Hickman, "Habermas's Unresolved Dualism: *Zweckrationalität* as *Idée Fixe*" (org. from 1996) in Lewis Edwin Hahn, ed., *Perspectives on Habermas* (Chicago and La Salle, Illinois, Open Court, 2000), also included with variations in Hickman, *Pragmatism as Post-postmodernism* (New York, Fordham University Press, 2007) as "Classical

Pragmatism and Communicative Action: Jürgen Habermas". Matthew Festenstein, *Pragmatism and Political Theory. From Dewey to Rorty* (Chicago, The University of Chicago Press, 1997). Scott Bartlett, "Discursive Democracy and a Democratic Way of Life", also in Hahn, ed., *Perspectives on Habermas.* Sandra Rosenthal, "Habermas, Dewey and the Democratic Self', in Mitchell Aboulafia, Myra Bookman, and Catherine Jemp, *Habermas and Pragmatism* (London, Routledge, 2002). Also John R. Shook, "Deliberative Democracy and Moral Pluralism. Dewey vs. Rawls and Habermas", and Paul Thompson, "Pragmatism, Practical and Discourse Ethics", both in John Ryder and Kristina Wilkoszewska, eds., *Deconstruction and Reconstruction.* Studies in Pragmatism and Values, The Central European Pragmatism Forum, Vol. 2 (Amsterddam-NY, Rodopi, 2004).

10."The Ethics of Democracy" (1888), in *The Early works of John Dewey, 1882-1895* (Carbondale: Southern Illinois University Press, 1969-1972), Vol. 1, p. 241.

11. Ibid., pp. 234, 240.

12. "The problem of Truth" (1911), in *The Middle Works of John Dewey*, Vol. 6, p. 16 . See my "A Grammar of Experience: John Dewey and Cultural Struggles", in Maura Striano and Stefano Oliveiro, eds., *John Dewey's Thought between Psychology, Philosophy and Pedagogy* (Napoli, Fridericiana, 2009, forthcoming) on how this text modifies partially Honneth's interpretation of Dewey's ideas about social integration and structure of labour.

13. *Democracy and Education,* p. 93.

14. Ibid., pp. 7, 8.

15. *Reconstruction in Philosophy* (1920), in *The Middle Works of John Dewey*, Vol. 12., p.186.

16. Ibid., pp. 197-198.

17. *The Public and Its Problems,* p. 36.

18. See Jürgen Habermas, *Between Facts and Norms: Contributions to a Discoursive Theory of Law and Democracy* (Cambridge, Mass. MIT Press, 1992), chapter VII, "Deliberative Politics: A Procedural Concept of Democracy."

19. *Human Nature and Human Conduct. An Introduction to Social Pshychology* (1922), in *The Middle Works of John Dewey*, Vol. 14, p. 43.

20. "Context and Thought" (1931), in *The Later Works of John Dewey*, Vol. 6, pp 3-21. See also "Qualitative Thought" (1930), in *Ibid.*, Vol. 5, pp. 243-262.

21. On Honneth's criticism of this position (defended by Habermas in *Between Norms and Facts*) see again, "Democracy as Reflexive Cooperation: John Dewey and the Theory of Democracy Today", *Ibid.*, pp. 779-780. Compare with his critique of Habermas's functionalism in "Reification and Recognition: A New Look at An Old Idea". See also Hans

Joas's criticism of Habermas's functionalism in chapter 6 of *Pragmatism and Social Theory*.

22. Axel Honneth, "The Other of Justice", in Stephen K. White, ed., *The Cambridge Companion to Habermas* (Cambridge University Press, 1995, p. 304). See broader relations between ethics, alterity and neopragmatism in Leszek Koczanowiz, "Sources of Solidarity: Is Non-Foundational Ethics Irrational?", in John Ryder and Kristina Wilkoszewska, eds., *Deconstruction and Reconstruction*, pp. 69-88.

23. *The Later Works of John Dewey*, Vol. 13, p. 303.

24. Ibid., Vol. 14, p. 228.

25. "Philosophy and Democracy", in *The Middle Works of John Dewey*, Vol. 11, p. 53. See a broader and exhaustive analysis of this point in Stefen Neubert, "Dewey's Pluralism Reconsidered" and Kersten Reich, "Democracy and Education", both in Jim Garrison, ed., *Reconstruction Democracy, Recontextualizing Dewey* (State University of New York Press, 2008), pp. 89-117 and 55-88.

26. James Gouinlock, *John Dewey's Philosophy of Value* (New York, Humanities Press, 1972), p. 239.

27. *Reconstruction in Philosophy*, *Ibid*., p. 181. On the variety of virtues and capabilities that democracy requires see Gregory Pappas, *John Dewey's Ethics. Democracy as Experience* (Bloomington and Indianapolis, Indiana University Press, 2008).

28. "Half-Hearted Naturalism" (1927), in *The Later Works of John Dewey*, Vol. 3, pp.75-76.

29. See Honneth's remarks on the *historical* dimension of recognition in Nancy Fraser and Axel Honneth, *Redistribution or Recognition. A Political-Philosophical Exchange* (London, Verso, 2003), p. 246.

30. See on a similar interpretation of Dewey, Sidney Hook, Introduction to John Dewey, *Experience and Nature*, in *The Later Works of John Dewey*, Vol. 1. p. xiv.

31. *Art as Experience* (1934), in *The Later Works of John Dewey*, Vol. 10, p. 51.

32. See Robert Westbrook, *John Dewey and American Democracy* (Ithaca, Cornell University Press, 1991). See also Martin Jay, *Songs of Experience. Modern American and European Variations on a Universal Theme* (Berkeley, University of California Press, 2005), p. 166-67.

33. *The Public and Its Problems*, in *Ibid*., pp. 322-23

34. *Individualism, Old and New* (1930), in *The Later Works of John Dewey*, Vol. 5. p. 84.

35. *Ibid*., pp. 82-83.

36. *Freedom and Culture*, in *Ibid*., p. 176.

37. On Dewey's own notion of community see James Campbell, "Dewey's Conception of Community" in Larry Hickman, ed., *Reading Dewey: Interpretations for a Postmodern Generation*, (Bloomington and

Indianapolis, Indiana University Press, 1998), pp. 23-42. On a broader context, see Richard Bernstein, "Community in the Pragmatic Tradition", in Morris Dickstein, ed., *The Revival of Pragmatism. New essays on Social Thought, Law, and Culture* (Durham, North Carolina, Duke University Press), pp. 141-156.

38. See as an example of this misunderstanding of Dewey Raymond Geuss' reply to Honneth in Martin Jay, ed., *Reification*, Ibid., p. 124.

39. See on this Han Joas's comparison between American Pragmatism and German thinkers in *Pragmatism and Social Action*, chapter, 4. Joas based his valuation on Heidegger on Carl Friedrich Gethmann's works.

40. On social visibility see Avishai Margalit, *The Decent Society* (Cambridge, Mass., Harvard University Press, 1996), chapter 6. See also Theodore George: "From Invisibility to Intimacy: Honneth, Gadamer and the Recognition of the Other" (Texas A&M, Philosophy Dpt., forthcoming).

41. I would like to thank Mariana Alessandri (Penn State University) and Alicia García (Universidad de Barcelona) for having read carefully a draft of this paper. Antonio Gómez's (Univ. Carlos III de Madrid) comments have also been of great value. This paper has been written with the support of a Research Project sponsored by Ministerio de Ciencia e Innovación, Gobierno de España. *Esfera pública, conflicto de valores y experiencia social. Una perspectiva pragmatista* (FFI2008-03310/FISO).

Eight

DEWEY'S VALUE THEORY AND THE ANALYTIC TRADITION OF MORAL PHILOSOPHY

Ángel M. Faerna

Although pragmatist philosophers in general, and John Dewey in particular, were mostly concerned with human improvement and the promotion of the habits of thought and judgment that are more likely to result in new and better forms of individual and social life, it is striking how little their views have been taken into account among contemporary moral philosophers. Two facts should be considered in this regard. First, that contemporary moral philosophy has knowingly avoided every approach to ethics that could be seen to involve normative commitments. On the basis of this assumption —and this is the second fact—, moral philosophers have limited themselves to surveying moral language in order to attain a purely descriptive, meta-ethical analysis of moral concepts and value judgments. These two facts would account for the relative marginalization of Dewey's contribution, seeing as (1) Dewey explicitly denied that linguistic analysis could actually throw any light on the real content and import of moral evaluations, and (2) he was persuaded that a philosophical clarification of ethical subject-matters would result in some kind of "moral progress."

Now, I believe that Dewey was right in challenging these two central tenets of the analytical approach to ethics, i.e., that moral language is the key issue in any philosophical account of the subject, and that in adopting a meta-ethical point of view philosophers are evading any prescriptive form of discourse. I will tackle both these questions separately. The first section of this paper deals with Dewey's discussion of "emotive meaning" —the notion that originated the idea that moral expressions belong to a distinctive semantic category and should not be treated as real propositions— in his 1939 essay "Theory of Valuation"[1] My point will be that Dewey's arguments against that notion, judged with hindsight, were grounded on a better understanding of meaning and on a more adequate insight into the nature of language. In the second section, Dewey's concept of an "empirical ethics" will be examined in order to show that (1) it satisfies the methodological requirements of a meta-ethical viewpoint, but that (2) it also involves a normative commitment to attain justified moral valuations.

1. Emotive Meaning

There is a sense in which Dewey's "Theory of Valuation" should not be taken as a central reference when assessing the real character and scope of Dewey's contribution to ethics. In this work, "values" valuations are taken mainly as an outcome of intellectual operations and processes, and not so much as a qualitative aspect of experience. It is not the distinctive character of certain values as moral that concerns Dewey here, but the essential unity he claims that exists between value judgments —whether moral or not— and matter-of-fact judgments as regards the methods by which they are posed, tested, and verified. Therefore, upon reading "Theory of Valuation" one could get the erroneous impression that Dewey should be labelled as a straightforward intellectualist in ethics, which he certainly was not, and that his approach to moral problems deprived these problems of their vital import and equated them to a sheer calculus of instrumental utility. As Gregory Pappas has quite rightly observed:

> To take Dewey's theory of value, as it appears in such places as *Theory of Valuation* [...], as central or as good summaries of Dewey's mature moral thought is a mistake. [For they] leave out or do not do justice to what is explicitly recognized in his 1930 essay 'Three Independent Factors in Morals': that there are at least three distinct and incommensurable qualities designated by moral value [i.e., good, duty, and virtue]."[2]

But this, of course, does not mean that "Theory of Valuation" is a loose end or even a deviation in Dewey's ethical views. Far from that, it re-states one of the central themes of his whole philosophical framework: that is, the continuity between the world of facts and the realm of values and, in more general terms, the close link that philosophy should help to reinforce between our intellectual abilities and our practical (moral, political, prudential...) decisions, or, in short, between knowledge and action. This certainly is the spirit with which we should read some statements in "Theory of Valuation" such as, for instance "not only is science itself *a* value [...] but it is the supreme means of the valid determination of all valuations in all aspects of human and social life,"[3] which a careless reader could take for the most blatant declaration of scientism and of moral short-sightedness.

Now then, why in his "Theory of Valuation" did Dewey blur the distinction between moral and non-moral judgments of value, as well as between value judgments and matter-of-fact judgments, after having disclosed the variety of independent qualitative aspects of moral values that preclude any attempt to reduce them to non-moral elements of experience in his 1930 essay ("Three Independent Factors in Morals")? Moreover, why is the "Theory of Valuation" mainly focused on one of those qualitative aspects of moral experience (the good, or the ends), leaving out the others (duty, virtue) he had already discerned?

It is precisely the context of the latter work that provides the answer to such questions. Pappas[4] is right once again when he hints that the polemical character of "Theory of Valuation" is a key factor. Let's recall that an important event had taken place between 1930 and 1939, one that was going to change America's philosophical bent and would seal the fate of Pragmatism: namely, the arrival of European neo-positivists to the United States. We should also remember that the "Theory of Valuation" was published as volume II, number 4, of the International Encyclopaedia of Unified Science, edited at the time by Otto Neurath. This strongly determined the purpose of the essay as well as the line of argument chosen by Dewey. On the one hand, Dewey wished to criticize the emotive theory of values proposed by Ayer and the Vienna Circle a few years earlier. But, on the other, he tried to do so in a way that could be understood as well as eventually accepted by the positivists themselves. There was, I believe, a sympathetic impulse or, at least, a non-contentious mood in Dewey's efforts to persuade positivist philosophers that value-talk could be perfectly meaningful —a tenet that Positivism had declared "metaphysical" in the derogatory sense in which the positivists used this word. This was why he placed so much stress (possibly too much) on the "scientific" spirit of his ethics, and why he was so eager to prove that judgments of value are not all that different from judgments of fact.

The polemical context of "Theory of Valuation" is also important to explain the fact that it only concentrates on "the good" or what is praised and desired as an end, ignoring the other qualitative aspects of moral values. For it was such terms as "good", "bad", "desirable", etc., that had been the object of minute examination by the positivists, as well as by Moore and Russell before them. This also forced Dewey to account for the meaning of these terms or, in other words, to adopt the language-oriented standpoint that analytic philosophy had imposed on ethics by that time. It is no surprise, then, that Dewey phrases the question which "Theory of Valuation" was meant to answer as "the problem of the possibility of genuine propositions about the direction of human affairs."[5]

Let us concentrate now on this discussion about the meaning of moral terms and, consequently, about the possibility of moral propositions. According to Ayer's standard version of emotivism, there can be no value-propositions since propositions only state facts and values are not facts. Hence, to use an example that Dewey quotes from Ayer:

> If I say to someone, 'You acted wrongly in stealing that money,' I am not *stating* anything more than if I had simply said 'You stole that money'.[6]

The first sentence ("You acted wrongly in stealing that money") should be analysed as containing the proposition "You stole that money" plus an extra element, the term "wrong", similar to a prosodic mark, which provides information about a certain feeling in the speaker. Thus, the sentence says that a theft took place and shows how the speaker feels about it. In the more sophisticated version of emotivism that Stevenson would develop a few years later, this alleged double function of the sentence was to be respectively re-phrased in terms of "descriptive" versus "emotive" meaning. The "emotive meaning" or "expressive function" of value-propositions was assimilated by emotivist theorists to non-propositional linguistic utterances such as interjections or imperatives, and these in turn were assimilated in their function to natural manifestations of emotions such as blushing, smiling, weeping, and so on.

But, according to Dewey, there is a serious misunderstanding as to the nature of linguistic expressions or signs here. Admittedly, a natural or spontaneous smile expresses that the smiling person is in a certain state; indeed, any other organic manifestation of that person's body, such as the fact that the person sweats, coughs or sneezes, would be expressive or informative in the same way. This, however, is by no means linguistic. Organic reactions can be used by an observer to put forward causal hypotheses like "If she is weeping, it probably means that she did not win the match," or "If she sweats, it probably means that the room is overheated." The fact that the observer can arrive at the right hypothesis concerning the causes of the weeping or the sweating, taken as effects, is not a matter of communicative competence, but of empirical knowledge of how things work in the world. It is a wholly different question if weeping or smiling can also be used as a sign to say something and to convey a meaning. These purposive uses, as Dewey puts it, "exist in the medium of language."[7] In this case, the expressions are not effects or spontaneous reactions but instances of non-verbal communication intended to evoke a response in another person. So firstly, emotivism fails to make a distinction between "symbols" and mere "signals" (or "indices", if we adopt Peirce's terminology) with regard to these expressions.

Consequently, the analogy between value-judgments and emotive expressions should be referred, if at all, to the symbolic or quasi-linguistic use of such expressions and not to their primary, non-linguistic function. No evaluation is involved in a smile or an instance of weeping whenever these appear as purely organic reactions, because organic reactions as such do not presuppose a subject is even aware that there is something wrong (or right) with the surrounding situation, or even thinks that something should be done about it. However, when smiling or weeping can be said to have an evaluative or estimative import, they clearly possess meaning and the conclusion the emotivist is trying to draw from the proposed analogy does not follow.

But the emotivist still has an opportunity to make his point by distinguishing between a descriptive aspect and an emotive aspect within meaning. In making a proposition one refers to some fact, and this referent contributes descriptive meaning and cognitive content to the proposition. However, the utterance will usually include other elements that provide information about the speaker's emotions and attitudes concerning the fact described, and these elements are said to constitute its emotive meaning, which is non-cognitive and accounts for the persuasive or prescriptive intentions of the speaker.

Now, to make sense of this distinction some linguistic signs need to be identified as performing the specific function of meaning emotions and attitudes; otherwise, all that we have is the rather trivial remark that whatever a person says or means can betray his attitudes and influence other people's behaviour. For the emotivist, that is precisely the function fulfilled by typically moral terms like "good", "bad", "right", or "wrong". In support of this analysis, it is alleged that other special words, like interjections, admittedly play this very role in language.

Dewey, however, denies that the semantic role of a term within a sentence can be specified without reference to what he calls the "existential context" of the utterance. In order to have any meaning at all, not only value terms but also interjections and even quasi-linguistic signs like smiles or weeping depend on particular identifications and discriminations that should be made within the situation in which they occur (real-life situations, by the way, usually exist within a long time-span, so that the identifications and discriminations in question may be fairly complex and demand a great deal of empirical knowledge to be substantiated; and all this is "cognitive labour"). Only the existential context can tell whether the word or sign means an emotion and, if it does, what kind of emotion it is —for obviously they can be very different, some of them connected with the very same term or sound or gesture. But, in such a case, all these words would have a referent, namely, the particular emotion or attitude they stand for. The idea that emotive meaning is defined by opposition to descriptive meaning, and that meaning is attached to words and terms independently of their context, constitutes emotivism's second mistake.

Perhaps the emotivist may concede that emotive meaning refers to emotions, as long as it is recognized that emotions are not facts. This, of course, was the initial claim made by positivists, but their efforts to state that claim on a semantic basis seem to have failed up to this point. For Dewey it was clear that, on a psychological basis, emotions or feelings are facts because he thought of them in behavioural terms, as parts of the total organic state of an individual involved in a situation. Indeed, Dewey accused positivism of having relied on an:

> [...] alleged psychological theory which is couched in mentalistic terms, or in terms of alleged states of an inner consciousness or something of that sort.[8]

If you set out with a definition of value judgments in terms of unobservable ("metaphysical") entities like mental states, it is no wonder that you are soon led to the conclusion that value judgments are not empirical propositions at all but metaphysical pseudo-propositions; though if a different psychology is employed, the ensuing conclusion may also be different. Dewey's point, in any case, is no less semantic than psychological. Mental states cannot be the meaning of any word or sign, for if this were the case:

> [...] there cannot be any assurance that any given word when used by two different persons even refers to the same thing, since the thing is not open to common observation and description.[9]

Therefore, emotivism's third mistake consists in ignoring the fact that meaning—all meanings— are essentially public.

Now then, what about the persuasive and prescriptive function value judgments typically perform? In saying that valuations are genuine propositions and therefore have empirical meaning and cognitive content, was Dewey denying that they serve to direct or influence other people's behaviour? Certainly not. In a critical review of Stevenson's *Ethics and Language*, which appeared in 1945, he clearly stated that "of ethical sentences as ordinarily used, it may be said, I believe, that their *entire* use and function is directive or 'practical'."[10]Therefore, for Dewey it is one thing to understand the function performed by a sentence as it is used, and a wholly different matter to apprehend its meaning as a sign. Or to put it in another way, a sentence taken as a means to some purpose should not be confused with the sentence as a statement of a meaning. Dewey suggested that Stevenson fell victim to this confusion:

> I get the impression that Stevenson is influenced at times in connection with the 'meaning' of moral judgments, by that ambiguity in which meaning has the sense of both design or purpose and that which a sign indicates.[11]

It is interesting to note that John Searle would denounce this very same mistake in the philosophy of moral language of his own time almost twenty years later:

> To point at the association between 'good' and commendation is not to answer the question 'What does 'good' mean?' but the question 'What is it to call something good?' Thus the old philosophical question 'What is it for something to be good?' has been confused with the question 'What is it to call something good?' The assumption that these two questions are the same is one of the most

> common mistakes in contemporary philosophy; it is a mistake arising from a particular interpretation of the view that the meaning of a word is its use.[12]

This does not mean that Dewey (not to say Searle) believed that the meaning of a term can subsist apart from its uses. As we have seen, quite to the contrary, Dewey stressed the importance of the "existential context" of utterances in order to determine the semantic import of terms; and the "existential context" firstly includes the purposes and expectations the speaker associates to his own use of signs. The point is simply that facts concerning the use and function of signs do not answer in themselves the question of what kind of object a particular sign stands for. The failure to realize this point can be considered as emotivism's fourth mistake.

Let me summarize the conclusions of this section in a way that emphasizes the fact that the conception of meaning that Dewey put forward at least as far back as 1939 anticipated several important traits of the post-positivist philosophy of language.

(1) In pointing to the difference between signals and symbols, Dewey focuses on the pragmatic aspect of meaning. It is the speaker's communicative intention when uttering certain sounds that renders such sounds meaningful. When meaning is so closely linked to the purposive uses of language, that is, to performing characteristic acts ("speech acts"), it can no longer be maintained that "You did wrong in stealing that money" means the same as "You stole that money."

(2) When Dewey demonstrates that non-descriptive expressions, such as interjections or imperatives, should not be treated as non-propositional, he avoids the "descriptivist fallacy" which is characteristic of positivist theories of meaning. To assume that meaning only has to do with describing or picturing facts is a form of reductionism which ignores the many uses of language.

(3) When Dewey states that meaning should be analysed by referring to "existential situations," he stresses the continuity between linguistic and non-linguistic behaviour. Language is embedded within a whole practical intercourse among individuals within a shared situation. Consequently, meaning belongs to social practice and is essentially inter-subjective.

(4) Lastly, Dewey warned against the temptation of trying to solve — or dissolve— significant philosophical problems by means of purely linguistic analysis. Concerning moral issues in particular, positivism's errors prove that misconceptions about meaning can cloud over a clear understanding of ethical debates and conflicts of value. However, an improved conception of meaning as can be found in later developments in the analytic philosophy of moral language does not in itself resolve all the problems. As Dewey put it:

> The conclusion is that verbal usage gives us little help. Indeed, when it is used to give direction to the discussion, it proves confusing. The most that reference to linguistic expressions can do at the outset is to point out certain problems. These problems may be used to delimit the topic under discussion.[13]

2. Empirical Ethics

The analytic tradition of moral philosophy has been mainly concerned with a specific set of problems that is usually labelled under the name "meta-ethics." The discussion revolved around such topics as the meaning of moral terms, the status of moral sentences, the nature of moral disagreements, or the possibility of moral knowledge. Little was said by professional philosophers about what particular objects deserved the moral terms "good" or "bad", which particular assertions should be made in moral sentences, who is right when a specific moral disagreement arises, or what we know (if anything) in the realm of moral values. Most theorists insisted that their meta-ethical claims were neutral as regards these ethical questions. Moreover, the fact that those meta-ethical claims need not vary as we consider different or even incompatible ethical answers would count as evidence for their acceptability. Intuitionism, naturalism, emotivism, or prescriptivism professed to be just descriptions —at a sophisticated and logically apt level— of moral discourse, not moral discourses themselves.

Now, when it comes to the task of placing Dewey's ethics within this theoretical landscape the situation appears to be puzzling. On one hand, Dewey could be labelled as a "meta-ethical" theorist since his entire work on ethics was devoted to attaining a clear understanding of what notions, operations and activities are ingredients in the construction of any moral perspective (though for him the discursive or linguistic aspect of morals was, as we have seen, of secondary importance). On the other, however, we feel that Dewey's theory is clearly committed to finding some answers, albeit very general ones, to those ethical questions to the detriment of others. Of course this remark is not novel, for it lies at the root of two criticisms that Dewey's approach to ethics typically received in the past. From meta-ethical quarters, Dewey was accused of smuggling a great deal of prescriptive views into his descriptive analyses. Other commentators, in turn, dismissed those very views as hopelessly vague and incomplete as to their final import in terms of definite ends and substantial values.

A simple analogy will show that both criticisms missed the point. Consider the set of problems that usually goes under the name "philosophy of science." There is a sense in which these problems should be properly referred to as "meta-scientific," for they mirror almost point by point the meta-ethical questions mentioned above: the meaning of scientific terms (whether theoretical or observational), the status of scientific statements, the nature of scientific disagreement (i.e., theory-conflict), or the possibility (epistemic status) of scientific knowledge. To this extent, it is not the task of philosophers of science to tell us what objects in particular should be designated by scientific terms, what the content of scientific sentences should be, which particular scientific theory must prevail over its rivals, or what we know (if anything) about the world of facts. In a sense, once again, philosophy of science is neutral in relation to these scientific problems, as it is just a description of scientific discourse and not a scientific discourse itself.

In another sense, however, philosophy of science is —or purports to be— overtly prescriptive. In describing the notions, operations and activities which characterize the scientific endeavour, it states at the same time a normative distinction between science and pseudoscience. The conditions imposed on scientific inquiries by philosophers of science must be general enough so as to allow for alternative theories explaining the same phenomena. However, if these conditions are to have any philosophical significance —and not, for instance, a merely sociological interest—, they must also exclude bogus science. Conversely, such conditions must be sensitive to the merits that render any theory a scientific accomplishment, but cannot be so demanding (and pretentious) as to prejudge which of them is more likely to be true.

I do not think this analogy is unfaithful to the spirit of Dewey's ethics. Far from it, it captures the idea —which Dewey expressed at a very early stage— that it is not the task of ethics to tell us "what to do, but *how to decide* what to do."[14] In other words, meta-ethics is a methodology, and methodology is no less prescriptive than descriptive. In calling some types of inquiry "science" or a particular assertion "scientific", we are ascribing a special value to them according to certain standards, which are not scientific themselves. In exactly the same way, every time Dewey calls some types of inquiry "ethics" or a particular assertion "ethical", he is endowing them with a special, non-ethical value. Again, we are not betraying the spirit of Dewey's thought if we assume that the standards and values involved are the same in both cases: they amount to the statement that such inquiries and assertions stem from experience, have been submitted to systematic processes of intelligent elaboration, and are subject to an ongoing assessment through experimental testing and control.

This normative component in Dewey's use of the word "ethics" is one of the features that differentiated his approach from that of the analytic moral philosophers. It created the false impression that Dewey was going beyond the limits of intellectual analysis, while falling short at the same time of a straightforward, full-blown moral theory. But the question, of course, is whether the analysis can dispense with that normative component without losing its whole point, as would be the case if we attempted to develop a philosophy of science in which the distinction science/pseudoscience were not applicable.

When criticizing Stevenson's *Ethics and Language*, Dewey wrote:

> There can be no doubt that sentences *claiming* to be ethical often use an extra-cognitive 'emotive' factor to influence conduct. [...] As far as non-cognitive, extra-cognitive, factors enter into the subject-matter or content of sentences purporting to be legitimately ethical, those sentences are by just that much deprived of the properties sentences should have in order to be genuinely *ethical*.[15]

More than thirty years later, Stevenson's reply was:

> I was taking ethical sentences just as I found them in common use, regardless of whether their use was 'genuinely' ethical, and regardless (accordingly) of whether they were supported by reasons.[16]

Thus, it is evident that while the sheer claim to be ethical does not fulfil Dewey's requirements for a sentence to be such, for Stevenson (and I believe that he can be taken on this point as a representative of a whole family of analytic moral philosophers) it is the common use of language that determines whether a sentence counts as a moral statement—moral philosophy therefore becomes a province of the philosophy of language.

The problem with a demarcation criterion based on the common use of sentences, or of terms, is that it tends to ignore the contexts within which such sentences and terms acquire their specifically moral meaning and express specifically moral valuations. When people use a "moral term" —instead of any other commendatory idiom, or a simple imperative— in order to induce this or that behaviour, they are not just saying that the behaviour should be accomplished. If they mean what they say (i.e., if they are not just trying to be emphatic), they are actually investing it with a distinct value. What they mean is that, no matter what other considerations could speak in favour of an alternative course of action, the behaviour in question must be preferred.

As a matter of fact, we are involved in many different kinds of behaviour, we are subject to very different obligations, and we pursue ends of different sorts. Typically moral questions such as "How should I behave?", "What obligations should I recognize?", or "What ends should I pursue?" are not questions about what to do but about what it is better to do. Therefore, in answering them we are invoking some kind of implicit standard upon which a special value has been stamped. And, by the very nature of the case, this value attribution can be wrong. I am not referring here to moral wrongness; it may be wrong simply because there are no reasons to call upon in order to affirm the moral preference of one course of action over the others. Eventually, it may turn out that there are no genuine ethical sentences at all and all our preferences have to rest upon custom, prejudice, authority, or any of the like. In such a case, our use of a moral vocabulary would have then proven to be as empty and superstitious as the talk of shamans and theologians.

This would be an important and instructive conclusion indeed, both for philosophical and vital purposes, but it cannot be obtained by simply looking into how people actually use "moral terms." Over and above that, we need some kind of normative standard of what would count as a genuine ethical sentence in order to differentiate ethical and pseudo-ethical sentences within the common use of these terms, or even to draw the final conclusion that all ethics is bogus. If we suppress this normative component shared by concepts like "ethics" and "science" (and I think "art" could be added to the same list), we leave out the essential aspect of what people mean when they use such vocabularies, and we preclude in advance the very questions upon which ethical or scientific (or aesthetic) issues seem to hinge, i. e., how to decide what to do or believe (or cherish).

When Stevenson and his philosophical partners were accused of spreading ethical scepticism and moral indifference, they protested that their analyses were being misconstrued, and pleaded "not guilty." As a matter of fact, I believe they were indeed innocent, in both senses of the word. I would happily concede that their intention was not to corrupt our moral souls or to scorn our moral lives. But they should have foreseen such hostile reactions as a natural consequence (alas!) of people's common use of moral terms. For if we are told that moral prescriptions have nothing to do with some standards that happen to be more creditable (for whatever reasons) than persuasion, influence, or emotional outbreaks, from then on we will avoid using such terms, or if we use them, we will no longer mean what we say.

After this short discussion, I feel we are now in a position to fully grasp Dewey's idea of an "empirical (or experimental) ethics" in a rather simple and straightforward way. As we have seen, the central question that any meta-ethical theory must address should be: according to what standards are we to administrate the use of those terms conventionally referred to as "typically moral" by means of which we adjudicate different value-claims? Note that this question (1) presupposes our use of the vocabulary of morals, but (2) cannot be answered by a simple description of that use. What it demands is a methodological decision about how to proceed in order to reach justified, reliable or successful value-judgments. As we bring this question to the foreground of moral philosophy, the problem of meaning becomes increasingly irrelevant. This is because what is preserved of moral vocabulary is merely its "structural function" as the *ultima ratio* within a system of valuations taking place at different levels. Eventually, we could even drop that vocabulary altogether if we so wished. As a matter of fact, Dewey was not very much interested in it. As James Gouinlock so rightly observed in his classic study on Dewey's philosophy of value, "in most discourse of moral import Dewey prefers that moral language not be used at all."[17]

Dewey's methodological proposal then is simply that we should adopt an empirical standard when dealing with substantial ethical problems like which particular objects deserve the moral terms "good" or "bad", what particular assertions should be made in moral sentences, who is right when a particular moral disagreement arises, or what we know in the realm of moral values. What I would like to stress here is that, according to the preceding discussion, this proposal cannot be interpreted as a moral choice in itself nor can it therefore be discussed in moral terms. To say, for instance, that such an empirical ethics denies "the very essence" of moral values and obligations, or that it threatens "the true meaning" of moral life simply does not make any sense at all. These objections, as Dewey repeatedly denounced, invoke a highly esoteric, intellectually non-clarified concept of ethics. They resemble the criticisms that dismissed scientific knowledge for not delving deep enough into "the real nature" of the world. It was not a scientific bias, but a real concern for the relevance and factual standing of our knowledge, that once led us to adopt empirical standards with regard to the meta-scientific question "How to decide what to believe?" Similarly, it is not a particular moral bias, but a real concern for the relevance and factual standing of our value judgments which should now lead us to adopt no less empirical standards with regard to the meta-ethical question "How to decide what to do?"

Dewey's empirical ethics is therefore not committed to a preconceived interpretation of what good is, but rather to a definite interpretation of what experience is. This point exceeds the limits of the present discussion, so I will confine myself to the final remark that Dewey's methodological thesis requires that experience be interpreted as already containing all the materials upon which our value judgments should be built, contrasted, and tested. Value cannot descend on experience from outside or from above, for in such a case, what is observable and predictable would never be the *ultima ratio* that lends support and reliability to our value judgments. In other words, there must be "experienced goods," as there are other experienced traits of the world that supply us with the materials to gain a reliable knowledge about it. These raw goods, as we could call them, form the subject matter of an empirical ethics and cleanse the expression "moral experience" of its transcendent or supernal excrescences. The task of ethics, according to Dewey, is then to procure as much growth and balance as possible to those experienced goods through an intelligent, open-ended process of reconstruction.

NOTES

1. John Dewey, *Theory of Valuation* (1939), in *The Later Works of John Dewey, 1925-1953* (Carbondale: Southern Illinois University Press, 1981-1991), Vol. 13, pp. 189-251.

2. Gregory F. Pappas, *John Dewey's Ethics. Democracy as Experience* (Bloomington and Indianapolis: Indiana University Press, 2008), p. 111.

3. Dewey, *Theory of Valuation*, p. 250.

4. Pappas, *John Dewey's Ethics*, p. 111, n. 11.

5. Dewey, *Theory of Valuation*, p. 193.

6. Ibid., p. 196.

7. Ibid., p. 198.

8. Ibid., p. 199.

9. Ibid., p. 200.

10. John Dewey, "Ethical Subject-Matter and Language" (1945), in *The Later Works*, vol. 15, pp. 127-140; p. 137.

11. Ibid., p. 129.

12. John R. Searle, "Meaning and Speech Acts", *Philosophical Review*, LXXI (1962), pp. 423-432; p. 429.

13. Dewey, *Theory of Valuation*, p. 196.

14. John Dewey, "Teaching Ethics in the High School" (1893), in *The Early Works of John Dewey, 1882-1898* (Carbondale: Southern Illinois University Press, 1967-1972), Vol. 4, pp. 54-61; p. 56.

15. Dewey, "Ethical Subject-Matter and Language", p. 129, n. 3, and p. 137.

16. Charles L. Stevenson, "Introduction" (1978) to *The Middle Works of John Dewey, 1899-1924* (Carbondale: Southern Illinois University Press, 1976-1983), Vol. 5, pp. ix-xxxiv; p. xxiii, n. 19.

17. James Gouinlock, *John Dewey's Philosophy of Value* (New York: Humanities Press, 1972), p. 196.

Nine

SANTAYANA'S CRITIQUE OF DEWEY'S PHILOSOPHY: PRAGMATIC MORALISM AND THE POLITICIZATION OF PHILOSOPHY[1]

Matthew Caleb Flamm

> "...if anyone says that the visible splendor of one moment helps to produce that of another, he does not seem ever to have seen the light."
>
> "...the dominance of the foreground is in all Dewey's traditions: it is the soul of transcendentalism and also of empiricism; it is the soul of moralism and of that kind of religion which summons the universe to vindicate human notions of justice or to subserve the interests of mankind or of some special nation or civilization. In America the dominance of the foreground is...emphasized by the prevalent absorption in business life and in home affections, and by a general feeling that anything ancient, foreign, or theoretical cannot be of much consequence. Pragmatism may be regarded as a synthesis of all these ways of making the foreground dominant: the most close-reefed of philosophical craft, most tightly hugging appearance, use, and relevance to practice today and here, least drawn by the lure of speculative distances."[2]

I want here to establish two, interrelated things. First, I want to highlight a significant point of affinity between Santayana and Dewey that both they and subsequent scholars have overlooked. The affinity is as follows: each thinker diagnoses, with different conceptual approaches and uniquely distinct contributions to the problem, the key mistake of traditional philosophy to be its confused account of nature, a confusion entailing an idealistic fallacy. Second, I intend to argue in support of Santayana's rather than Dewey's solution to this mutually diagnosed problem. The difference between the two solutions was made obvious in the much discussed "dominance of the foreground" critique Santayana provided in his 1925 review of *Experience and Nature*. Santayana's argument that Dewey's philosophy remains too much in the foreground involved his views that philosophy without cosmology is insufficiently naturalistic, and that ontology offers the best means to avoid metaphysical extravagances of the past. I shall examine these views as they transact with Dewey's philosophy, and in my conclusion offer suggestions on their use and application to a contemporary strand of pragmatism.

1. The Published Exchange

The second epigraph passage from Santayana's 1925 review of Dewey's *Experience and Nature* is critically sweeping. Santayana places Dewey and his foreground-hugging pragmatism within a long tradition of human-interest evangelists. Dewey's pragmatism is perceived by Santayana to encapsulate narrow features of the American character; it leaves little to the imagination except action for its own sake and a distracted insistence on the exclusively practical significance of ideas. Santayana describes the "foreground" as the "immediate experience of things," and as "only the dream which accompanies our action."[3] Such language could not impress Dewey, who wrote in the first chapter of the very work Santayana was reviewing in his criticism: "...experience...is no infinitesimally thin layer or foreground of nature, but...penetrates into it, reaching down into its depths..."[4] In his second (1939) reply to Santayana's critique Dewey charged that Santayana's dream-notion of experience is a mistaken product of "the traditional 'mentalistic' view of British psychology," one wrongly severing the mind from its environing conditions.[5]

The larger issue regarding which the two famously traded barbs is naturalism. Santayana accused Dewey of being a half-hearted naturalist.[6] He sensed the deficiency in Dewey's key concepts. Dewey's talk of "events," "affairs," and "situations" conveyed the belief that "human affairs" are "projections, continuations, complications, of the nature which exists in the physical and pre-human world."[7] He used the concepts, in other words, to bridge the alleged gap between human experience and nature. But Dewey's bridge, to Santayana, was a hedge. Such things as events, affairs, and situations revealed the essentially *ethical* nerve of Dewey's thinking.[8] Santayana charged that the *experience* of which these concepts are said by Dewey to be constituted turns out, partly due to its intentionally vague character, to be the social world, the world of politics and convention. For Santayana, *affairs* and *situations* reside forever in the moral foreground, human abstractions wholly distinct from deeper natural realities, from the material conditions funding human experience.

The idea that his world was insufficiently naturalistic and restricted to the moral-political foreground bemused Dewey. The misreading on this point was accounted for as coming from Santayana's mistaken impression that Dewey held nothing but the "immediate" to be real. Dewey conceded to holding that "everything which is *experienced* has immediacy," and that "natural existences" have immediacy in their "unique and brutal [existential] particularity,"[9] but added that this is what he thinks of as *traits* of existence, not existence writ whole. Traits of existence, Dewey protested, should not be confused with existence *exclusively*.[10]

In fact, Dewey detects that very confusion in Santayana's philosophy. To Dewey the purported deeper existential, material realities Santayana insists upon as standing outside of immediacy repeats a tradition of philosophies that estrange humans from nature, that persist in a concept of nature that is wholly "negative and antithetic," devoid of human traits. If his naturalism is half-hearted, then, Dewey replies, Santayana's is broken-backed, a "kneeling before the unknowable and an adjuration of all that is human."[11]

Those strongly loyal to one or the other naturalist lean of Santayana or Dewey undoubtedly walk away with the confidence that theirs was the victor in the charged exchange. As David Dilworth characterizes the end result: "Each won in his own way by privileging the hermeneutical circle of his own thought."[12] To add to reader's frustrations, for all of the difference of opinion displayed, the published exchange seems not to bring to light what scholars have helpfully found to be deeper differences. As Paul Forster has observed, "their impasse is far more intricate and perplexing than either of them suggests in their published contributions...."[13] Forster perceives the deeper impasse to consist in an unbridgeable difference on the *starting point* of philosophic criticism: "Santayana rejects Dewey's naturalism because he thinks transcendental criticism shows that it presumes too much, while Dewey dismisses Santayana's transcendental criticism as indefensible in light of the very science that Santayana finds philosophically dubitable."[14] In similar fashion, John R. Shook has argued that the division can be found in the thinker's opposed attitudes towards the possibilities of an "empiricist naturalism."[15] Santayana, Shook shows, rejects immediate empiricism, which is the basis from which Dewey establishes a functionalist, "transactional and ecological theory of meaning" on which his empiricist-naturalism is founded: once again, never the two shall meet.

Stepping back from these principled points of impasse, there is an important sense in which each thinker can be said to have failed to adequately encounter the other. Certainly credence may be given to each of their broad criticisms. There are respects in which Dewey's vision of nature is (sometimes cringingly) blind to the significance of extra-human realities, and in which Santayana's (sometimes gleefully) demeans human perspectives. But each was too-eager to catch the other in a straw-built lie. If, as Santayana suggests, all that Dewey's careful critical examination of humans, nature, institutions, science, and culture amounts to is little more than a subjectivist fallacy,[16] it is difficult to account for the powerful and pervasive reception of his thinking among scientists and philosophical realists.[17] Conversely, if, as Dewey charges, Santayana's view of nature is merely typical of a dehumanizing tradition in Western philosophy, one is confounded to discover the praise his work receives as an exemplary instance of appreciation for the human condition and as a source for reconciling humans to their deepest experiences of suffering and loss. It seems to me that the published exchange was as much an evasion as an encounter, and if we wish to understand it more fully we have to consult the larger issues looming.

2. Essence, Cosmology, and *The* Philosophic Fallacy

A further, but underappreciated feature of the Santayana-Dewey exchange is the curious lack of acknowledgement of key points of kinship. For anyone reading, with equal degrees of appreciation and respect, books as rich as Dewey's *Experience and Nature* and Santayana's *Scepticism and Animal Faith*, books published within two years of each other, the published exchange displays too little acknowledgment of the closeness of some of the author's respective views. Compare the following passages from the two discussing the understanding of "essence" as found in Ancient Greek philosophy:

> (Dewey): "Greek thinkers...distinguished patterns...of consummatory uses of speech, and the meanings that were discovered to be indispensable to communication were treated as final and ultimate in nature itself. Essences were hypostatized into original and constitutive forms of all existence."[18]
> (And Santayana): "Thus [my notion of] essence, while confirming Platonic logic in the ideal status which it assigns to the terms of discourse... destroys the illusions of Platonism, because it shows that essences, being non-existent and omnimodal, can exercise no domination over matter [i.e; cannot be hypostatized], but themselves come to light in nature or in thought only as material exigencies may call them forth and select them."[19]

Each thinker recognizes here the tendency of Ancient philosophers to hypostatize essences, to confuse the settled results of discourse and thought for the natural realities from which they arise. They both attribute the Ancient's cart-before-horse mistake to an insufficient appreciation of the determinations made upon essences by realities of originating existential influence. Though they speak independently of one another on this issue, and in preferred modes of speech and vocabulary, their shared critique can be summarized as an identification of the idealist fallacy: a philosophic privileging of ideal over naturally existing realities.

The crucial difference on this point lies in the respective solutions proposed to the problem, and it is here that the importance of Santayana's foreground critique of Dewey looms: if Dewey's philosophy remains in the foreground it begs to be said what *background* it lacks. Dewey's philosophy, Santayana takes occasion to note in his critique, lacks *cosmology*.[20] This is a point which deserves more attention than it has received.

To say that Dewey's philosophy lacks cosmology is not to say that it fails to speculate about the character or nature of the world as such. In fact Dewey initiated such speculation when he invoked the phrase Santayana chose to single out as the most problematic: "naturalistic metaphysics."[21] Importantly, the context in which Dewey used the phrase performed the negative work of fending off Platonic or idealistic metaphysics, of avoiding in other words the idealistic fallacy just identified. Dewey characterized the appeal of traditional Platonic metaphysics as one biased by "the [a]esthetic character of logical coherence rather than its tested coherence with fact..."[22] The positive side of Dewey's endorsement of a naturalistic metaphysics entailed its identification of "the generic traits of existence," which he forthrightly claimed turn out to be "evidential of the character of the world."[23] So Dewey's naturalistic metaphysics was an attempt to provide an account of existence that avoids the idealistic fallacy, the Platonic mistake of form-fitting existential realities with essential definitions and meanings.

Moreover, speculating about the character of the world, we find Dewey continually warning, should not succumb to cosmology: that is, giving an account of the world in its alleged totality. Dewey protests cosmology throughout his writings almost pathologically; he speaks of it as an outmoded artifact, an Ancient philosophic penchant for "supreme science"[24]; cosmology is for Dewey nothing but an "animistic teleology" that has thankfully been "disintegrated" in modern times[25]; he associates cosmology with theology as twin enemies of the progress of science[26]; with the (again, outmoded Ancient) belief in the superiority of "Ideal being"[27]; and most strikingly, in *Experience and Nature,* he proclaims the appearance of cosmology as signal of having committed "*the* philosophic fallacy."[28]

Dewey's identification of "*the* philosophic fallacy" gives explicit understanding of his rejection of cosmology. The fallacy involves a critique of the problem of universals vexing historical schools of metaphysicians. Larry Hickman nicely summarizes the argument:

> Dewey's novel solution of [the] traditional problem [of so-called "universals"] was to relocate generality. Realists located generality in existent things: for Platonists, those things were supernatural; for Aristotelians, they were natural. Nominalists denied generality altogether…Conceptualists located the *grounds* for generality in things, but maintained that the activity of the human intellect is required to construct the classes into which they are sorted on the basis of their naturally occurring essential properties…It was Dewey's radical proposal that generality has to do with productive activities undertaken in inference, not with things or events in terms of the status prior to inquiry and as existential.[29]

Hickman goes on to summarize that in Dewey's view, those who solve the problem of universals by appeal to "things" or "events" existing independently of or prior to inquiries undertaken, commit "*the* philosophic fallacy." The fallacy is multiply described by Dewey as "converting" "eventual functions into antecedent existence[s],"[30] a conversion that Hickman, following Dewey, claims opens the way to "gratuitous metaphysical entities and misleading hypostatizing of all sorts."[31] It is this fallacious conversion of eventual functions into antecedent existences that Dewey has in mind as the error of cosmology. *Any* attempt, for Dewey, to give an account of the world in its alleged totality is an instance of the philosophic fallacy.[32] And here is where we find the crucial departure of Santayana from Dewey, because while he shares Dewey's reservations about the *idealistic* fallacy involved in traditional metaphysics, he does not equate it with grandiose cosmological speculation. Santayana does not follow Dewey in equating cosmology with "gratuitous hypostatizing."

If, as Dewey holds, providing an account of the universe in its totality only amounts to "gratuitous hypostatizing" we can easily see why he was not disposed to Santayana's systematic ontology. Santayana's fourfold realms of being, with matter among them as the existential component of his ontological system, can be said from a certain angle to be a concession to cosmology. The realms are said by Santayana to be "qualities of reality" not to be confused as being "parts of the cosmos."[33] Yet it is clear from the responses Dewey has to his philosophy and criticism that he finds Santayana's ontological propensities objectionable on the same ground that cosmology is objectionable.

To briefly elaborate, I offer an unpublished critique Dewey gives in a 1927 letter he wrote to Herbert Schneider. Schneider had then recently visited Santayana in Rome and was engaged in ongoing exchanges with he and Charles Strong about the doctrine of essence and related issues. After receiving what he describes as "scraps" from Schneider containing overviews of Santayana's views about essence, Dewey replied:

> Santayan [sic] seems to be wholly wrapped up in an initail premiss [sic, read: "initial premise"] which he takes to be an unquestioned fact but which seems to be a historical survival. Namely, that there are such things as data of immediate of [sic] or in immediate experience. Its [sic] an inheritence [sic] from subjectivistic psychology. If there were any such things, they would be the kind of things S calls essences; he shows his acumen in that. This isnt [sic] argument of course; simply assertion. But I dont [sic] see how you can argue about facts...[34]

Dewey goes on to list five interesting objections to what he observes of Santayana's views. Among the objections is one that illustrates Dewey's reading of Santayana's ontology as unwarranted cosmology. Dewey objects: "Some descriptive matter of fact statement of the organic conditions under which essences come out of the sky and are present is called for, not just as a desiderated addendum but intrinsically."[35] Dewey understands Santayana's appeal to essence—the realm of being that constitutes his account of the foreground—as "coming out of the sky" and as a "desiderated addendum." In other words, Santayana is guilty in his ontological propensities of committing the very same philosophic fallacy of which cosmologists of old are guilty. One discovers that the shared critique of metaphysicians of the past on the notion of essence observed at the opening of this section ends in the passages excerpted. Dewey groups Santayana among gratuitous metaphysicians of the past and is not convinced by his numerous attempts to establish that his essences, unlike the Ancient's, are not capable of hypostatization.

This aids in understanding the dissonance between Santayana and Dewey on the troubled controversy over foreground, and more generally provides the opportunity for appreciating what is at stake in their different views of cosmology. It introduces the need for an assessment of their respective readings of historical philosophies on the question of essence (entailing the problem of universals that Dewey diagnosed in his identification of *the* philosophic fallacy), and more particularly their opposed views of the value of or need for cosmology. I propose in the next section that these questions come down to the question of the need for ontology, which for Santayana allows space for the necessary "background" lacking in Dewey's philosophy. Philosophy that lacks such background leads to the politicization of thought that one finds in contemporary pragmatism.

3. Ontology or Why One Needs to Appreciate the "Background"

Santayana's foreground-privileging critique of Dewey, David Dilworth has recently claimed, anticipates "the entire gamut of transcendentalist, phenomenological, pragmatic, and historicist styles that flourished in Dewey's day and continue to flourish in today's academy."[36] I think it is appropriate to term the trend Dilworth has in mind here the *politicization* of philosophy, especially as it developed after 1950. If Dilworth is correct, as I think he is, Santayana's critique of Dewey is much more historically sweeping than sheer attention to opposed philosophic principles appreciates.

Santayana called in his day the politicizing one finds in contemporary philosophy "moralism."[37] He identified two forms of "moralism." The first, "moralism proper," according to Santayana, is a species of Kantianism, where "the categorical imperative of an absolute reason or duty determining right judgment and conduct" is asserted.[38] Kant's philosophy "conceived that duty was something absolute and not a method of securing whatever goods of all sorts are attainable by action."[39] The shortcomings of such absolutism of duty in the philosophy of Kant have been much discussed by a wide variety of philosophers, including Deweyan pragmatists, yet the tendency persists among contemporary ethicists striving for exclusively "normative" moral theories. Such theorists follow the absolutism in Kant in so far as they hold out for the possibility of establishing standardized ethical norms that are capable of guiding conduct, as though conduct is so exclusively and neatly capable, or even in need of, guidance by norms.

But Santayana identifies a second form of moralism, one that implicates Deweyan pragmatists as well. He calls the second form of moralism "a principle of cosmology and religion [asserting] the actual dominance of reason or goodness over the universe at large."[40] It is revealing that the very same cosmology (and religion) that Dewey suspected lurked behind troubled philosophies of old would be invoked by Santayana to implicate pragmatism. This would of course be amusing news both to Dewey and to Deweyan pragmatists, who if nothing else view their position as opposed to any form of cosmic absolutism so described. What can Santayana mean in the charge?

Santayana suspected that Dewey protested cosmic absolutism too much, and failed to admit the commitments made in his consistent appeals to the self-sufficing character of empirical "traits" or appearances. This is a subtle but important aspect of Dewey's thinking that merits deeper consideration. Dewey consistently emphasized the creaturely environment, so it is one of the appropriate legacies of his thinking that it is situated among contemporary philosophies of ecology.[41] Dewey's discussion of environment entailed the notion of growth, of an organism's relative desire, need, and ability to adequately adjust to its surroundings.[42] Growth occurs in any organic situation when a phase of equilibrium or harmony comes about, when the desire stemming from inadequate adjustment to the creaturely environment is, for the time being, sated.

Dewey's speculation about the natural scope of the creaturely environment does not stop at the empirical ground-level of biological adjustment and growth, as he writes: "There is in nature, even below the level of life, something more than mere flux and change. Form is arrived at whenever a stable, even though moving equilibrium is reached."[43] Form itself is owed to a stage of equilibrium *in* nature, a temporary settlement of previously unsettled and destabilized features of organic life. It is in these modes of speculation that Santayana interpreted a hedge in Dewey's naturalism, a tendency for the erstwhile "natural facts" he spoke of to become "so strangely unseizable and perplexing."[44]

To say, as Dewey consistently said, that experience reaches down and down into the depths of nature without necessarily committing individuals to any idea of its total character was for Santayana, if not disingenuous, inconsolable with active animal life. Santayana tries to puzzle out the problem in his own way in a dense section of his critique where he considers Dewey's use of the term "event." He summarizes Dewey's view that events are "natural waves, pulsations of being, each of which, without any interruption in its material inheritance and fertility, forms a unit of higher order."[45] The view commits Dewey, Santayana claims, to the idea that mind in such instances, that is, conceived as a part of what Dewey understands to be events, "possesses a hypostatic spiritual existence." In other words, Dewey's understanding of events commits him to a view of mind that goes beyond behaviorism, one that attributes to mind a being separable from material conditions. The problem is, Santayana charges, Dewey only commits himself to this idea of mind begrudgingly, and hastens to cover it up in other parts of his thinking. The idea of a substantial mind apart from material conditions to which Dewey sometimes concedes when speaking of aesthetic contemplation is rejected, Santayana contends, when Dewey turns to what he calls "knowledge of acquaintance."[46]

What Santayana is getting at in this overlooked prelude to the foreground-privileging charge is that Dewey refuses to admit the undeniable superficiality of such things as "events and affairs."[47] To so admit of course would be a reversal of Dewey's overarching aim, which is to vindicate the notion of experience against traditional, historical slurs. For human experience to get its due, Dewey strives to argue, it must not in any respect be viewed as superficial, but, Santayana asks here: what does this do to one's conception of nature and consciousness? What below or beyond-the-surface character to experience is committed to that does not at the same time risk belittling those vital phenomena of philosophic import? This is the challenge Santayana presents to Dewey when he suggests: "[Dewey's understanding of] events, situations and histories hold all facts and all persons in solution."[48] In fact, events and situations pass into the empyrean of remembered experience and that ephemeral character cannot (should not) be denied.

Now recall Dewey's phrase regarding form that its emergence is due to something "in nature, even below the level of life." Why risk in this context speculation about extra-human, natural sources of unity in experience but deny it in other contexts? Cosmology, Santayana agrees with Dewey, goes wrong when it follows from hypostatizing existential elements of lived experience, when it becomes a game of identifying *the single* feature of life as met in action as the true root and source of such action; this, he agrees, commits the idealistic fallacy. But cosmology as speculation about the world in its totality, contra Dewey, is not wholly inadmissible. Such speculation is warranted where it aims to establish distinctions in being that parse out for understanding in lived experience an explanatory framework; not, it is important here to stress, for the purpose of *guiding everyday action* but of edifying the human perspective so that life as reflected upon broadly can be made more luminous and meaningful.

This is especially where Dewey and Santayana part ways; and where Santayana more generally parts with contemporary schools that from his perspective politicize philosophy. Dewey's pragmatism emphasizes the achievement of greater *control over* intelligent action, the scope of which, Santayana was right to criticize, is always in the foreground, or just-around-the-corner of every experiential context. Santayana's ontological categories reflect an appreciation for the role of the background, for the edifying possibilities that emerge as a result of taking a longer view of things, without insisting that any single component of that longer view be taken as exhaustively explanatory of life in the foreground.

4. Privilege of the Foreground in Dewey's Philosophy

Following the leads of Santayana, and extrapolations from Dilworth, one can relate the move of Dewey in rejecting traditional metaphysics with that of transcendentalists and phenomenologists, who bracket key presumptions of animal, natural existence. In Dewey's case, the bracketing occurs as a result of his overreaching critique of the problem of universals in traditional philosophy. The problem of universals in traditional metaphysics was, it should be remembered, an attempt to explain a very fundamental experience in natural life, the ascription of sameness or likeness to different events and entities across their various encounters. We forget sometimes from our positions of historical privilege that Plato would not have thought of himself as engaging in high metaphysics when in various dialectical modes he made Socrates demand of discussants a definition of a thing not reducible to the example given. To say that there must be something to the idea of "table," "piety" or "justice" that various appeals to individual cases fail to capture is not gratuitous metaphysics. It is asking a legitimate question about the nature of language, reference, and meaning. Similarly, Aristotle would have found it confusing at best to deem gratuitously metaphysical his speculation that first philosophy must be distinguished from the "so-called special sciences" in its being a "science which investigates being as being and the attributes which belong to this in virtue of its own nature."[49] Aristotle, like Plato with regard to experiential appeals, did not have as his main intent what Dewey characterizes to be an appeal to "things" or "events" existing independently of or prior to inquiries undertaken. Rather, they were interested in establishing a point at which talk of existences bottoms out, or, to put it in terms of Aristotle's understanding of essence, becomes "unanalyzable."[50]

What happened by the time of medieval scholasticism, when the problem of universals as conceived in Plato and Aristotle reached its well-known climax, is that the problem had became a sheer dialectical exercise; it had become a matter of proving or disproving the existence of divine presence. *This* was when the debate became "metaphysical" in the bad sense of the term to which both Santayana and Dewey object (and that William James well parodied in his squirrel anecdote at the beginning of his pragmatism lectures). By the time of scholasticism, neither realist nor nominalist could hope to win the debate save by privileging, to recall Dilworth's apt phrase, their preferred hermeneutical circle.

This history is insufficiently acknowledged in Dewey's, but not in Santayana's critique of traditional metaphysics. Dewey lumps the problem of universals in his catch-all category of "bad metaphysics," leaving its early origins unexplored. Santayana catches Dewey in this historical narrowness by calling the bluff of the philosophic fallacy; he helps one see how it has become a too-easy weapon of later generations of pragmatists against idolatries of many stripes:

> The typical philosopher's fallacy, in [Dewey's] eyes, has been the habit of hypostatizing the conclusions to which reflection may lead, and depicting them to be prior realities—the fallacy of dogmatism. These conclusions are in reality nothing but suggestions or, as Dewey calls them, "meanings" surrounding the passing experience in which, at some juncture, a person is immersed. They may be excellent in an instrumental capacity, if by their help instinctive action can be enlarged or adjusted more accurately to absent facts; but it would be sheer idolatry to regard them as realities or powers deeper than obvious objects, producing these objects and afterwards somehow revealing themselves, just as they are, to the thoughts of metaphysicians. Here is a rude blow dealt at dogma of every sort: God, matter, Platonic ideas, active spirits, and creative logics all seem to totter on their thrones; and if the blow could be effective, the endless battle of metaphysics would have to end for lack of combatants. (675)

Santayana acknowledges the importance of recognizing this fallacy, but as can be inferred by the word "would" in the last sentence he doubts it achieves Dewey's ultimate aim: to overthrow traditional metaphysics. In fact, Santayana argues here, the target of Dewey's critique is not so much metaphysics (he after all reserves room for his own "naturalistic" metaphysics), but rather "dogma of every sort," and *that* target, Santayana argues, is misguided.

Santayana is charging Dewey here with being overly zealous in his battles against fallacious metaphysics, because at the same time as he deftly articulates the fallacy itself, he fails to recognize his own metaphysical dogma: the unquestioned approval he displays for "the enterprise of life in all lay directions, in its technical and moral complexity...where individual initiative, although still demanded and prized, is quickly subjected to overwhelming democratic control...his inspiration is sheer fidelity to the task in hand and sympathy with the movement afoot... ."[51]

Again, the accusation here is severe but I think warranted given certain modes we find in Dewey's writing such as the following where he is thinking about the status of present living qualities:

> "This," whatever *this* may be, always implies a system of meanings focused at a point of stress, uncertainty, and need of regulation. It sums up a history, and at the same time opens a new page... a fulfillment and an opportunity...Every perception, or awareness, marks a "this," and every "this" being a consummation involves retention, and hence contains the capacity of remembering. Every "this" is transitive, momentarily becoming a "that." In its movement it is, therefore, conditioning of what is to come...[52]

Here one finds Dewey tangled in what Santayana thought of as the troubles resulting from a too-dominant preoccupation with the foreground. There is evidence here that Dewey's "inspiration is sheer fidelity to the task in hand and sympathy with the movement afoot." There is no *"this*" for Dewey, no present quality, which does not by definition dissolve into a final tradeoff between a dead past and pregnant future. The present is a "point of stress," in "need of regulation." The foreground for Dewey here is so present as to become incapable of presenting itself at all, prompting Santayana's disgruntled remark that Dewey's readers are "left with an uncomfortable suspicion that it is impossible to inspect anything for the first time."[53]

What is important to Dewey about present quality is not its retention of the past; the past is, after all retained in present quality and nothing more need be demanded of perception or awareness in regard to that aspect. Nor is the shining reality of the present as it is felt of much importance to Dewey because it is by nature inexorably transitive and incapable of being understood (as he revealingly characterizes, "whatever *this* may be"). What is important to Dewey rather is what present quality aids in one's *movement towards* what is to come.

For Santayana, it was very easy to negatively attribute such preoccupation with present movement and control of potential future outcomes with Dewey's Americanism and theory of democracy, which strives to "give individuals a personal interest in social relationships and control, and the habits of mind which secure social changes without introducing disorder."[54] I am uneasy with this negative association if only for the fact that I think it short-winded on the virtues of Dewey's democratic theory. But I remain convinced that it is a matter of great concern to explain the politicization, after Dewey's time, of philosophy; by this I mean the overwhelming predominance of social and political concerns in philosophic discourse, exemplified in a certain dominant strand of post-Deweyan pragmatism.

5. Conclusion: The Politicization of Philosophy in Post-Deweyan Pragmatism

Exclusive fidelity to the forward-moving character of life takes curious forms in contemporary pragmatism, forms that sometimes reduce the entire doctrine to a political stump speech for progressive liberals. In one context Richard Shusterman accepts without appreciative analysis the characterization of pragmatism as "forward-looking philosophy," and praises as a "radical" example of such philosophy the work of John J. Stuhr. Shusterman praises, reminds us that "Change...is inevitable, so instead of letting it happen to us passively we should actively embrace and shape it by willfully striving to change our lives and thought into directions that we determine, rather than just going with the flow of previous directions."[55] The foreground here is not just dominant: it has romantically reached a place of untouchable privilege. Philosophy, in the style of Shusterman and Stuhr, is a style of politics that is also a curious form of psychotherapy, of identifying sources of individual anxiety and suffering and awakening persons to possibilities of healing and growth.

Richard Rorty played a leading role in encouraging this moral-"therapeutic" notion of pragmatism.[56] He ushered in the idea of Deweyan pragmatism as departing with tradition (along with Continental philosophy) in its disavowal of "appeal[s] to commonly shared criteria [that have the] ability to decide issues of ultimate significance for our lives."[57] This narrow, therapeutic account of pragmatism is unfortunate. If all that the entire corpus of Dewey, Peirce, and James can offer contemporary philosophers like Shusterman and Stuhr is a disavowal of consensus about ultimate meanings and a warning to "get with the times," to shake off the stubborn yolk of dead pasts, one might just as well deliver the message by making pragmatists a footnote to the early songs of Bob Dylan. It is a very legitimate question, one that diagnostically leads back to Santayana's foreground-privileging critique of Dewey: how does it come about that tired inspirational slogans like "change is inevitable" and "don't just go with the flow" (none of which would have impressed Dewey), become the core message of pragmatism? Space limitations forbid a sufficient examination of this phenomenon, but some necessary historical points of focus are in order.

Rorty is preeminently the figure responsible both for reinvigorating, and at the same time politicizing pragmatism after Dewey. Intellectually daring in a time when philosophy had become almost exclusively an affair of logic-chopping and argument-refuting, Rorty spearheaded in the late seventies a return to pragmatism, among other philosophic orientations, as a means of bridging the growing chasm between philosophic discourse and social life. As Neil Gross accurately portrays, Rorty's invocation and embracement of Dewey was due, biographically, to his political patriotism and his "sense that Dewey [like him]... viewed philosophy through a sociological lens."[58] Rorty's contribution to the reinvigoration of interest in Deweyan pragmatism, it is well known, is alternatively embraced or violently resisted where he has been perceived as laying to waste time-honored dogmas persisting in academic philosophy. Rorty has, in other words, been received as a "radical" in the sense, clearly, that pragmatists such as Shusterman and Stuhr are trying to emulate.

The sad fact is that pragmatism under this influence struggles to be taken seriously as a *philosophy*. Rorty, uninterested in legitimating anything as a philosophy, least of all pragmatism, contented himself with playing ironically with competing vocabularies and with the idea that pragmatists believe:

> ...that the best hope for philosophy is not to practice Philosophy. They think it will not help to say something true to think about Truth, nor will it help to act well to think about Goodness, nor will it help to be rational to think about Rationality.[59]

Contemporary pragmatists such as Stuhr echo Rorty in this and think it misguided to even worry whether pragmatism ever gets taken seriously as a philosophy. To me this is a deep problem for those truly interested in recovering the core insights of the sort found in John Dewey's *Experience and Nature.*

Thinking along these lines, Santayana's critique of Dewey is more challenging and potentially offensive than strict consultation of the published exchange can reveal, especially by limited attention to the "half-hearted naturalism" dig. His foreground criticism was in a deeper sense a backhanded way of saying that Dewey's thinking is not sufficiently broad enough to be considered *a philosophy*. One can surmise this consulting Santayana's writings beyond the critique, in his various attempts to characterize philosophy and to assess its relative worth. Throughout his thinking he associated philosophy with religion in its ultimate scope: "Philosophy and religion are nothing if not ultimate; it is their business to deal with general principles and final aims"[60]; "...what is philosophy...except religion liberated from groundless fear or anxiety, that is to say from superstition, and also from rage at honest illusions?"[61] And he strongly contrasted the enterprises of politics and philosophy as emancipating activities: "It is not politics that can bring true liberty to the soul; that must be achieved, if at all, by philosophy...".[62] More poignantly, Santayana saw the worth of a philosopher's moral ideal as lying in its ability to both conceive of morality naturalistically, and to appreciate the breadth of moral forms as they range across political, social, and existential lines:

> ...in considering the moral ideal of any philosopher, two questions arise. First, does he, like Spinoza, understand the natural basis of

> morality, or is he confused and superstitious on the subject? Second, how human and representative is his sense for the good, and how far, by his disposition or sympathetic intelligence, does he appreciate all the types of excellence towards with life may be directed?[63]

Bringing these characterizations into contact with Santayana's criticisms it is clear that he thought Dewey's thinking failed both of these evaluative criteria, in large part for the reasons explored in this examination.

What, after all, is to be said of the *ultimate scope* Santayana understands it to be the business of philosophers to establish? Very little from the standpoint of many contemporary, at least Rortian, pragmatists. Their aim is to hold in skeptical abeyance all claims of ultimate significance and meaning. But there is a cost, a tradeoff that Santayana presciently diagnosed of Dewey's pragmatism. Not holding to significant account the general direction of modern life, viewing it as self-justifying as a basis for morality, for conceiving means as at least equal to ends, for privileging intelligent methods and processes over ideal ends and preferred outcomes; and not least, politicizing pragmatist doctrine to such a degree as to render altogether questionable its philosophic legitimacy: these are the undeniable legacies of the Rortian strand of Deweyan pragmatism. The moralism Santayana suspected of Dewey's pragmatism may have only now made its full appearance.

NOTES

1. This article is a revised version of a same-titled article appearing in (full length) *Limbo*, Núm. 29, 2009, pp. 21-38, ISSN: 0210-1602, and in condensed length in *Overheard in Seville: Bulletin of the Santayana Society* (forthcoming, October, 2009). The author wishes to extend his gratitude to participants, including Larry Hickman, in the "American and European Values" Dewey conference held in Opole, Poland, in June 2009. After delivering an early version of this paper the author received valuable feedback from participants in post-talk discussions.

2. Santayana, George. "Dewey's Naturalistic Metaphysics,"*The Journal of Philosophy.* December 3, 1925. Vol. XXII, No. 25: 678; 679. All page references in the present essay are from the *Journal of Philosophy* version, and cited in the following notes as "DNM," followed by page number.

3. DNM, 684.

4. *Experience and Nature*, LW 1: 11.

5. John Dewey. *The Library of Living Philosophers: The Philosophy of John Dewey.* Edited by Paul A. Schilpp. New York: Tudor Publishing Company, 1939, 1951: 530-531.

6. "...his naturalism is half-hearted and short-winded." DNM, 680.

7. John Dewey. " 'Half-Hearted Naturalism'," *The Journal of Philosophy,* Vol. XXIV, NO. 3, February 3, 1927: 58. All page references in the present essay are from this *Journal of Philosophy* version, and cited in the following notes as "HHN," followed by page number.

8. "The words 'affairs' and 'situations' in their intentional vagueness, express very well the ethical nerve of this philosophy..." DNM, 681.

9. HHN, 60.

10. HHN, 61.

11. HHN, 58.

12. David Dilworth. "Santayana's Review of Dewey's *Experience and Nature:* Pivotal Expression of a Philosophy of Nature and Vivacious Spirit." *Overheard in Seville: Bulletin of the Santayana Society,* No. 21, Fall 2003 (http://www.math.uwaterloo.ca/~kerrlaws/Santayana/Bulletin/y03/dilworth.pdf) Accessed as of 3/25/09.

13. Paul Forster. "Animal Faith or Natural Knowledge? Why Dewey and Santayana Can't Agree About Philosophy." *Under Any Sky: Contemporary Readings of George Santayana.* Cambridge Scholars Publishing, 2007: 46.

14. Forster, 60.

15. John R. Shook. "The Possibility of an Empiricist Naturalist: Dewey and Santayana." *Overheard in Seville: Bulletin of the Santayana Society.* No. 21, Fall 2003: http://www.math.uwaterloo.ca/~kerrlaws/ Santayana/ Bulletin/y03/shook.pdf.

16. Or to what Edmund Husserl elsewhere diagnosed as "psychologism," a misconceived elaboration of the Protagorean maxim about man being the measure of all things.

17. Also, and more paradoxically given the associations of progressive education with lowering standards, it is difficult to understand Dewey's influence on facets of educational practice that display a striking appreciation for standardized methods, established technique, and historical authority.

18. *Experience and Nature,* LW 1: 145.

19. Santayana, George. *Scepticism and Animal Faith.* New York: Dover Publications, 1923, 1955: 78-80.

20. DNM, 678.

21. LW 1: 73.

22. LW 1: 72.

23. LW 1: 62.

24. MW 2: 192.

25. MW 4: 34.

26. MW 4: 36.

27. MW 12: 119.

28. LW 1: 51.

29, Larry Hickman. *John Dewey's Pragmatic Technology*. Indiana University Press, 1992: 129.

30. Sidney Hook's characterization at LW 12: xxi; Dewey's association and identification of *"the* philosophic fallacy" is at LW 1: 34.

31. Hickman, 125.

32. As Sidney Hook helpfully puts it in the introduction to Dewey's 1915 *Essays*: "Dewey argues that this conclusion [cosmology from the standpoint of biological evolution], which admittedly could never be established by any scientific inquiry, depends upon an unacceptable conception of metaphysics that regards the world as a whole or in its totality as a legitimate subject of inquiry." (MW 8: xi)

33. George Santayana. *The Realms of Being*. New York, Charles Scribner's Sons: 1942: 183.

34. Letter from John Dewey to Herbert W. Schneider, dated Jan 6, 1927, courtesy of Special Collections, Morris Library, Southern Illinois University at Carbondale.

35. The fuller excerpt of Dewey's elaboration: "1.): In the first place 'consciousness' goes; except as a name for essences made existent by the conditions of the organism. Its [sic] as flat a denial of consciousness as any behaviorist ever made the entities called essences taking its place. 2.) There is no ground for any distinction between psychics and physical, since that is a matter of existences. The essences must be neutral, and there is no way by which animal faith can refer some to physical and some to psychical existence, for these terms have no meaning. 3.) Some descriptive matter of fact statement of the organic conditions under which essences come out of the sky and are present is called for, not just as a desiderated addendum but intrinsically. 4.) I can see how, or admit that, animal faith might make a reference of essences to existence in general, to an X thing-in-itself, but how existnece [sic] can be qualitatively and numerically discriminated by them I dont [sic] see. 5.) Its [sic] perhaps a case of the same point but in his comment S seems to me to use 'walk' in two senses. Does a walk have an essence or is [it?] an essence? If the former it is presumably an act which is existentially discriminated and hence neither a datum nor reached by animal faith from a datum, If the latter we don't take the same act twice but only the same essence; and it may be doubted whether we can literally take the same walk, in the sense of performing the same act twice, since identity is a matter of essence. In other words, I think the argument skips back and forth between essence and existence at will."

36. David Dilworth. "Santayana's Review of Dewey's *Experience and Nature:* Pivotal Expression of a Philosophy of Nature and Vivacious Spirit." *Overheard in Seville: Bulletin of the Santayana Society,* No. 21, Fall 2003 (http://www.math.uwaterloo.ca/~kerrlaws/Santayana/Bulletin/y03/dilworth.pdf) accessed as of 3/25/09.

37. My translation of Santayana's "moralism" into "politicization" is an attempt to facilitate its more direct application to examples of contemporary philosophic discourse. Identity politics in social and political philosophy is a particular instance of politicization in philosophy, where claims are frequently made as to the priority of an individual's political status over other considerations, such as (for example) those having to with inherited biological, or even cultural features that determine their lives. I would add too however that privileging "the political" in any way from a philosophical point of view tacitly rejects other broader strategies of criticism, such as those pursuing the same questions from an ontological perspective. The latter concern is at the center of the present examination.

38. George Santayana. "Apologia Pro Mente Sua." *The Library of Living Philosophers: The Philosophy of George Santayana.* Schilpp, Paul Arthur, editor. New York: Tudor Publishing Company, 1951: 502.

39. George Santayana. *Egotism in German Philosophy.* Charles Scribner's Sons, 1915: 107.

40. PGS, "Apologia," 502.

41. For example in Neil W. Brown's *The World in Which We Occur: John Dewey, Pragmatist Ecology, and American Ecological Writing in the Twentieth Century.* University of Alabama Press, 2007.

42. "Every need, say hunger for fresh air or food, is a lack that denotes at least a temporary absence of adequate adjustment with surroundings. But it is also a demand, a reaching out into the environment to make good the lack and to restore adjustment by building at least a temporary equilibrium." (*Art as Experience,* LW 10: 19).

43. Ibid, 20.

44. DNM, 678.

45. DNM, 677.

46. Dewey discusses this most extensively in *Experience and Nature,* LW 1: 248.

47. An example that Santayana undoubtedly has in mind: "It follows that theories which identify knowledge with acquaintance, recognition, definition and classification give evidence, all the better for being wholly unintended, that we know not just events but events-with-meanings." (LW 1: 249)

48. DNM, 675.

49. *Metaphysics,* Book IV, chapter 1, lines 21-25 (translation from *The Basic Works of Aristotle* edited by Richard McKeon).

50. Ibid, Book II, chapter 2, lines 17-25.

51. DNM, 675-676.

52. *Experience and Nature,* LW 1: 264.

53. DNM, 678.

54. *Democracy and Education,* MW 9: 105.

55. Richard Shusterman. "Making Sense and Changing Lives: Directions in Contemporary Pragmatism." *The Journal of Speculative Philosophy*. Vol. 19, No. 1, 2005:67

56. "I think of pragmatism as primarily therapeutic philosophy—therapy conducted on certain mind-sets created by previous philosophers. In so far as reading pragmatism frees you up from various old habits and convictions, it does it in the same way that a startling new literary text does. It makes you think, 'Gee, I never knew you could look at it that way before!' But therapy isn't the same as providing criteria, or a theory." Richard Rorty (quoted), with E.P. Ragg. "Words or Worlds Apart? The Consequences of Pragmatism for Literary Studies: An Interview with Richard Rorty." *Philosophy and Literature*. Vol. 26, No. 2, October 2002: 373.

57. Richard Rorty. "Pragmatism Without Method," *Objectivity, Relativism, and Truth*. Vol. 1 of *Philosophical Papers*. Cambridge: CUP, 1991: 75.

58. Neil Gross. "Richard Rorty's Pragmatism: A Case Study in the Sociology of Ideas." *Theory and Society*. Vol. 32, No. 1 (February 2003) :121.

59. Richard Rorty. *Consequences of Pragmatism*. University of Minnesota Press, 1982: Introduction xv.

60. George Santayana. *Interpretations of Poetry and Religion*. Charles Scribners's Sons, 1900: 208 ("The Poetry of Barbarism").

61. *Dominations and Powers: Reflections on Liberty, Society, and Government*. New York, Charles Scribner's Sons: 1951: 285.

62. *Soliloquies in England*. Charles Scribner's Sons, 1922: 184.

63. George Santayana. *Persons and Places*. New York: Charles Scribner's Sons, 1981: 245.

Ten

DEWEY'S THEORY OF VALUES

Hugh McDonald

> "Value in the sense of *good* is inherently connected with that which promotes, furthers, assists, a course of activity, and value in the sense of *right* is inherently connected with that which is needed, required, in the maintenance of a course of activity..."[1]

There has been much confusion in the literature about John Dewey's theory of values. Some have denied that Dewey has a notion of intrinsic value.[2] Dewey has also been taken as a conativist by a number of authors, despite his sustained critique of conativism in "Theory of Valuation."[3] Moreover, Donald Lee has argued that the value theories of the pragmatists are "similar enough" to warrant common treatment. In this paper I will examine Dewey's theory of value in detail, in the hope of clarifying its main outline and thereby distinguishing it both from conative theories and from that of C.I. Lewis, another well-known pragmatist.

1. Theory of Value

A. Dewey's Value Vocabulary

I will begin with a brief analysis of Dewey's value vocabulary, including value, valuing, and valuation. On the whole, Dewey is in agreement with those who see in value a concept of wide scope. Dewey believes "value" can belong to any object and he considers "value" as a potential quality of anything. He remarks that value is not confined to a "peculiar class of things. Anything under the sun may come into possession of what is named by 'value' as its adjective."[4] Also, all "deliberate...planned human conduct, personal and collective, seems to be influenced, if not controlled, by estimates of value or worth of ends to be attained."[5] This primarily concerns activities, but, as a term with wide applicability, can include any things or objects of value, actions, situations, etc., as well as different kinds of goods. Value can belong also to ends in view, that is, future objects or projected conditions, including ideals.[6] It is a property, quality or character of that which it describes, e.g. a good tool.

For Dewey, "good is always found in a present growth in significance in activity." The emphasis on activity separates Dewey decisively from subjective theories of value, as activity is public and in the world. Although desire and interest have a role in the formation of value, they are not mentioned in the definition of values. Despite some texts that seem to endorse the identification of value and desire or interest, Dewey ultimately rejects this definition as uncritical. But more, he rejects the approach initiated by the Austrian school of values in which value is grounded in the psychological states of a subject or valuer. The emotional and the conative are included in valuing but value is not ultimately an emotional state.

Dewey connects valuing more with action. Valuing is primarily acting in the form of prizing and caring for while value activity as a whole includes valuation and valuing. Valuing more narrowly consists in "prizing…honoring, regarding highly, and appraising in the sense of putting a value upon, assigning value to."[7] Valuing is more narrowly defined in order to mark off the activities of value in its primary significance as a positive direction of change. As such activity is often necessary to bring what is prized about or into existence, value as a quality may be the consequence or result of valuing.[8] It may even include the value of all of the elements of the situation, each of which has some place in the activity as a whole.

The process of bringing this about involves valuation, evaluation and judgments of value. These terms refer to certain crucial elements of the value situation. "The good of the situation has to be discovered, projected and attained on the basis of the exact defect and trouble to be rectified."[9] Valuation is one form of valuing,[10] which he identifies with "warranted desire," i.e. desires actually formed from the many impulses generated by the problematic situation.[11] Valuation only applies to impulses transformed by deliberation and evaluation[12] due to some problem that needs to be resolved "by means of changing existing conditions." Valuation is used "when it is necessary to bring something into existence which is lacking or to conserve in existence something which is…menaced...."[13] Bringing into existence is a result or consequence and thus involves means and ends—including evaluation of and choice between different means and ends. Dewey distinguishes valuation in the hope of making clear the difference between the value or valuing of present goods, and activity to bring about valued future goods, that will help in a problematic situation. But the process of valuation itself may carry value.

What I want to emphasize is that value is not equivalent to either valuing or valuation. Value is a character of something, whether an object, an activity, or even a feeling. Although valuing may bring about a valued result, the result can bear inherent value. These activities are distinct from the value that attaches to the activity of valuing as such, as complex as this may sound, as well as to the "bearer" of value, whether an activity or an object, although each may be involved in the process. Value, then, has the widest scope, while valuing and valuation are progressively smaller in scope, and confined to activity or behavior.

Valuing includes appraisals, prizings, value judgments and other "ratings." When something is prized, value is assigned to it: Dewey calls this an appraisal, a term which has the same root as prizing. According to Dewey, prizing emphasizes the "personal" aspect, including any emotions involved in valuing. Appraisal is concerned with "rating," i.e. "a relational property of objects so that an intellectual aspect is uppermost…."[14] He compares this to the distinction between 'estimate' and 'esteem.' It is similar to evaluation as it is commonly used, that is, judgments and comparisons of value, or rating objects, activities etc. in terms of their value. This distinction will prove to be important in Dewey's overall theory; in this context, its distinction as a mode of valuing from prizing should be noted. However, prizing, and rating or appraising are both activities or involve action.

"What is 'valuable' is the object of a certain kind of activity." The objective is an object of value in some sense because of the activity: activity is basic to valuation. There is a sense of activity attached to prizing, an active quality rather than a quality of a thing. The evidence for this active prizing is caring for something. "The measure of the value a person attaches to a given end is not what he *says* about its preciousness but the care he devotes to obtaining and using the *means* without which it cannot be attained."[15] Activity is a sign or criterion of actual valuation, which involves an active pursuit of value not passive contemplation. Activity is public and observable; prizing something is revealed in actually caring for something, not in mere lip service. Without this caring as a context of support, valued objects would exist in conditions adverse to their maintenance; with it they are supported in existence.

Evaluation is a species of valuing, viz. appraisal, or putting a value on.[16] Impulses are evaluated before they become ends in view. The result is warranted desires or the desirable, which *should* be desired or valued and is an element in ends in view. Thus we make evaluations of proposed values, ends in view in a problematic situation, which would seemingly resolve a problematic situation. We then act to bring about such results as an act of valuing. The consequence is new goods that have value. Evaluation involves both a comparison of different lines of action and a critical appraisal of them. They are evaluated in an external relation to each other and to circumstances. It involves weighing alternative means as well as evaluating the outcome in relation to expectations, i.e. the end in relation to the results. Dewey considered the objection that evaluation properly only applies to means, and argues that ends should also be evaluated. The ends are appraised in the same evaluation with the means. The process of evaluating alternatives ends with a judgment.[17]

Dewey rules out any "private, introspective view of the field of values,"[18] which would exclude almost all psychologistic and subjective approaches as too narrow, thereby differentiating his position from that of C. I. Lewis. He views such theories as having methodological shortcomings *qua* private: they are not observable, and thus not subject to scientific treatment. Also, science itself is a form of conduct for Dewey, within the field of values.

He also rules out value as exclusively attaching to an "agency or agent." By this he means a subject, a valuer, but Dewey also argues against "self-contained," and "short-span acts," in favor of considering "the whole span of life." A holistic element is introduced by this qualification, the whole span of life as the main point of reference for value as well as an implicit critique of the value of short term ends, i.e. as valuable in themselves *without reference to a larger whole*.

Dewey also rules out value in the sense of an independent entity, whether in the sense of "the good" or the economic sense of "a good." "Value is an adjectival word, naming that which is a trait, property, qualification of some thing."[19] Dewey thereby retains the Aristotelian *pros hen* relation of good as adjectival or a property of something, as referring back to a first in judgments, although he rejects the idea that it is a property of "being." Value is adjectival and characterizes a wide field of conduct or behavior as well as good things. In the former sense, value attaches to things or objects, e.g. objects of desire, and "to value" is derivative. However, as verbal, it is connected to action. Value as a perceived quality is connected more with some quality of an object.

Values and morals share the entire field of the practical and both involve situations in which better or worse outcomes are one factor, but they are not identical. Value is wider in scope than morals since although both are concerned with improving human life and the quantity of good in it, morals are attached to activity, especially activity in a social setting, i.e. mostly to conduct.[20] Moral goods are a specific "kind of values," viz. those that "make a difference to the self, as determining what one will be, instead of what one will have."[21] Morals are also "social," involving the recognition of others and the requirements of a life in a social setting. Since collective conduct is social, there is a social aspect to values.

I noted above that for Dewey value is "adjectival," that is a property or character of something. In what sense is value a quality of activity? Dewey characterizes value as "directions of change in the quality of experience." As experience is behaviorally defined, value is a property characterizing directions of change in the quality of behavior, broadly conceived. Value is "whatever is taken to have rightful authority in the direction of conduct."[22] Conduct is good or bad, i.e. an activity that can lead in a positive or negative direction. If it is rightly directed it aims at good outcomes and value is the "rightful authority" or warrant for directing it in a specific direction. Conduct as a form of behavior involves activity and the value of the activity as its direction is a property that attaches to the direction of the activity.

Why is a "change" in direction indicated? Both conduct and activity as a whole may go in a good or a bad direction. By starting a new activity, the organism can change direction, whether for the better or the worse. There is another sense of change however, which involves consideration of the situation in which value questions arise. Dewey connects value with "selection-rejection." Selection of a direction for activity involves a process of evaluation and deliberation.[23] Choices must be made between alternatives and some selected and others rejected. Value is connected with this process, resolving problematic situations. "Good consists in the meaning that is experienced to belong to an activity when conflict and entanglement of various incompatible impulses and habits terminate in a unified, orderly release in action."[24] One alternative is selected from the many, often conflicting, possibilities and is the basis of acting. This process marks a change in direction or outlook that is projected as good. Or rather, it is enough to do away with a bad or problematic situation with a change in direction that will resolve the problematic situation. This change in direction for the better marks all such changes as identically good. "Good is the same in quality wherever it is found, whether in some other self or in one's own."[25]

B. Instrumental, Inherent and Intrinsic

What is the relation of such a meliorist change in direction to what is prized or cared for? What is prized is based in some respect on past experiences in which agents have made a connection between objects helpful in the positive direction of change. They are valuable as tools for improvement and as such they are cared for, i.e. active attention is made for their preservation. I noted above that Dewey favors the good of activity as primary and that of objects as secondary. The relation of prizing and caring for objects reflects this evaluation, as objects that are prized or cared for involve the species of activity characterized by a positive change in direction. Objects are the objective of the activities of prizing and caring for; their value involves the activity of prizing and caring for in some respect. Dewey does not deny the so-called objectivity of the value of objects, then, but incorporates it into a larger, activity-oriented theory in which they are instruments of activity.

Culminations of experience are described in terms of "satisfactions" and "enjoyments." Enjoyments are values "in and of themselves." Humans "form purposes, strive for the realization of ends" because "they believe these ends have an intrinsic value of their own; they are good, satisfactory."[26] Further, as Gouinlock stressed, such satisfactions can be "consummations" of an overall process in which an initially problematic situation is resolved.[27] However, Dewey refines satisfaction or enjoyments in accord with his more action-oriented and critical theory of value. He states that "there is no value except where there is a satisfaction, but there have to be certain conditions fulfilled to transform a satisfaction into a value."[28] Firstly, satisfaction is not mere passive contentment, as it involves a prior activity to bring it about and as a "consummation" of this activity is attached to the activity itself as part of the process. Secondly, although the satisfaction of a desire may bring enjoyment of itself, it may not fit well into overall values and the activity requisite to them.[29] Candy is dandy but may cause obesity, tooth decay and malnutrition. It may have immediate but not ultimate value. Satisfactions, then, have to be evaluated in order to determine whether they are truly or ultimately satisfactory. Dewey argues that there must be a "degree of regulation of valued enjoyments." This is measurable by the effort that is made to control the conditions of the occurrence of the valued enjoyments, i.e. prizing and caring for or valuing. If valued enjoyments result in interest and a desire to care for them, they will continue to be sources of satisfaction in an ongoing process, in which as satisfactions and enjoyments are attained, new ones are generated. Intrinsic values in the form of consummations of a process of activity are a part of value as a whole then. Humans value such consummations and will pursue them. Enhancing such values may be a factor in evaluating alternative courses of action. However, they are not the only consideration, nor the foundation of the process.

Intrinsic value has two meanings in Dewey, then. One is the larger, "adjectival" sense of a quality of an action or an object, a value inherent in an activity or intrinsic to an object. In the narrower sense, intrinsic value is attached to culminations of experience in the form of satisfactions and enjoyments reached as part of an ongoing process. The latter can be specific content of the former. Experience of satisfactions is the specific content of activities valued in themselves. Lewis distinguishes these as inherent and intrinsic value, and I believe that Dewey would concur. Dewey's approach to value is to stress activity over objects of activity, and experience is a form of activity, since it is tied to behavior, habit and other actions. The experience of satisfaction or enjoyment is within such activities, not removed from them in a Cartesian world of the subject. Experience and satisfaction are moments in a larger process of experience as an ongoing activity.

Objects of value that we prize and care for can also have a value that is inherent to them. Such objects can be natural but more often are made, "artificial" objects. Such objects are the goal of the process of making or caring for. New ones are brought about through activity that has the object as its objective or goal, and has a value of its own. "Goods" as objects, i.e. qualities of prized objects, are derivative from our activity of bringing them about. Combined, such activities bring about objects as their objective, but this is grounded in evaluation as the basis of the object. The object is, initially, an end in view or goal. Activity is primary or basic to bringing about prized objects as an objective, including the care rendered for more natural objects, e.g. that of farmers for their crop. "Objectivity" of value is revalued by this view, since it reflects activity and the public space of action more than the perceiving subject.[30]

C. The Highest Good

In the tradition, intrinsic value and the highest good have often been equated, but they are distinct. While the highest good requires a notion of intrinsic value, the reverse is not true, for there can be a multitude of intrinsically valuable items without any one of them forming the highest one. Aristotle's argument that something can be an end of a process or a "final cause" but also a means to higher ends contains a tension that Dewey noticed. For in his treatment of each situation as unique and as of equal value,[31] he is following the logic of intrinsic value, even if he avoids such language. If something is of intrinsic value, its value cannot derive from a higher value, for then it would be *ex*trinsic. It would derive from the higher value, not itself. Now Dewey is careful not to argue too strongly for the merits of intrinsic value, for he wishes to preserve the external relatedness and thus meaning of good: to situations, consequences, further activities, etc. This includes its relations to longer-term goals. Thus I would submit that his de-emphasis of intrinsic value is due to his emphasis on relations and connections. He does not deny inherent qualities, although even these involve a relation of a quality to an activity or thing, but argues that these have relations to overall goals. The latter are not fixed, transcendent, etc. but are "stable" or endure from situation to situation. This would include the life of the organism that encounters situations, its growth processes, etc., in short natural elements of a living organism.

The relation of the short-term in time to the long-term in time constitutes some of the meaning of the situation. Thus if "highest good" means a fixed, transcendent, singular or isolated end, Dewey does not have such a theory. If it means that there are overall goals which mark a longer-term good and which regulate short-term good, in a relation established by deliberation, then Dewey does incorporate a highest good. If growth or value for life is viewed as a "highest good," it is a major transformation of the model. For both growth and life are marked by change in which the decision over which they are partly regulative marks a change in the life or growth themselves of the agent. In either case, Dewey's notion of overall value is not supreme over all other considerations in deliberation and evaluation.

There is also a notion of hierarchy, or rather of degree in Dewey. That is, different possibilities for resolving a problematic situation may present themselves and they must be ranked in order of value by such criteria as which best meets the circumstances or will best contribute to overall growth. The choice is made of a "preference out of competing preferences." Some prizings are clearly better, and in many cases this is quite evident, e.g. in the case of the health of the organism. Again, certain means to ends may be better as more efficacious, which Dewey argues can be scientifically tested.[32] However, this does not involve an *a priori* hierarchy in any conflict of goods, for the choice is in terms of the unique situation: it is competitive rather than hierarchical, temporary, not eternal. As Dewey puts it: "The better is the good; the best is not better than the good but is simply the discovered good. Comparative and superlative degrees are only paths to the positive degree of action. The worse or evil is a rejected good. Until it is rejected it is a competing good."[33] Moreover, these are valuings, not value as such.

Dewey, somewhat reluctantly, concedes that ideas of general ends or goods develop, but argues that they are based on induction from empirical inquiries, not *a priori* standards.[34] For example, a "general idea of health as an end and a good (value)"[35] is formed on the evidence provided by examination of the occurrence of disease as a disturbing situation and the means of overcoming it. End, good and value are equated in this interesting text. Not only are good and value more or less equivalent terms for Dewey, but ends can be good or valuable. This is the first of many such texts providing evidence that Dewey's critique of intrinsic value represents a refinement of the notion, rather than its abandonment. The model or conception formed is an inductive generality, however, not an absolute standard. Dewey denies that health is an "absolute end in itself," as it is one of many natural goods that have a place in a larger whole. Similarly the value of "learning," or knowledge, which plays a very important role as an instrument of deliberation, is recognized as a general end. The role of general values is as "tools of inquiry into the individual case." Such general notions may be useful for classifying the situation and thus suggesting further lines of inquiry and methods by connecting the instant situation with similar experiences. "They are tools of insight; their value is in promoting an individualized response in the individual situation."[36] Like objects, they are instrumental for further activity, as aids in removing the causes of problems.

General ends may also constitute ideals. In a sense, every valued end is an ideal as something that is wanted and does not exist.[37] Ideals arise from an actual situation, but are not a mere projection. For in them is projected in a "securer and wider form some good" which was previously experienced in a precarious way. They are general ends of some kind. However, ideals in the strict sense are the higher values.[38] Dewey made this clear in his critical comments about the idea of material goods as sufficient. If too much of a premium is put upon such goods, "goods that are more ideal, [a]esthetic, intellectual values…are forced into subordination."[39] Ideals, then, are generic, general ends representing certain higher values. In sum, Dewey advances a "practical idealism" in which the ideal of continuous development is tied to concrete conditions for growth:

> Practicable idealism is found only in a fulfillment, a consumption which is a replenishing, growth, renewal of mind and body. Harmony of social interests is found in widespread sharing of activities significant in themselves.[40]

Dewey argues that the end as an ongoing process rather than a fixed terminus changes the time orientation of the activity. Ends are connected with the future. However, activity takes place in the present, and Dewey argues that his theory places emphasis on the present, where it belongs. Dewey regards it as "tragic" that the present is sacrificed to a remote future. He argues that "good, happiness is found in the present meaning of activity…." and that sacrificing the present to the future has bad consequences. Value as a quality, is to be found in present experiences, "for distinction, quality, is a matter of present meaning."[41] "The genuine heart of reasonableness (and of goodness of conduct) lies in effective mastery of the conditions which now enter into action."[42] In dealing with a problematic situation the stress ought to be on the present. The location of value in present activity is part of Dewey's critique and transformation of the means-ends relation

2. Bringing About

Valuing as primarily connected with activity is a complicated process. I will analyze some of the elements or aspects of this process below, but it should be noted first that each of these elements has its own value as well as the value it contributes to the whole process. Or, perhaps better, its value is the "value for," the instrumental value it has in the process of activity. Its instrumental value is in accordance with the wide scope value terms have in Dewey's view: each of the elements may be said to come under the scope of value as instruments or elements in the value situation taken as a whole. The good is primarily instrumental in just this relation, i.e. as an element in a larger whole to which it serves an instrumental function, whether an ideal, a general end, an object or an activity. "Value in the sense of good is inherently connected with that which promotes, furthers, assists a course of activity…."[43]

How do desires in valuation differ from facts? Dewey argues that valuations differ from other facts in that the presence of desire for some object, an end-in-view that will resolve the problematic situation for the better, involves a lack. The object which will fulfill the need or satisfy the desire is not itself a fact. Rather, it must be "made" a fact or brought into existence. "Because valuations in the sense of prizing and caring for occur only when it is necessary to bring something into existence which is lacking, or to conserve in existence something...valuation involves desiring." Desire is the root of technical changes, which bring about new objects. This involves a whole process, which is missed by more contemplative theories of value and touches on the relation of values and facts in a brilliant way, for which Dewey has not, as far as I know, received any credit. Valuation is the source of a whole class of facts as their ground. Desire, a felt lack, motivates the search for means and ends to fill the lack. Creativity of some type distinguishes valuation.

Activity aims at preserving a good, caring for or prizing what is unstable; or creating more goods, bringing about a superior resolution to a problematic situation. Bringing something into existence to satisfy a lack can be referred to as "actualization," by contrast with potential existence: what has actually been brought about. "Intelligent action is not concerned with the bare consequences of the thing known, but with consequences to be brought into existence by action conditioned on the knowledge."[44] Thus actualization is the special form of activity connected with the value situation which generally involves action as a means. Actualization is the means to a special but crucial class of facts, namely, created facts, facts in the original sense of deeds, those that make life worthwhile.[45] As a form of action, it is knowable both as itself observable and by its consequences. Moreover, Dewey argues that this does not subordinate morals to facts, which would separate ideal standards from customs. He argues that "morality resides not in perception of fact, but in the use made of its perception." By being known, facts have changed context, as they have entered a "context of foresight and judgment of better and worse."[46] In a sense facts are derived from morals for Dewey as what is brought into existence, deeds or facts in the original sense, is regulated by morals.

In turn, objects brought into existence, such as tools, are used as instruments for further activity. The good that attaches to objects is as instruments that function.[47] Tools function more or less well. Those that function properly or well are prized. They must be cared for or may break or disintegrate and must then be replaced, starting the process over again. Thus the relation of activity and objects, to which value attaches in the primary and secondary sense, is also part of an ongoing process. It includes the culmination or consummation of experience, i.e. intrinsically valued experiences, in ongoing events in the natural environment. However, the process is not based on or grounded in intrinsic value.

To transform a wish into a means requires a study of the conditions which make the existing situation possible, i.e. a study or judgment connecting an end-in-view to some means which will effect the end-in-view. Dewey views the connection of means to ends as one of the central problems of philosophy. In particular it is the "interaction" of judgment of ends with knowledge of means. The process by which they are brought together Dewey calls, following the tradition, deliberation. Deliberation includes evaluation and judgment, but is not equivalent to them. The role of deliberation is to consider "various alternative desires (and hence end-values) in terms of conditions that are means…." In short, the relation of means to ends is parallel to that of cause and effect, and the causal sequence is a means to a consequence that will effect an end-in-view. Deliberation includes consideration of which alternative means will effect the end-goal.

However, it should be pointed out that the end-in-view is formed distinctly, apart from consideration of cause and effect, although in relation to it. Deliberation includes tying ends-in-view to cause and effect, but is distinct.

Dewey has separated the constitution of ends-in-view from cause and effect, an implicit qualification of Aristotle. As deliberation involves consideration of the means-end relation, a traditional office of reason, Dewey notes the instrumental role of reason in deliberation. This includes the "narrow" role of making a logical connection between a means and a fixed end-in-view and a "wide" role. The wide role of reason "regards the end-in-view in deliberation as tentative and permits…encourages the coming into view of consequences which will transform it and create a new purpose and plan."[48] Not just the immediate means and end but further connections and meaning are sought. The use of general principles which will aid in resolution of the problematic situation and create links to other, like circumstances may also be involved.[49]

Evaluation and judgments of value may precede the activity of actualization, the bringing into existence of new values. It is in this respect that the title of one of the chapters of *The Quest for Certainty*, "the Construction of Good," can be read. Good can be constructed as an object, which is actualized as a result, as activity that improves the problematic situation and thus, increases the good. The good is constructed as a project over a lifetime, as the judgments that collectively result in continuing growth.

3. Conclusion

Above all, traditional morals have ignored actual conditions and full consideration of consequences. Dewey traces this all the way back to Aristotle, who he believes avoided inquiry into actual conditions and suitable means.[50] The Aristotelian attribution of value or good to being is the beginning of the separation of value from action, which should be its primary locus. This relation accounts for what Dewey perceives as the lack of progress in value theory as compared with natural science. The standards and ideals of value lie "outside actual valuations," as they are tied to singular, fixed ends. Consideration of consequences has been evaded in favor of establishing relations to fixed ideals. The result has been the modern separation of will from actions, of feelings from consequences. Values in the modern period are not based on action, what men do, but on a relation to the knower. The theoretical has been valorized at the expense of the practical; the ideal at the expense of the actual. Modern thought has accentuated the divide between ultimate values and natural objects and goods. By separating morals from science, it has completely subjectivized value.[51]Dewey's theory of value is that the good consists in a positive direction of change. Thus it is tied primarily to action, the agency of change for the better. The good that attaches to objects is as instruments that function. Objects are prized and cared for in an ongoing process, including intrinsically valuable experiences of satisfaction and enjoyment. These are part of a larger whole in the form of a life in which growth in a positive direction is the overall inclusive value. Thus experiences of intrinsic value in life are within a naturalistic context: Dewey presents a consistent naturalism. He has thereby clarified value theory as well as expanded it by making some important distinctions, especially between value, valuing and valuation.

The end-in-view or objective is the instrument of activity aimed at bringing new results about. This delimits the sphere of value as unique by virtue of actualization: of bringing about new facts, objects, objectives and results. Value is primarily connected with action in the sense of valuing. Action is the actual agency of bringing about value in making and virtue.[52] Value is a guide or instrument for action at the same time that action incorporates value into objects by bringing them about for the sake of rectifying a problematic situation. The value characterizing the action as a positive direction of change is the cause of the value of the object actualized. Practice is basic as the cause of the object of value: the good as cause in making. It is the condition of improvement of the problematic situation: meliorism. Value is the focus of change marking the novel, the gateway to the new world of the future. Value is regulative over what will come about and thus of destiny. But value is within the world for it cannot go beyond the potentialities of what is naturally possible. What is brought about will be within the natural world.

In summary, Dewey's treatment of the means-end continuum argues that ends can always be instrumental. As ends can later be means, the distinction of ends and means is not absolute but relative to circumstances. "Actual consequences, that is, effects which have happened in the past, become possible future consequences of acts still to be performed."[53] What is an end in one situation may be a means in another. This point was accepted by some figures in the tradition, of course, but Dewey's emphasis of this feature allows him to criticize transcendent ends as incompatible with it. "In a strict sense, an end-in-view is a means in present action; present action is not a means to a remote end."[54] In this sense, Dewey can rightly call his position "instrumentalism," as ends are instrumental both as means to or motive for some present objective and, when achieved as consequences, as possible future means.[55] Even general ends and notions of general value are conceived of as instrumental in a situation. The value is in relation to a situation and partly defined in terms of that situation. Growth of an organism or the value of health can be in a different direction in one situation than in another.

Further, he has developed a plausible account of the relation of is and ought, of fact and value in which they are not in separate universes. The ought or ideational is brought into existence or actuality through activity. The valuable is the ground of new facts in the form of new technologies and the objects made possible by these. The latter are prized and cared for, reflecting past values. Values, in turn, reflect a critical evaluation of desires, a natural fact. In sum, Dewey has incorporated the conative theory, as desire may ultimately be what moves one to action in the problematic situation. However, it is not a conative theory of value, in that desire must be critically transformed by other elements in the value situation. Conation is a condition but not sufficient for valuation.

Dewey did not adopt conativism so much as transform it. He stressed intelligent deliberation about forming desires to give them warrant: the "desirable." Thus there is a normative element in what is a warranted desire. As far as comparing him to the other pragmatists, he rejects the subjective turn that C. I. Lewis took in value theory. The stress is on action, not feeling. Warrant comes from successful adaptation of organism to the environment in a problematic situation far more than momentary satisfactions, although he does not reject the latter. His emphasis is on valuing over value.

The office of philosophy in this context is to revise traditional judgments of value and to project "ideas about values which might be the basis of a new integration of human conduct."[56] Conduct is to be based on ideas about values in this vision: a consequentialist model.

NOTES

1. "Theory of Valuation," from the *International Encyclopedia of Unified Science* (Chicago: Univ. of Chicago Press, Vol. II, #4, 1939) 437.

2. Beardsley, M., "Intrinsic Value," *Philosophy and Phenomenological Research,* 26, 1965, p. 6 ff. However, this is a misreading, as Gouinlock, *inter alia*, has extensively documented. (For Gouinlock's view see *John Dewey's Philosophy of Value* (New York: Humanities Press, 1972). One need only read the first and last chapters of *Experience and Natu*re to realize that Dewey did not reject intrinsic value, only foundational grounding in intrinsic value.

3. See, inter alia, Mitchell, E.T., "Dewey's Theory of Valuation," Ethics, LV, 7/1945, 287-97; and Vivas, E., *The Moral Life and the Ethical Life*, Univ. of Chicago Press, 1950, p. 108. Cf. Bennett, J., "Beyond Good and Evil: a Critique of Richard Taylor's 'Moral Voluntarism'," *Journal of Value Inquiry*, XII, 4, 1978, 313-319, in which Dewey, along with James, is associated with the theory that good consists in something "being desired." Eric Katz makes a similar charge. Holmes, in "John Dewey's Moral Philosophy in Contemporary Perspective" (*Review of Metaphysics*, XX, 1, 1966, 42-70, p. 55) argues that Dewey's early vocabulary is conativist.

4. "The Field of Value," in Lepley, R., ed., *Value, a Cooperative Inquiry* (New York: Columbia Univ. Press, 1949) ch. 3, p. 66 ff., p. 67.

5. "Theory of Valuation," sect. 1, p. 382.

6. Ibid., 413.

7. Ibid., sect. 1, p. 385. Cf. sect. 8, p. 438.

8. The terms value and valuing are used in these senses in Dewey's last word on value, "The Field of 'Value'," in Lepley, 1949, p. 65-68. These terms do not always occur in texts of the middle period, but clarify Dewey's meaning for his theory of value as a whole.

9. *Reconstruction in Philosophy* (Boston: Beacon, 1920, 1972). p. 169.

10. Although some texts treat them as equivalent. See "Theory of Valuation," p. 385.

11. "Theory of Valuation," p. 404, 409, 435 and 440. 'Valuation' is not used in "The Field of Value," (1949); Dewey wants to reduce the separation of valuation and valuing in the later work. However, the element of warranted desire and its object remains. This is an important text as Dewey's last published work on value; the "Theory of Valuation" was not his final word.

12. "Theory of Valuation," 414.

13. Ibid., p. 395, 397.

14. Ibid., sect. 1, p. 385.

15. Ibid., sect. 4, p. 407. This text includes the relation of ends and means in valuing. Ch. Peirce's original definition of pragmatism in which

meaning is the totality of conceivable practical consequences. For Dewey, if valuing does not result in practical consequences it is not valuing; this involves activity in the form of caring for.

16. Ibid, ("Theory of Valuation,") 400.

17. Ibid., sect. 4, p. 403 ff.

18. "The Field of Value," ch. 3, p. 66.

19. Ibid., ch. 3, p. 66.

20. For a contrary view see Gouinlock, J., *John Dewey's Philosophy of Value*, ch. 3, sect. 1. Although Gouinlock's reading of Dewey on this issue is plausible, it does not do justice to Dewey's analysis of the distinctively moral, esp. in Part IV of *Human Nature and Conduct* (New York: Modern Library, 1922/1957) where he characterizes "morality as social"); and as connected with activity, rather than bearers of value. Morality is connected with all experience organically but it is still distinctive. A tool may be valuable but is not as such moral. Gouinlock's view that Dewey denies the "dualism" of moral and non-moral value is true only in the sense that moral values are within natural processes. Moral values are still distinctive as a kind from other natural values.

21. Dewey, John, and Tufts, J., *Ethics* (New York: Henry Holt, 1908, 1932 revision) p. 302. I agree with Caspary that this "tension" is resolved and self-development is seen within the wider perspective of social ethics.

22. The first quote was from *Reconstruction in Philosophy*, ch. 7, p. 177; the latter from *The Quest for Certainty, A Study of the Relation of Knowledge and Action* (New York: Capricorn/ Putnam, 1929b), ch. 10, p. 256. Dewey notes that it is not important that value be considered exclusively as a character of the direction of action so long as the distinction is kept and the relation of values to the direction of conduct is noted.

23. The process of deliberation will be treated below.

24. *Human Nature and Conduct*, III, 5, p. 196.

25. Ibid., IV, 1, p. 269. This is difficult to reconcile with another text which states exactly the contrary, viz., "in quality, the good is never twice alike. It never copies itself…It is unique in every presentation." (ibid., III, 5, p. 197). As the former text is later in the same book it may be his final word. The latter text may represent an ontological point, that the good as the occurrence of an event of experienced good is never identical, as it is a unique occurrence. This is indicated by the third sentence. The point may also be that the good does not "copy itself," i.e. it is not an identical transcendent cause which forms copies of itself, quasi-Platonic style. Finally, Dewey may simply be emphasizing the contextual relation of good to specific circumstances in one text, and the formal identity of good in the later one. However, these explanations may be inadequate.

26. *Reconstruction in Philosophy*, ch. 7, p. 167.

27. Values as consummations of experience are stressed especially in *Experience and Nature* (New York: Dover, 1929a/1958) *passim*. (For Gouinlock's view see *John Dewey's Philosophy of Value*, ch. 3, p. 125.)

28. *The Quest for Certainty*, ch. 10, p. 268.

29. I am indebted to Gouinlock, *John Dewey's Philosophy of Value* on this point, which clarifies the relation of immediate and ultimate satisfactions. However, Dewey is admittedly of two minds on this score since he both rejects immediate gratification for long-term consequences and also argues against deferred enjoyment. Dewey argues against "self-contained," and "short-span acts," in favor of considering "the whole span of life." However, he also regards it as "tragic" that the present is sacrificed to a remote future. He argues that "good, happiness is found in the present meaning of activity…." and that sacrificing the present to the future has bad consequences. Value as a quality, is to be found in present experiences, "for distinction, quality, is a matter of present meaning." (*Human Nature and Conduct*, III, 9, p. 250).

30. This holds for C.I. Lewis as well. See *An Analysis of Knowledge and Valuation*, Introduction, ch. 1.

31. For Dewey's argument that a good in one situation is as good as any good in any other situation see *Reconstruction in Philosophy*, ch. 7, p. 176.

32. "Theory of Valuation," sect. 4, p. 402.

33. *Human Nature and Conduct*, IV, 1, p. 257.

34. "Theory of Valuation," sect. 6, p. 427. The argument is probably aimed against Kant. It constitutes a denial of *a priori* categories of value, and thus is an implicit argument against Scheler.

35. Ibid. A further point is that such generality may be a by-product of the experience of good as singular.

36. *Reconstruction in Philosophy*, ch. 7, p. 169.

37. *Human Nature and Conduct*, III, 8, p. 239. Dewey argues in this section that although ends are ideals, they are not ideal in common sense terms, as ideals for the latter includes "the quality of the plan proposed," i.e. evaluations of means.

38. *Experience and Nature*, ch. 1, p. 32 ff.

39. Dewey, John, and Tufts, J., *Ethics* 1932 revision, p. 229. As this is an early text, it may not represent Dewey's more mature view. Later texts will emphasize the reintegration of material activities with intellectual and aesthetic.

40. *Human Nature and Conduct*, III, 9, p. 251.

41. Ibid., p. 250. Dewey argues that, in general, economics does not provide a good model of valuation, as it is a socially contingent form and, on the contrary, exhibits in extreme degree the separation of present activity as a means from future ends in production, where drudge work is without meaning. "Production apart from fulfillment, becomes purely a matter of quantity." (ibid., III, 5, p. 205). This "empties present activity of meaning by making it a mere instrumentality." (ibid., III, 9, p. 252)

42. Ibid., I, 4, p. 63.

43. "Theory of Valuation," sect. 8, p. 437.

44. *Human Nature and Conduct*, IV, 2, p. 275.

45. As morals had their origin in social customs, i.e. "specific empirical facts," Dewey rejects any supernatural origin (ibid., IV, 2, p. 271.)

46. (ibid., p. 274) Dewey notes that the analogy of morals to functions "uproots the causes which have made morals subjective and 'individualistic'." (*Human Nature and Conduct*, I, 1, p. 18).

47. He accuses relational theories of separating means from ends ("Theory of Valuation," sect. 5, p. 416-17).

48. *Human Nature and Conduct*, sect. 5, p. 200.

49. Dewey argues that because of the novelty of the problematic situation, general principles may be needed to avoid "hopeless confusion" (ibid., sect. 7, p. 225). Habit also is a stabilizing factor.

50. *The Quest for Certainty*, ch. 1, p. 17.

51. Ibid., ch,. 3, p. 49 ff. Dewey notes the wide range of theories in the modern period, from *a priori* to emotive ("Theory of Valuation," sect. 1). However, they alike subjectivize value. In his view, the main division in the modern period is between idealist and psychological theories (*The Quest for Certainty*, ch. 10, p. 256 ff.).

52. Again, use of the term "ontological" is misleading, since Dewey opposed the view of value as an attribute of "fixed being." A better term would be that activity is the bearer or agency of value.

53. *Human Nature and Conduct*, III, 6, p. 209.

54. Ibid., p. 210.

55. Cf. "Theory of Valuation," sect. 8, p. 443, where he calls for a "theory of valuation as an effective instrumentality."

56. *Experience and Nature*, p. 46.

SCIENCE AND LOGIC

Eleven

SCIENTIFIC CONCEPTS IN DEWEY'S PEDAGOGY

Larry Hickman

Philosophically and etymologically speaking, a concept (Latin *conceptus*) is an offspring of the mind. As we know, concepts have in one way or another been the focus of philosophical discussion at least since the time of Socrates and Plato in Greek antiquity. From Avicenna and Aquinas in the Arab and Latin middle ages; to Descartes and Leibniz in the modern period; to 19th and 20th century logicians such as Gottlob Frege, Pragmatists such as William James and John Dewey, and postmodernists such as Jacques Derrida and Felix Guattari; to 21st century logicians and cognitive scientists too numerous to name, discussions of the nature of concepts have been associated with psychology, aesthetics, ethics, the philosophy of mind, and logic. A satisfactory theory of concepts is, of course, particularly pertinent to philosophy of education.

About seventy-five years ago John Dewey wrote that "With respect to teaching there is no more important topic than the question of the way in which genuine concepts are formed."[1] In the following pages I will discuss what these "genuine" concepts are and what part they play in education. I believe that Dewey's understanding of these matters ensures that his work will continue to be highly relevant to 21st century debates concerning the future of education, long after celebrations of his 150th birthday (during 2009) have been completed and preparations are under way to celebrate his 200th birthday (in 2059).

Dewey defines the genuine, or logical concept as a "factor or status in a reflective situation; it is always a predicate of judgment, in use in interpreting and developing the logical subject, or datum of perception."[2]

One of the first things to notice about this remark is that it brackets (or dissolves) the traditional metaphysical/epistemological debates about the status of concepts. Dewey simply says that a concept is a factor or status within a process. This is a significant move because there have been periods during which logicians and historians of logic have debated the ontological status of concepts without any real resolution, and at the expense of what Dewey thought was their real import, namely, their role as functional aspects of processes of inquiry.

Even a cursory examination of histories of logic, such as those by William and Martha Kneale and I. M. Bochenski, for example, or a review of the debates about the foundations of logic that flourished from the 1870s well into the 1930s and '40s, reveals an enormous literature dedicated to debates about the metaphysics of concepts. The three main Scholastic traditions from the 13th to the 16th centuries, for example, which were inspired by Thomas, Scotus, and Ockham, featured heated debates about these matters, and, perhaps more interestingly, debates about the higher level concepts that they termed "second intentions". (It is now clear that the late Scholastics not only anticipated some of the main features of the debates among logicians of the late 19th and early 20th centuries, but even anticipated the treatments of higher order predicates developed by Frege and Russell.)

Logicians might debate *ad infinitum* whether concepts are abstract entities or only names that denote classes taken in their extensional sense. But it would be strange indeed if they could not at the very least agree that there are processes of inquiry and that those processes have functional aspects.

The fact is that Dewey was not interested in discussions about the ontological status of concepts. He thought that such discussions obscured the real issues: function and transaction–that is to say, how do we make our lives and the lives of others more richly meaningful. In other words, he was interested in improving processes of inquiry.

It might be objected that Dewey's instrumentalism conceals his own implicit metaphysical commitments. But this objection does not do justice to the texts, especially the treatments of concepts in his logics of 1916 and 1938. His accounts start from processes of inquiry and what happens within those processes. He denies the primacy of the subject/object distinction on which traditional metaphysical accounts of concepts are based. To start with such a distinction, he argued, would be to commit the "philosophers' fallacy", that is, the taking of the result of inquiry as if it had existed prior to inquiry.

But what about the fact that concepts are abstractions? Does this not introduce questions about the nature of "abstract entities" and therefore metaphysical/epistemological considerations? In his 1938 *Logic*, Dewey was clear about how he avoids this traditional problem:

> [U]nless conceptual subject-matter is interpreted solely and wholly on the ground of the function it performs in the conduct of inquiry, [the] difference in dimensions between the conceptual and the existential creates a basic philosophic problem. For the only possible alternative interpretations are either the (highly unsatisfactory) view that the conceptions are mere devices of practical convenience, or that in some fashion or other they are descriptive of something actually existing in the material dealt with.

> From the standpoint of the *function* that conceptual subject-matters actually serve in inquiry, the problem does not need to be "solved"; it simply does not exist.[3]

Dewey's account thus avoids traditional metaphysical/epistemological quicksand by starting with processes of actual inquiry. Reflecting on meanings of experienced objects and events and seeking to enrich those meanings, concepts are treated both as discriminations that enrich prior experiences and as tools or instruments capable of enriching future experiences.

The *locus classicus* of this position is his 1916 lecture to the Columbia philosophy club, entitled "Logical Objects." In that lecture he invited his audience to get past the ontological debates among those who have held that concepts are variously a) physical properties abstracted, b) mental existences, or c) "metaphysical" entities, that is, a *tertium quid*. He dissolved the metaphysical aspects of the problem at hand by instrumentalizing, functionalizing, and naturalizing concepts. Here is Dewey: "Consider the bare possibility that tools and works of art give the key to the question in hand: that works and tools of art are precisely the sought-for alternative to physical, psychical, and metaphysical entities." Further, "[concepts] are simply prior natural things reshaped for the sake of entering effectively into some type of behavior"[4]. How, more specifically, does this de-ontologizing of concepts–this functionalizing and naturalizing of them–cash out? Dewey thought that there are "invalid" concepts, that is, concepts that are not genuine. A concept fails to be genuine from the point of view of logic when it has not been properly formed and can therefore not provide much (or anything) in the way of real work. (Dewey does in fact use the term "invalid" to characterize some ideas. See, for example, "Theory of Valuation".[5] Some invalid concepts have been formed on the basis of simple induction or enumeration. Some may have been abstracted as a part of a "dialectic of concepts" and then just left floating or dangling with no way of getting back to proper application to existential affairs. At one point he calls such orphan concepts "lost souls."[6] In this respect, at least, Dewey is well within the tradition of the British Empiricists who criticized much of the Scholastic metaphysical apparatus as devoid of meaning. John Locke, and Jonathan Edwards, among others, criticized invalid concepts, calling for a discourse of naked words.[7]

Dewey's discussion of abstractions in his 1938 *Logic* further supports the claim that he has bracketed the traditional metaphysical/epistemological disputes about the status of concepts. He distinguishes between what Charles Peirce termed "precision," which is abstraction by means of selective discrimination, and "hypostatic abstraction," in which the predicate of a subject functions as the subject for further predication. For the educator, the introduction and teaching of new concepts is primarily a matter of abstraction by means of precision.

What this means in educational practice is that concepts must be constructed as a part of the learning process, not just given ready-made from some external source. They are "prior natural things reshaped for the sake of entering effectively into some type of behavior."[8] As reshaped or reconstructed, they constitute a system of attributes that has to be held together by some determining principle. Viewed from the inside, the principle must control its own instances in order to insure that they are maintained as a well-integrated whole. Viewed from the outside, the principle specifies its own limits in such a way that it excludes everything else.[9]

At this point it is worth recalling Dewey's definition of the logical concept as a "factor or status in a reflective situation; it is always a predicate of judgment, in use in interpreting and developing the logical subject, or datum of perception."[10] Nominal abstractions are simply not adequate substitutes for scientific concepts. It is certainly true that the concept "color," for example, may be nominally abstracted from several variously colored blocks. This may be an appropriate way to teach a child how to use a name–in this case the generic term "color."

But genuine or robust learning involves much more than getting control of the usage of nominal concepts. It involves a scientific understanding of situations–in this case an understanding of the concept color that goes well beyond simple inductive or taxonomic exercises. Dewey tells us that in order to construct a scientific concept of color, there must be a shift from the empirical exercise that I have just described (nominal abstraction from individual colored blocks in a box) to one that is experimental in the sense of being able to serve as an instrument for the control of future situations.

If we want a scientific concept of color, for example, then it must be constructed as a part of discovery and analysis of light waves, whose various frequencies constitute the various colors of the spectrum.[11] This is the "determining principle." It is in this way that a jumble of unordered color experiences, red, blue, and yellow blocks in a box, for example, become parts of a color system. And this is a genuine or scientific concept in the sense that it can in turn serve as a tool for further investigation. Genuine or scientific concepts thus point in two directions: they express the results of experimentation and they are able to be of use in further inquiries. Nominal concepts are limited; they cannot adequately perform these roles.

In terms of educational philosophy, it is instructive to contrast Dewey's view of how "scientific" concepts are formed to what Margolis and Laurence (*Stanford Encyclopedia of Philosophy*) term the "classical" view of concepts. It is, they write:

> worth pausing to see what's so appealing about classical or definitional structure [of concepts]. Much of its appeal comes from

> the way it offers unified treatments of concept acquisition, categorization, and reference determination. In each case, the crucial work is being done by the very same components. Concept acquisition can be understood as a process in which new complex concepts are created by assembling their definitional constituents. Categorization can be understood as a psychological process in which a complex concept is matched to a target item by checking to see if each and every one of its definitional constituents applies to the target. And reference determination, we've already seen, is a matter of whether the definitional constituents do apply to the target.[12]

Moreover:

> the classical theory receives further motivation through its connection with a philosophical method that goes back to antiquity and that continues to exert its influence over contemporary thought. This is the method of conceptual analysis. Paradigmatic conceptual analyses offer definitions of concepts that are to be tested against potential counterexamples that are identified via thought experiments. Conceptual analysis is supposed to be a distinctively a priori activity that many take to be the essence of philosophy. To the extent that paradigmatic conceptual analyses are available and successful, this will convey support for the classical theory. Conversely, if the definitions aren't there to be discovered, this would seem to put in jeopardy a venerable view of what philosophy is and how philosophical investigations ought to proceed.[13]

The "classic view" of concepts, on this account, involves conceptual analysis. Acquisition involves assembly of definitional constituents, categorization involves matching definitional constituents, and reference involves application of definitional components. The complex concept "bachelor," for example, is presumably created by assembling putative definitional constituents such as "unmarried" and "male," "holder of a baccalaureate degree," and so on. Then the definitional constituents are checked to see if each and every one of them applies to the target, John Doe (thus eliminating "holders of baccalaureate degrees," for example, as a definitional constituent–since John Doe may not hold a baccalaureate degree and since there are female holders of such degrees). If the refined or revised definitional constituents correspond, then we are entitled to the propositional belief that "John Doe is a bachelor," that is, that the definitional constituents apply to John Doe.

Translating this "classical view" of concept formation into the language of education, a child who is learning to use the concept "dog" would first assemble possible definitional constituents such as color, size, shape, number of legs, quantity and quality of hair, foods eaten, and so on, by considering a number of dogs.[14] The child would then categorize, eliminating irrelevant constituents, such as size and color, but retaining traits such as four-legged, domesticated, and so on. Finally, the revised definitional constituents would be used to refer to a target, Fido, as well as other dogs.

Dewey understood concept-formation (or what the classical view terms "concept acquisition") quite differently. He thought that the classical account did not present an adequate picture of the actual processes of learning. His own account emphasizes expansion of significance and meaning by means of experimentation, rather than assembling, categorizing, and referring. To continue with the example already discussed, he thought that a child begins in the middle of things, with whatever is for the child, the current significance of one dog. That significance (which is relatively inchoate with respect to the eventual concept) is carried over into new experiences in terms of behavioral expectations of various sorts. The individual may call a horse, for example, a "big dog" and then be corrected by a teacher or other adult. The child's concept of "dog" is thus refined, not by thought experiments or *a priori* conceptual analysis, but by objective responses to behavior based on interest and activity. The type of abstraction in play in this example is discrimination based adjustments of attention to some parts of experience and neglect of other parts of experience. This is more or less what Charles Peirce meant by "precision" and what Dewey thought is the essence of concept-formation in education.

Dewey thus presented a model that is quite different from the "classic" model of concepts. Instead of acquisition being a matter of induction, that is, extraction of a common meaning from several "ready-made" objects, such as several dogs, his account of concept-formation involves adjustment through time on the basis of behavioral responses to an idea that is initially vague but becomes increasingly more definite through trial and error at first, and then, gradually, through more systematic investigation. "By these processes, his [or her] idea gets body, steadiness, distinction; it becomes a concept."[15]

A genuine or scientific concept is never completely fixed or finished. It is dynamic. It is a well ordered arrangement of previously unordered data in what Dewey terms "a dynamic continuity of existence."[16] Our use of concepts must thus incorporate a strong sense of fallibilism, that is, a profound appreciation of the manner in which changing circumstances and new data are capable of undermining some of our most entrenched and cherished beliefs. At the same time however, and this is a point that some fallibility of the post-modernist variety tend to miss, it is important to take the term "final" in its logical meaning. As Dewey puts the matter, "final" means "'final' for that case." But he also points out that "apart from a condition of tension between a person and environing conditions there is . . . no occasion for evocation of desire for something else...."[17] Infinite re-interpretability or re-description of all concepts is itself an invalid concept. This is because concepts tend to standardize knowledge as habits standardize behavior. There are concepts and habits that are capable of lasting a lifetime, or many lifetimes.

An excellent example of the superiority of Dewey's account of concept-formation in learning comes not from Dewey's own work, but from physicist Richard Feynman.[18] He reports looking at a science textbook used in the first grade. He finds three pictures. The first one is a picture of a wind-up toy dog. The second one is a picture of a real dog. The third one is a picture of a motor bike. Each picture has the same caption: "What makes it move?" Then he picks up the teachers' handbook. The teacher is supposed to tell the children that it is energy that makes the toy dog, the real dog, and the motor bike move.

If at this point the teacher employs the classical account of concept acquisition, this person will lead the children to acquire the concept "energy" by first assembling putative definitional constituents such as "capacity," "physical system," and "work," then categorizing or matching definitional constituents, and finally referencing definitional components. The primary activities on the parts of the children will be conceptual analysis, thought experiments, and, of course, memorization.

If the teacher utilizes the "blocks in a box" method, then the lesson will be even briefer: all three examples fit the nominal definition of "energy." We agree that all three things: the toy dog, the real dog, and the motor bike, run on "energy."

Nobel Prize-winner Feynman, however, didn't think that providing the answer in the teachers' handbook was a very good way to learn about science. Nor, more particularly, was it a good way to form a concept of energy. It would be like Dewey's example of attempting to teach the concept of color by pointing to colored blocks in a box. Energy would be taught as nothing more than a nominal concept, badly formed and incapable of performing any useful work.

Nor would a genuine concept of energy be formed as a result of conceptual analysis. A genuine concept of energy would be the result of experimental activity that sharpens vague experiences and brings them to a useful conclusion. A genuine concept of energy would be able to perform various functions, such as getting additional data about some problem that we want to solve. It would serve as a key that opens the door to lots of new ideas and new data. A genuine concept of energy would not be produced by asking the children to memorize a definition and then telling them that all three cases–the toy dog, the real dog, and the motor bike–fit the definition. Feynman thought that that method of concept formation would be an excellent way to destroy any interest in learning about science.

So Feynman's approach to concept-formation was very much like Dewey's. The reason that the toy dog moves is that you wind up the spring, which begins to unwind when you release the crank. As the spring unwinds, it runs the gears, and makes the toy dog's legs move, and so on. That experiment is admittedly a bit vague, but it has the potential of leading to discussion of how the ability to do work gets stored, how it is measured, and how it is transmitted. New connections are made and the concept becomes sharper. As Dewey puts it, it by this type of process that the children's idea "gets body, steadiness, distinction; it becomes a concept."[19]

Feynman and Dewey are clear: it is better for the children to take the toy dog apart and see how it works than to memorize the definition of "energy." Taking the toy dog apart is a good way to teach and learn science. Getting the children to memorize or analyze the definition of "energy" for a test is not.

If inductive generalization and conceptual analysis are not adequate to the formation of a scientific concept of color, neither are such methods adequate to the formation of moral or political concepts. This is a particularly important point, given the charges of positivism, scientism, and even moral nihilism that continue to be leveled against Dewey's educational theories. The collection of data relating to moral values and practices across human populations, and abstraction from such data of common elements that are then treated as "universal" in the sense of "fixed," does not, in Dewey's view, constitute a scientifically useful concept in the field of morals.

Scientific concepts that involve moral behavior, and therefore concepts that activate and are active in the process of learning, must be more than abstractions from enumerated data (or definitional constituents): they must have historical or genetic depth and the potential for instrumental involvement with future situations. Their formation must involve processes by means of which what is morally valued is analyzed in concrete, emerging contexts and sharpened by being determined to be morally valuable within and across those contexts. As we know, this view has put Dewey at odds with those who view education as simply or primarily the "transmission of knowledge and values." These people are easy enough to find on the Internet. It has also put him at odds with "natural law" theorists.

Dewey's view that genuine concepts are plastic (but not arbitrary) is easily applicable to current debates about moral values. One such debate which is a matter of considerable concern in the United States as I write, and which serves as an excellent example of his view of the generation of scientific concepts in the context of moral debates, involves the presence (either by biological parentage, adoption, or foster care) of children in households of same-sex couples. According to the year 2000 census, there were 250,000 children in the United States being raised by same-sex parents.[20] Of course this matter is much more complex now than it was a decade ago; as I write, six states recognize same-sex marriage, and several more states make provisions for civil unions between same-sex partners. But on November 4, 2008, "Arkansas voters approved a measure to ban anyone 'co-habituating outside of a valid marriage' from being foster parents or adopting children."[21] (It is interesting to note the breadth–whether intended or not–of this statute: it prohibits adoption or foster care by non-married heterosexual couples as well.)

My suggestion is that the Arkansas statute and similar prohibitions are for the most part based either on religious authority or on inductive generalizations of the type I have just discussed. In the former case, religious-based concepts of successful (or legitimate) families tend to be both attenuated and ossified: they take as its determining principle cultural practices that are most often antique and inflexible with respect to the demands of changed and changing social circumstances. In the latter case, inductive generalization usually becomes circular and thus impotent, since the determining principle is selected in advance in a manner that insures that only cases of "correctly defined" (read "narrowly selected") families are selected as instances of the generic concept "legitimate family." Even if we were able to find within such narrowly selected human populations some invariant practice regarding children in same-sex households, however, we would still not have gained any insight into the scientific significance of that practice.

The problem is that in both cases, arguments from religious authority and inductive generalization, there is a failure of scientific concept-formation. In one case good science is most often trumped by adherence to religious authority, and in the other what we have is not so much good science, but faulty application of information gained through sampling. First, there is a failure to properly identify instances of what constitutes a successful, nurturing (read legitimate) family situation because the choice of the determining principle has been artificially restricted. Beyond that, however, there is a failure to supplement those narrowly selected instances of successful traditional family arrangements with non-traditional arrangements of the same kind, that is, with non-traditional familial arrangements in which children are loved and nurtured in ways that promote their healthy maturation. Finally, there is failure to locate instances of successful family relationships of both types–traditional and non-traditional alike–within a wider conceptual system, such as a system of laws.

This pattern of identifying, supplementing, and locating is another way of describing the learning process: identifying (starting with) a particular dog, supplementing behavioral responses, and then locating the refined concept in a larger context.

Scientific significance is thus not based on the authority of sacred scriptures or other received traditions, nor is it a result of inductive generalization. It is instead based on experimentation with relevant data and linked to the possibility of control of future experience. Dewey argued that scientific significance, and therefore the possibility of control of future experience (read learning), is disclosed only through knowledge of historical facts and their permutation through time, that is, their evolutionary development.

In the example under discussion, this would require a careful assessment of successful child-rearing by same-sex couples; comparison of that data with similar data derived from analyses of child-rearing by traditional heterosexual couples; and analyses of collateral issues such as those that are economic and legal. The result would be the formation of a "genuine" concept of successful (legitimate) families that would be capable of serving as the basis for further experimentation that would, in turn, be likely to provide the basis for legal and economic reforms. A "genuine" concept of a successful or legitimate family is thus able to contribute to the possibility of successful control of future experience in ways that other types of concepts cannot.

On matters of this nature, Dewey was clear: "[t]he essence of moral struggle and of moral progress lies . . .precisely in that region where sections of society, or groups of individuals, are becoming conscious of the necessity of ideals of a higher and more generalized order than those recognized in the past. To fix upon that which has been believed everywhere, and at all times, 'as the essential content of the moral law, ' would give practical morality a tremendous set-back."[22]

This is not to say that concepts do not standardize our knowledge, for they do. They are instruments that serve to anchor our thinking in a constantly changing world. As I have already suggested, they aid identification of objects and events, their supplementation with objects and events of the same kind, and their location or placement within a wider system of objects and events.

In Dewey's view, simple pedagogy of transmission of standardized knowledge, although for the most part easily and efficiently accomplished, simply does not contribute much to robust educational practice. It leaves most learners instead with concepts that tend to be nominal and remote from their own experience. For learning to occur, there must be a generalization and extension of learned concepts. What Dewey terms a "genuine" or "logical" or "scientific" concept thus becomes "a working tool of further apprehensions, an instrument of understanding other things."[23]

The debate between the two main positions I have described–one involving nominal concepts and the other what Dewey termed "genuine" concepts, has enormous implications for education. In one case we have transmission of values. In the other we have active experimentation with values and their reconstruction.

I will conclude with another example of an attempt at genuine concept-formation. In her op-ed piece in the *New York Times* June 5, 2009, former Israeli foreign minister Tzipi Livni calls for the development of what Dewey would probably have called a "genuine concept" of democracy. Like Dewey, she rejects the idea that mere participation in elections constitutes or defines a democracy. As we know from the history of Central and Eastern Europe, there are "one time" democrats, as she puts it, "ready to use the democratic system against itself." So she calls for a "universal code for participation in democratic elections." She continues: "This would require every party running for office to renounce violence, pursue its aims by peaceful means and commit to binding laws and international agreements. This code should be adopted by international institutions, like the United Nations, as well as regional bodies. It would guide elections monitors and individual nations in deciding whether to accord parties the stamp of democratic legitimacy, and signal to voters that electing an undemocratic party would have negative international consequences for their country."[24]

In my view Livni presents a particularly interesting case to which Dewey's criteria for "genuine concepts" can be applied, and which has considerable pedagogical value. First, Livni has identified a concept of democracy–participation in the voting process–that has been advanced by some, including the recent Bush administration. Second, she has supplemented the concept with additional provisions, such as renouncing violence and abiding by international agreements. And third, she has located the concept within a larger legal and political framework. Finally, her conception of democracy is treated not as an end in itself, but as a tool for further development: it is offered as a tool for alerting voters about the international ramifications of their actions and, by implication, further refining the concept.

During the few pages allotted to me I have discussed what Dewey characterized as "genuine" concepts and I have distinguished them from those that are "nominal" and those that are "analytic." I have suggested that the consequences for education of these distinctions are considerable, and I have provided two examples, one at length and the other in embryo, in which I have attempted to explore some of those consequences. In all of this, my point has been to emphasize the importance of one of the central insights of Dewey's philosophy of education.

NOTES

1. John Dewey, "Santayana's Orthodoxy" (1934), in *The Collected Works of John Dewey: The Later Works, 1925-1953*, vol. 9, ed. Jo Ann Boydston (Carbondale: Southern Illinois University Press,1986), p. 240.

2. John Dewey, *Studies in Logical Theory* (1903), in *The Collected Works of John Dewey: The Middle Works, 1899-1924*, vol. 2, ed. Jo Ann Boydston (Carbondale: Southern Illinois University Press,1976), p. 359, note 10.

3. John Dewey, *Logic: The Theory of Inquiry* (1938), in *The Collected Works of John Dewey: The Later Works, 1925-1953*, vol. 12, ed. Jo Ann Boydston (Carbondale: Southern Illinois University Press,1986), p. 462.

4. John Dewey, "Logical Objects" (1916), in *The Collected Works of John Dewey: The Middle Works, 1899-1924*, vol. 10, ed. Jo Ann Boydston (Carbondale: Southern Illinois University Press,1980), p. 92.

5. John Dewey, *Theory of Valuation* (1939), in *The Collected Works of John Dewey: The Later Works, 1925-1953*, vol. 13, ed. Jo Ann Boydston (Carbondale: Southern Illinois University Press,1988), p. 230.

6. Dewey, "Logical Objects," p. 93.

7. John Locke, *An Essay Concerning Human Understanding*, vol. 1, book 2, chapter 11, section 8, ed. Alexander C. Fraser (New York: Dover Publications, 1959).

8. Dewey, "Logical Objects," p. 92.

9. Dewey, *Studies in Logical Theory*, p. 345.

10. Ibid, p. 359, note 10.

11.Ibid, p. 345.

12. http://plato.stanford.edu/entries/concepts/#OntCon.

13. Ibid.

14. John Dewey, *How We Think* (1933), in *The Collected Works of John Dewey: The Later Works, 1925-1953*, vol. 8, ed. Jo Ann Boydston (Carbondale: Southern Illinois University Press,1986), p. 240.

15. Ibid, p. 141.

16. Dewey, *Studies in Logical Theory*, p. 345.

17. Dewey, *Theory of Valuation*, p. 231.

18.Richard P. Feynman, *The Pleasure of Finding Things Out* (Cambridge, Mass.: Perseus Books, 1999), p. 178.

19. Dewey, *How We Think*, p. 141.

20. Dan Savage, "The Gay Child Left Behind," *New York Times*, February 17, 2005, p. 27.

21. Wikipedia, http://en.wikipedia.org/wiki/LGBT_adoption Retrieved 06.14.09.

22. Dewey, *Studies in Logical Theory*, p. 19.

23. Dewey, *How We Think*, p. 242.

24. Tzipi Livni, "Democracy's Price of Admission," *New York Times*, June 5, 2009, p. A21.

Twelve

DEWEY'S SCIENCE: A TRANSACTIVE MODEL OF RESEARCH PROCESSES

Philipp Dorstewitz

1. Introduction

Dewey's supporters cannot complain about the recognition his work has received around the world. Dewey's impact is greatest on those aspects of our common life that he himself deemed most important: primary and secondary education. Yet, it is astonishing how little attention other fields have given to Dewey's work, where his contribution is equally ground-breaking. With a few notable exceptions, mainstream philosophy of science has ignored Dewey as a predecessor of many topical debates in the field. Dewey's conceptions of "experience" and "inquiry" were not only at the very heart of his philosophy; they were also highly innovative contributions and could still be tapped as an important resource for contemporary theory of science. Since Dewey's death in 1952, philosophers of science have debated how we can live up to the methodological rigor of empiricism without losing the realist notion that our theories are actually about something natural; they have discussed ways to acknowledge the fallible character of all knowledge while at the same time believing in the scientific method as a warrant for improvement. Many wondered how we can do justice to the historical nature of science as a social institution while still crediting scientific research with a rational or intelligent method. During the last thirty years, cognitive scientists began to argue that experience is a productive and not merely receptive process, and philosophers of science saw that experiments create facts rather than just observing or registering them. Since the 1960's new attention has been given to the idea that science is not only a bystander in society and politics, but an important player, and itself a field of ideological altercations. Questions regarding objectivity and value-neutrality of scientific research have recently been revived. Finally, attention is currently being given to science in the making, or "how the ship gets into the bottle", together with debates on the instrumental character of theories and scientific models. Dewey[1] asked most of these questions more than fifty years ago and came up with innovative answers that today's philosophers of science could benefit from.

Curd and Cover's[2] collection, *Philosophy of Science – the central issues*, comprises some fifty representative contributions. Dewey is only marginally mentioned in three articles (by Carl G. Hempel, Larry Laudan, and Ian Hacking). Hacking once titled a subheading “On not reading Dewey”[3], under which he owned up to the confession, “I once tried valiantly to read Dewey, but it did not click. He goes on and on.” Hacking made the admission: “My own view, that realism is more a matter of intervention in the world, than of representing it in words and thought, surely owes much to Dewey”[4]. Yet later he qualifies: “*Owes*. Not in the sense that Dewey incited me to think that way, but in the sense that, when I looked back over the history of philosophy, I recognised that Dewey had been there before me”[5].

Admittedly, reading Dewey can be like swimming in porridge[6], but Dewey scholars such as Richard Bernstein[7], Larry Hickman[8], John R. Shook[9], David Hildebrand[10] and many others have done a thorough job of clarifying the complex concepts of “inquiry” and “experience”, and have made Dewey’s convoluted work fertile ground. Dewey’s style is no longer an excuse for ignoring what he has to offer to contemporary debates.

My aim here is not to trace Dewey’s footprints in supposedly newly discovered territory. I will discuss four fields where I believe that Dewey’s theory of inquiry has an important contribution to make to contemporary post-Kuhnian debates.

The first concerns a new determination of the elements of scientific inquiry and their relations. Many theorists of the late 20th century have claimed that we cannot consistently separate the working of conceptual, intellectual, and empirical phases of research; that we must reorganize our understanding of tasks like “formulating hypotheses”, “theoretical reasoning”, and “experiment” as mutually dependent elements within scientific practice. The Deweyan model of inquiry that I will present here redefines the relation between theory and practice and between “representing” and “intervening” as Ian Hacking argued[11]. The present model will introduce the elements “hypothesis”, “reasoning”, and “experiment” as concurrent and mutually dependent activity streams. These centre on the imaginative production of warranted beliefs and normative orientations.

The second field where Dewey can make a contribution is the question of how scientific research methodology relates to practical circumstances. Post-Kuhnian philosophers rejected Poppers’ separation between logical and historical accounts of scientific research. They insisted on looking at the history of science as episodes of discovery with an open outcome. Theorists of science have recently focused on scientific practice as a process situated in concrete places like offices, laboratories, zoos or botanical gardens, and gave great attention to the economic, social, instrumental, and political conditions of research as part of its epistemic situation. Dewey’s theory of inquiry is surprisingly apt at accommodating this sciences-studies perspective within a logical and methodological account of scientific research. At the same time, he grounds the origin of inquiry and possible criteria for its successful conclusion in natural qualities, and not in arbitrary conventions.

The notion of a "community of inquiry" refers to the third field where Dewey has a contribution to offer to contemporary debates. Much thought has been given to the social nature of scientific research, as e.g. the formation of research networks. I will point out how Dewey's transactive notion of agency can account for achieving coordination and collaboration in initially dispersed research environments. Dewey's transactive notion of communication helps to overcome problems arriving from Kuhn's notion of incommensurable paradigms.

The focus here is not on a theoretical treatment of these three problem-complexes, nor do I offer a systematic analysis of Dewey's contribution to their solution. Using a model, I attempt to create a synthetic perspective on the Deweyan research process, which identifies and contextualizes its elements and collates them with several recent debates in the philosophy of science. I thereby hope to make Dewey's theory of inquiry a bit more accessible to mainstream philosophy of science.

The model of scientific research proposed here is an adaptation of a process-model of policy-making and planning, the "transactive rationality model", which Prof. Shyama Kuruvilla and I developed together[12]. The isomorphic structure of these two models implies a great semantic overlap between the elements of scientific research and policy making. In the present article I will concentrate on modeling scientific inquiry, and this task will not leave me with room to discuss the relation between this model and its original formulation as a policy and planning model. For those who are familiar with the "transactive rationality model", the caveat must suffice that I do not believe a strict one-to-one match between the elements of both models is possible. One of the defining characteristics of a transactive model is its flexible internal structure and the heuristic nature of its elements. It depends on the application in concrete contexts what elements may be distinguished and how exactly they come to be related.

2. Reciprocity in the Learning process

In "Logic: Theory of Inquiry", Dewey spells out his pattern of inquiry in five steps. Some have interpreted this as a procedural logic of sequenced steps or stages in the research process.

1. Indeterminate Situation: Punctuated Equilibirum
2. Institution of a Problem: Recognizing a Problematic situation
3. Determination of "problem-solutions": Formulating a hypothesis
4. Reasoning: transforming symbols and arriving at a judgment
5. Operational character of Facts-Meanings: Controlled experiment

I will argue that a linear reading of these elements as subsequent steps is mistaken. Ernest Nagel in his "Introduction to Dewey's Logic" proffers such a linear procedural interpretation: "As Dewey saw the pattern of inquiry, it is a succession of distinguishable stages. ... This pattern obviously includes, but goes beyond, what has been called the "hypothetico-deductive method"[13]. Dewey himself gives rise to this misunderstanding by making the claim that "The pattern [of inquiry] is serial or sequential"[14]. However, in his explanation Dewey only refers to a serial or sequential order of the rough "macro" pattern of inquiry as presented in figure 2 (in a later section). I will introduce a model of Dewey's pattern of inquiry that represents the elements as concurrent and interdependent activity-modes, which together form the organic whole of a learning-process (see fig. 1).

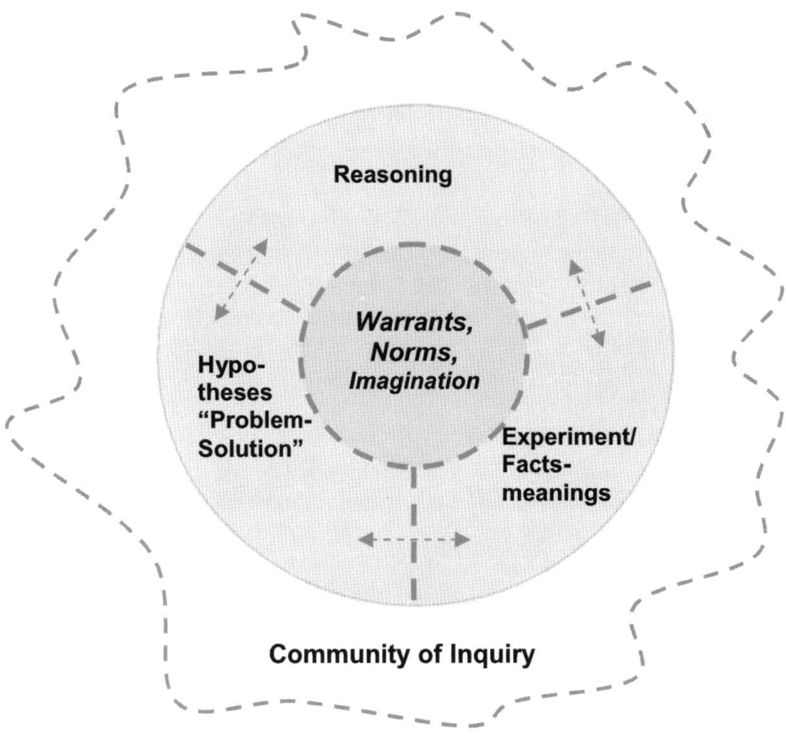

Figure 1: The process of scientific research

Core and periphery refer to themes that Dewey did not explicitly include among the five elements of the pattern of inquiry that he distinguished in chapter six of his *Logic: theory of inquiry*. However, these fields address themes that are prevalent throughout Dewey's work, and should be included in a Deweyan model of inquiry.

The idea of organizing a research process as a non-linear interaction between different activity modes can best be illustrated by the task of writing an academic paper. It is indeed possible to pursue a strict procedural logic such as: 1) define a topic, 2) formulate a hypothesis, 3) collect data, 4) analyse data, and 5) write up the paper. Yet this is only one way of going about the task. It may be better to start by drafting a conclusion as a means of formulating and sharpening a hypothesis. Often data-collection, analysis and writing build parallel streams in a process. In my own experience, the tasks of conceptualizing, structuring and framing a topic become more relevant and fertile at the end of the writing process.

Below I will analyze the internal structure of the proposed inquiry model. I will argue that Dewey's own definitions suggest such a parallel and interdependent ordering of the elements in his pattern of inquiry. Only if we see the elements of Dewey's pattern as interactive and interdependent parallel activities are we able to understand inquiry as an organic whole and a learning-process.

Hypothesis: Dewey uses the hyphenated word "problem-solution" to indicate that a hypothesis is both an issue within and the conclusion of inquiry. A hypothesis organizes research by guiding and organizing experimental activity – on this account, "hypothesis" can rightfully claim its position at the beginning of inquiry. At the same time, research aims at formulating new and viable hypotheses, and in this sense "hypothesis" belongs at the end of the inquiry process. The hyphenated notion "problem-solution" indicates this wide stretch that a hypothesis stage has to perform.

In the above model, "hypothesis" is present at all stages of the process as an element, aspect or activity mode of inquiry. In early phases, "hypothesis" should mean a playful exploration of the situation in terms of concepts, frames, tentative problem-specifications and model-solutions. Despite its playful nature, the hypothesis is of great consequence, since "[t]o mistake the problem involved is to cause subsequent inquiry to be irrelevant or to go astray…. The way in which the problem is conceived decides what specific suggestions are entertained and which are dismissed."[15]

For Dewey, "problems" and "solutions" are changing functional distinctions in the creative process of structuring problematic situations. Their distinction is heuristic and they serve as guides or resources for our actions. Dewey often emphasizes the "thoroughly reciprocal character"[16] of means and ends in human conduct. "… [W]e know what the problem exactly is simultaneously with finding a way out and getting it resolved"[17]. Dewey further emphasizes the danger of fixing the hypotheses or ends of inquiry too early in the process. "If we assume, prematurely, that the problem involved is definite and clear, subsequent inquiry proceeds on the wrong track."[18]

Reasoning: Intelligent inquiry employs reasoning to settle definite (though tentative) judgements. For Dewey, reasoning means reflecting on the consequences of possible conceptualizations and ensuing experimental actions.

Reasoning determines further experimental activity and establishes criteria for its success and failure.

This intellectual aspect of research comprises the transformation of meanings by means of symbolic operations. It clarifies propositions, which translates to making them "more clearly *relevant* to the problem at hand"[19]. This also establishes the close tie between the segments "reasoning" and "hypothesis".

For Dewey, "reasoning" is much more closely connected to experience than traditional views would have it. Reasoning plays a constitutive role in all experimental activity. This notion of reasoning is of course incompatible with a linear sequential reading of pattern of inquiry. "The distinction and relation between observed data and directive ideas is a functional division of labour within inquiry"[20].

Some of the meanings that "reasoning" processes rely on must first be determined through experiments. In this sense, reasoning has no definite procedural position as it both guides experimental activity and is transformed by it.

Experiment: Dewey introduces the hyphenated word "fact-meanings" to point out the theory-ladenness of all experience. It corresponds with Kuhn's rejection of Carnap's "protocol sentences" and the idea that context and background understanding are crucial for the identification of basic empirical facts. At the same time, however, the notion of "fact-meanings" entails a naturalistic interpretation of *meaning* which also finds expression in Dewey's identification of *meanings* and *symbols*. Like symbols, facts have for Dewey an intentional character. They are not established by experience as "antecedent existences". Experience rather distinguishes or fixes facts as reference points for the organization of further actions. In this sense, all facts are originally meaningful because they are situated in a context, they are constituted by relations, and they refer to future transactions. Meanings are for Dewey not attributes attached to facts; they are partly constitutive of these facts. An increase in meaning, such as that brought about by an experiment, enriches facts with regard to the way they function in experience.

In this way, Dewey offers a promising solution to the problem of subjectivism, which ensues from Kuhn's rejection of a basic theory of neutral observation and his insistence that all experience is interpretation. For Dewey, facts are 'constructed' in experience. Still, this leaves them in place as facts in and of nature. Yet, this also means that facts are not self-sufficient: Being *evidence,* they function in a context of research activity. For Dewey, experimental facts are "operational"[21], i.e. experimental facts are intelligible only as *aspects*, *elements*, or *phases* within an inquiry process. They have a factual character insofar as they constitute elements in an experimental inquiry and learning-process. Observations are made or generated so as to give answers. Like technical products, they are designed and produced to function in a certain problematic context. If they function as 'designed,' they help to confirm a hypothesis and to settle the cognitive dissonances that initiated the inquiry. If we fail to observe or 'to bring about' facts, as our experimental setup was designed to do, we must conclude that there has been a failure in the technical procedure that we designed to produce a certain experience. This of course refers to possible faults in all three segments of the model. Besides faulty experimental equipment, the problem can reside with our conceptual tools, with our hypotheses, or with our reasoning procedures. In this sense, further problematic experimental observations bring us right back to the drawing-board, meaning to other sectors of the model.

Philosophers like Ian Hacking[22], Andrew Pickering[23] have recently made what Dewey calls the "operational character of facts-meaning" quite fashionable. Dewey summarizes this point as: "The experimental procedure is one that installs doing as the heart of knowing"[24]

The core: "warrants, norms, imagination: Science is not an algorithm; it cannot be identified with a logical procedure as positivists and critical rationalists have tried to establish. Science *employs* methods and principles, but it must not be defined by reference to a finite number of logical operations and procedures. In *Logic: Theory of Inquiry*, Dewey argues that scientific research is itself the source of logical forms. "[Logical] forms originate in operations of inquiry… primary inquiry is itself the *causa essendi* of the forms…"[25]. Deweyan inquiry is a situated enterprise. A scientific community must develop and adapt its own methods and tools, and must agree on their employment.

For a pragmatist, "methods and tools" would include deductive procedures and normative principles that guide and organize scientific practice. For example, scientists have to evaluate arguments and judgments not only by how cogently they follow from their premises or by how reliable the premises are; they must also ask whether the type of argument or method of reasoning applied is appropriate in a particular context. For example, they must ask whether a hypothetico-deductive method is suited to explain a migration-movement or an election result, or whether an ethnographer who studies witchcraft in the Bocage has any use for a perspective of disinterested objectivity. Scientists have to answer questions like 'what kind of data have to be investigated?', 'what would constitute a satisfactory explanation?', 'is a coefficient of 0.05 good enough to establish a significant result in a specific study?'; all of these referring to the question of whether the employed methods, norms and experiments yield a satisfactory solution to a *problematic situation* at hand.

"… just as in science the question of the advance of knowledge is the question of *what to do*, what experiments to perform, what apparatus to invent and use, what calculations to engage in, what branches of mathematics to employ or to perfect, so *the* problem of practice is what do we need to *know*, how shall we obtain that knowledge and how shall we apply it?"[26] Dewey defines inquiry as "the controlled and directed transformation of an indeterminate situation into one that is so determined in its constituent distinctions and relations as to convert the elements of the original situation into a unified whole"[27]. Besides explaining why

Dewey never received awards as an essayist, this quote deals with a widely held prejudice against the pragmatist movement. Inquiry is not an expedient way of making problems disappear, and a pragmatist does not judge the validity of scientific methods and logical norms by their utility. Inquiry tries to "unify", i.e. it aims at remedying or at 'making whole' a situation that was torn by contradictions and conflicts. Moreover, inquiry unifies by "determining". Conflicting theories of scientists must be so clarified as to yield complementary points of view; and incompatibilities between experiments and hypotheses must be so adjusted that beliefs and experiments work together in harmony. Dewey requires determining a situation in its "constituent distinctions and relations", which is more than finding some *modus vivendi* of living in problematic circumstances. It demands of the inquirer to explore and investigate it and to gain a deeper understanding of the context, which is the only warrant available for curing a disease rather than merely abating the symptoms.

By the same token, Dewey's translation of "truth" as "warranted assertibility" implies a rather restrictive criterion. Neither truth nor its warrants are timeless or necessarily valid. There may always be alternative ways of settling an indeterminate situation through inquiry. However, it is still the problematic quality of a situation that determines the worth of warrants.

Like Michael Polanyi[28], Dewey understands scientific inquiry as a craft, requiring the scientist to be an artisan. Scientists have to employ experience, knowledge of a particular situation, and imagination in order to transfer general principles to concrete contexts. "We know that some methods of inquiry are better than others in just the same way we know that some methods of surgery, farming, road-making, navigating, or what-not are better than others"[29].

Scientific inquiry must develop its own normative groundwork. It has help from established methods and principles, but in every research situation scientists must pass an implicit or explicit judgment on the applicability of norms, and must evaluate their worth in that particular context.

Recent work in philosophy of science has been sensitive to the normative and self-legislative character of scientific inquiry. Lorrain Daston and Peter Galison[30] present "objectivity" as a normative concept that underwent great transformations during the past three hundred years. The authors argue that "objectivity" should not be regarded as a timeless epistemic virtue or as a *gold-standard* for acting in a scientific manner. They parallel different instruments and methods used to produce scientific images with changing ideas of the purposes of these representations. Their story is about transformations of ethical standpoints that scientists obtain with respect to their work. For instance, the 18th century "sage, whose well-stocked memory synthesizes as lifetime of experience with skeletons or crystals or seashells into the type or that class of objects"[31] in contrast with the 19th century "ascetic", an "indefatigable worker whose strong will turns inward on itself to subdue the self into a passively registering machine"[32], or the

20th century expert who uses "trained judgement" to recognise patterns through the shroud of noisy data.

Steven Shapin and Simon Shaffer[33] investigate how scientific standards and methods were proposed as answers to social and political crises rather than merely as means of settling intellectual disputes. They interpreted Boyle's publically witnessed experiments and Hobbes' method of deductive reasoning as proposed means for manufacturing consent in situations of political disagreement or "civil war".

On such accounts, inquiry into the methods and normative principles of scientific inquiry itself is natively an ethical quest in a wider socio-political sense. Dewey understands scientific research as continuous with mundane problem-solving. Solutions to everyday problems create problems for further scientific inquiry. He therefore complains about "the highly technical and remote language of science [which] creates and maintains the feeling and idea of a complete gap. The paths of communication between common sense and science are as yet largely one-way lanes. Science takes its departure from common sense, but the return road into common sense is devious and blocked"[34]. The problem is that science is "…such a specialized mode of practice that it does not appear to be a mode of practice at all"[35]. Science is losing connection with its origin: the struggles of everyday problem-situations. This answers why methodological and logical questions of science were no longer regarded as having practical and ethical relevance in social and political spheres.

One of Dewey's deepest concerns is to reintegrate scientific inquiry and knowledge production into the moral life of citizens in a community. "The problem of restoring integration and cooperation between man's beliefs about the world in which he lives and his beliefs about values and purposes that should direct his conduct is the deepest problem of modern life."[36] Dewey seeks the solution to this problem in a conception of scientific inquiry that is at its heart a search for orientation in the world. "Orientation" refers equally to action-guiding knowledge and to situated normative principles, and so decreases the gap between "is" and "ought". For a pragmatist like Dewey, values are entirely empirical matters. "Why should we not proceed to employ our gains in science to improve our judgments about values, and to regulate our actions so as to make values more secure and more widely shared in existence?"[37]

At the core of a model of scientific inquiry we may expect to find a reference to "logic" or "scientific method" rather than "imagination", which may "have emotive appeal and evoke vivid pictorial associations, but which does not further our theoretical understanding of the phenomena under consideration."[38]

For Dewey, scientific inquiry is a matter of *reason* since it proceeds along the "method of intelligence". However, reason is not a form of "bloodless calculation"[39]. Emotions are part of the fabric of reason which he defines as "an effective relationship among desires rather than a thing opposed to desire"[40]. Hence, "feeling and intellect … have no separate existence"[41], and Dewey concludes that reason "depends upon a sensitive and proportionate emotional sensitiveness"[42]. This means that reason cannot be defined as a given faculty, but is a "laborious achievement", and intelligent inquiry must aim at transforming impulses until "a viable working harmony is established."[43] Dewey writes, "more 'passions,' not fewer is the answer… The man who would … cultivate intelligence will widen, not narrow, his life of strong impulses while aiming at their happy coincidence in operation."[44]

Imagination is a form of reasoning that involves and transforms our emotions. Dewey also calls it "dramatic rehearsal" because imagination uses thought experiments to explore situations and potential developments in *quasi* (i.e. imagined) lived experience. In a separate paper[45], I have distinguished nine dimensions and offices of imagination within reasoning. These include the ability to project current tendencies into the future, the intuitive and aesthetic grasp of a situation as a whole, and the ability to translate between two situations. The latter enables us to transfer methods, tools and normative principles and to apply them adequately in a given context.

Rehabilitating the status of emotions with respect to the traditional notion of dispassionate reason is a topic that psychologists and feminists have been discussing for several decades[46]. Appreciating the role of imagination in science seems to be an emerging trend[47].

3. Situated Research

Dewey's conception of inquiry relies on Peirce's "doubt-believe-scheme", which holds that scientific research begins with the breakdown of previous certainties. Beliefs are challenged and such challenges are of an existential nature: "doubt" means a loss of practical orientation and a growing disharmony between beliefs and practical conditions of action. Both Peirce and Dewey make the radical proposal that scientific inquiry has no logical starting-point save one given within scientific transactions themselves. Cartesian radical doubt does not suffice to initiate inquiry. A genuine question implies an acute crisis of a belief within its context. For Peirce, *doubt* as much as belief requires solid ground, and for Dewey, real doubt is marked by a disharmony between experimental observations and the concepts and theories employed to explain phenomena. Dewey identifies the point where "believe" yields to "doubt" as the transition from a habitual equilibrium to a situation of "indeterminacy" or "conflict". "Equilibrium" and "conflict" are terms that refer to the quality of experience within a situation. Such problematic experience is not limited to cognitive dissonances within a researcher's mind. For Dewey, "experience" comprises all transactions within a situation, i.e. all interactions between researchers and their natural and social environment. "Doubt" or "doubtful experience" does not necessarily refer to a researcher's state of mind, but can also

refer to conflict and disagreement among a peer-group or a community of scientists.
Dewey calls a situation "indeterminate" where such cognitive dissonances or conflicting beliefs are still implicit. As soon as "implicit" beliefs become articulated and acknowledged as meriting an active search for resolution, he uses the term "problematic situation" to indicate the beginning of "inquiry". Scientists can conclude inquiry only by adopting new beliefs which temporarily re-establish the equilibrium of experimental practices within the peer-community. This model (figure 2) provides the macro-context of the transactive inquiry model.

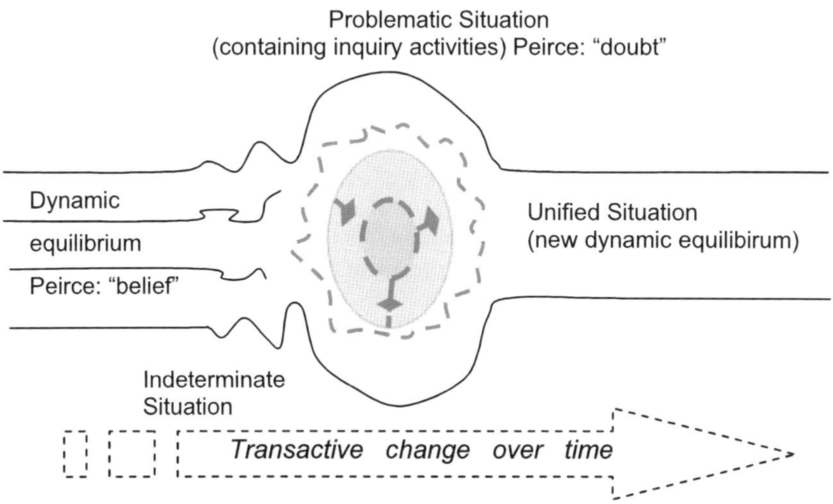

Figure 2: Peirce/Dewey Pattern of Inquiry

The notion of a pattern of inquiry is incompatible with the received view that a theory of the scientific method can safely neglect the origin of questions, problems or hypotheses. For example, Popper holds that 'science proper', i.e. the logical application of a scientific method or the "context of justification", begins only after a hypothesis has been proposed[48]. Any state of scientific puzzlement – any creative attempt at formulating theory and generating conceptual distinctions – is exported to a "context of discovery", which is deemed irrelevant for vindicating scientific conclusions. Dewey includes all phases of this "pattern of inquiry" in his "logic", which he defines as "theory of inquiry", and hence rejects treating "discovery" and "justification" as separate contexts.

A Pragmatist inquirer can rely neither on given research problems (hypotheses or questions) nor on experience as an antecedent category. For Dewey, experience does not confront us *sui generis* with *explananda;* and experience is not an independent source of facts for testing our theory either. "Pragmatism ... presents itself an extension of historical empiricism with this fundamental difference, that it does not insist on antecedent phenomena, but on consequent phenomena, not upon precedents but upon the possibilities of action. And this change in point of view is almost revolutionary in its consequences. An empiricism which is content with repeating facts already past has no place for possibility and for liberty"[49]. The fiction of given research problems and antecedent experience-data as natural starting-points for scientific research brings with it the tendency to reduce the scientific method to a rigid procedure or logical scheme.

Dewey's "pattern of inquiry" can be paralleled with Kuhn's notion of science as oscillating between "normal" and "revolutionary" phases. The move from a state of relative equilibrium to an indeterminate situation is well captured by Kuhn's notion of "anomalies" and the breakdown of a paradigm. The Title of Kuhn's *Structure of Scientific Revolutions*[50] diverts a bit from one of its greatest contributions: Kuhn provides a theory of *normal science*, which emphasizes that scientists manage to coordinate their work and agree on basic theories and concepts in order to "solve puzzles" or explore the full potential of a theoretical framework. Dewey does not provide a theory of normal science other than referring to phases of habit or equilibrium, but by doing so, he is at least able to accommodate Kuhn's ideas of "normal science", which Popper was not. Popper's falsification model can only accommodate a certain notion of revolutionary science, as it claims that theories must permanently remain at stake and under attack.

Perhaps Dewey has something to offer to Kuhn in return. Kuhn expressed legitimate skepticism about the over-optimistic view of science as "an enterprise of reasonable men, who use logic and facts in order to arrive at conclusion in ways that compel ascent without coercion"[51]. By Kuhn's account, scientists are socialized in scientific traditions or "paradigms", and these come furnished with criteria for criticizing and justifying claims. Kuhn holds that there is no neutral ground for 'logic and facts' to mediate in disputes between denizens of different paradigms. Standards of good and bad arguments are paradigm-dependent and observation is mediated by conceptual schemes, which makes the observer's expectations an important part of what will be perceived. This corresponds well with Dewey's view that an empirical scientific method cannot rely on antecedent phenomena because experience always includes the contribution of the observer.

Dewey's concept of experience has the power to reconcile what many saw as Kuhn's unbridgeable contradiction. On the one hand, Kuhn claims that a paradigm-shift does not merely change one's perspective or perception of a thing but actually changes the world: "… in the absence of some recourse to that hypothetical fixed nature that he saw differently, the principle of economy will urge us to say that after discovering oxygen Lavoisier worked in a different world"[52]. At the same time, Kuhn felt uncomfortable with such ontological conclusions which seemed to reduce nature or reality to a convention in scientific perception or description. Lacking a transactive conception of experience and nature, such as Dewey had developed, Kuhn deemed it necessary to soften some of his earlier claims that a changed paradigm should amount to ontological changes in the nature and reality of objects under investigation.

Dewey always rejected the distinction of reality *for us* and reality *in itself* wherever it was made to separate epistemological from ontological categories. Dewey saw all experiencing and knowing as a form of transaction *within* nature. Therefore, a change in scientific perception and description is not merely a change in representing the world, but constitutes a transformation of nature's very own transaction.

With a little help from Dewey, Kuhn could have maintained his thesis that Lavoisier worked in a different world after discovering oxygen, namely one that enabled new scientific experiments and a new organization of experience. At the same time he would not be pressed to reduce nature to a matter of convention.

4. Community of Inquiry

Kuhn captures the frustrating experience that occurs when our best arguments fail to reach our opponents – when we talk past one another. Perhaps Kuhn overshot his target with the notion of "paradigm incommensurability", and here Dewey could be of great help. It is not clear whether Kuhn ever settled upon a definition of his most successful concept. Besides a very useful notion of *paradigms* as 'examples' or as reference models for good scientific craftsmanship, he also uses the term to describe quasi unbridgeable intellectual, linguistic and ideological divides between groups and individuals, or quasi hermetic systems of scientific communication. Kuhn's paradigms often appear as community-based ethical standards for good scientific reasoning, in a way that invites Lakatos' rebuke of Kuhn's paradigms as "intellectual tribalism". The problem is that by Kuhn's account, meaningful scientific discourse between "incommensurable" paradigms should not only be difficult, but outright impossible.

To be sure, scientific communication often fails because of fundamental disagreements about basic concepts and background notions. People may differ even on basic observations since there is no theory-independent notion of experience. However, for Dewey, communication is always a creative process aimed at building a community. Scientific inquiry cannot presuppose a given scientific community or a common language. Dewey's transactive model of agency allows communities to form in response to problematic contexts. Successful communication is the sharing of experience, which Kuhn would translate as inhabiting a common paradigm.

However, for Dewey, such shared experience is an *achievement* and not an *antecedent* of communication. This means that Kuhn's paradigm-incommensurabilities are less daunting than they may appear at first glance. Where Kuhn implies a definition of communication as the use of common concepts and the application of shared conceptual schemes, Dewey defines communication as a creative struggle for understanding and the creation of "public symbols". By this latter account, incommensurability is not indicative of a permanent failure to communicate, but is rather the starting-point for creative attempts at building bridges. "Of all affairs communication is the most wonderful … that the fruit of communication should be participation, sharing, is a wonder by the side of which transubstantiation pales"[53]. However, this is not to leave the solution of Kuhn's problem to a miracle. On the contrary, communication and the production of common concepts and shared experience is hard transactive work. It is a type of work in which the scientific community resembles most closely a Deweyan democratic polity. Democracy as a method of intelligent cooperative inquiry requires "the improvement of the methods and conditions of debate, discussion and persuasion"[54]. For example, a substantial part of all social enterprises (including public policy, social planning and scientific research) is concerned with creating a community and with finding means of coordinating the activity of agents and networks.

Given this definition, how can we understand inquiry as a community-based enterprise? Stuhr complains, "Dewey talked confidently about the indeterminate situation that *we* experience, but … who is 'we'? … What happens if different persons or different groups … define different problems? What methods of framing are better than others? These kinds of questions point to real flaws in Dewey's *Logic*."[55] The answer to this puzzle is that the inquiring community is a transactive notion, i.e. a product of inquiry, but not its logical antecedent. Actor is for Dewey a "confused and confusing word; offering a primitive and usually deceptive organization for the complex behavioral transaction ... Under present postulation actor should always be taken as … as a trans-actor."[56]

Dewey defines "transaction" as a systems-point-of-view or a perspective on a problematic situation as a whole. Transactions comprise all processes and interactions that take place within such a situation, and make no a-priori distinction between human agency and nonhuman behaviour[57]. As the above quote indicates, the emergence of an intentionally coordinated agent (or actor) is the product of organizing activity. Hence, inquiry must always be concerned with a definition of the inquiring agency. A Deweyan transactive model of inquiry must remain open to transformations in the structure and composition of its form of agency or organisztion. This is symbolized by the grey shaded area framing the inquiry model with its flexible external borders (cf. fig.1).

This interpretation of Dewey's transactive theory of inquiry has some obvious parallels with Actor Network Theory (ANT), as developed by Callon, Latour and Law[58]. ANT emphasized the benefits of seeing humans, machines and environments as integrated parts of larger networks and processes. It also argued that such networks undergo transformations by creating coalitions, or through "resonances" in which a signal from one part of a network can trigger responses and activities in other parts[59].

Some critics felt unsatisfied with the permanently dispersed notion of agency in actor networks, which is to say that "actor-network analyses can be centered on any perspective or on multiple perspectives"[60].

Dewey's transactive theory of agency suffers less from this challenge because the production of a point of view or perspective that is able to organize experience and activities is the product of inquiry. Coordinated agency and the creation of a common perspective can be the creative product of inquiring activity within a problematic situation. Inquiry cannot presuppose a given or stable agent or 'inquirer', but inquiry can create forms of centralized coordination and vantage points.

Actor Network Theory was also criticized for its "culturally flat"[61] materialistic and mechanistic account of behaviour, which puts humans and machines on one level and leaves little room for culture as a determinant of behaviour. This criticism does not affect the present Deweyan inquiry model to the same extent, as the above discussion of the importance of communication and the formation of a shared experience has indicated.

Dewey's emphasis on "conjoint, communicated experience"[62] and "like-mindedness"[63] must not be mistaken for perfect agreement amongst all participants with regard to perceptions and their interpretation. Neither does shared experience imply identical meanings of concepts and symbols. Shared experience can only mean a certain harmony in linguistic and other commerce between interlocutors; it is a successful form of exchanging symbols and coordinating activity. Dewey makes clear that the aim of sharing experience in a democratic community does not imply the reduction of pluralism or diversity. On the contrary, shared experience becomes richer the more diverse perspectives enter common projects and deepen public discourses.

Peter Galison[64] contradicts Kuhn's notion of paradigm incommensurability by interpreting the meeting points of different paradigms as creative "trading-zones". For Galison, successful communication in such trading-zones is not limited to sharing opinions and perspectives. A seller and a buyer in a marketplace may have a very different idea about the use and value of the goods on offer; someone may sell headphones while another buys an ear-warmer, and still they come to terms. In the same sense, communication between scientific paradigms is possible where ideas, methods and concepts are transferred successfully or where a new language originates out of such trading-zones. Galison uses the example of biochemistry, which was established out of a trading zone between chemistry and biology and which created its own vocabulary, much as children of immigrant families sometimes create creoles that native speakers of neither original language are able to understand.

5. Conclusion and Acknowledgements

Dewey's theory of inquiry describes an integrated and organic learning-process. My aim is to provide a model Deweyan scientific inquiry that adequately accounts and relates its elements as parts of a dynamic whole. I am aware that this model is a rough as it stands. It offers itself for further refinement. I hope that this account of the scientific research process helps contemporary philosophers of science to recognize the relevance of Dewey's work for many of today's most prolific debates.

I am grateful to Prof. Shyama Kuruvilla not only for her helpful comments on this draft, but for years of collaboration in which the structure of this model matured as a "transactive rationality model" of policy processes. It is no great surprise that the structure of this policy model offers itself so readily to an account of scientific research, since it was originally derived from Dewey's account of inquiry. The "transactive rationality model" and the present inquiry model are structurally isomorphic. The policy model distinguishes "Define", "Design", and "Realize" as sectors surrounding a core of "Deliberation". Yet it would require a separate paper to see how precisely the elements of a Deweyan inquiry model relate to a model of rationality policy making. I am also grateful to the participants of the 2009 summer workshop on "American and European Values" in Opole, Poland for their constructive and helpful comments, and to Matt Brown, with whom I first discussed how to apply early versions of the *transactive rationality model* to the context of scientific inquiry.

NOTES

1. All reference to Dewey's work will follow the standard format and refer to: Dewey, J. (1996 [1882-1953]). The Collected Works 1882-1953. *Past Masters: The Collected Works of John Dewey*. J. A. Boydston and L. Hickman. London Amsterdam, Intelex, Southern Illinois University Press.

2. Curd, M. and J. A. Cover (1998). *Philosophy of science: the central issues*. New York; London, W.W. Norton.

3. Hacking, I. (2007). On not being a pragmatist: eight reasons and a cause. *New pragmatists*. C. J. Misak. Oxford, Clarendon Press. pp. 32-50

4. Hacking, I. (1983). *Representing and intervening: introductory topics in the philosophy of natural science*. Cambridge, Cambridge University Press. p.62

5. Hacking, I. (2007). On not being a pragmatist: eight reasons and a cause. *New pragmatists*. C. J. Misak. Oxford, Clarendon Press pp. 32-50

6. Alexander, T. M. (1987). *John Dewey's theory of art, experience, and nature: the horizons of feeling*. Albany, N.Y., State University of New York Press. p. 12

7. Bernstein, R. J. (1966). *John Dewey*. New York, Washington Square Press.

8. Hickman, L. (1990). *John Dewey's pragmatic technology*. Bloomington, Indiana University Press. Hickman, L. (1998). *Reading Dewey: interpretations for a postmodern generation*. Bloomington, Indiana University Press. Hickman, L. and T. M. Alexander (1998). *The essential Dewey*. Bloomington, Ind., Indiana University Press

9. Shook, J. R. (2000). *Dewey's Empirical Theory of Knowledge and Reality*. Nashville, Vanderbilt University Press.

10. Hildebrand, D. (2008). *Dewey: a beginner's guide*. Oxford, Oneworld.

11. Hacking, I. (1983). *Representing and intervening: introductory topics in the philosophy of natural science*. Cambridge, Cambridge University Press.

12. Dorstewitz, P. and S. Kuruvilla (2007). "Reviewing Rationality: A pragmatist perspective on policy & planning processes." *Management and Philosophy* 6(1): 91-117. Kuruvilla, S. and P. Dorstewitz (2008). "There is no 'point' in decision-making: a transactive model of rational public policy." *Policy Sciences* (Available at: http://www.springerlink.com/content/gp23hw6754p10511/).

13. Nagel, Ernest. Introduction to volume 12 "Logic: Theory of Inquiry" in *Dewey's Collected Works* LW 12.xv-xvi

14. Dewey, LW 12.41

15. Ibid. p. 112

16. Dewey, MW8.37

17. Dewey, LW8.201

18. Ibid.

19. Ibid. p. 115

20. Ibid. p. 509

21. Dewey, LW12.116

22. Hacking, I. (1983). *Representing and intervening: introductory topics in the philosophy of natural science*. Cambridge, Cambridge University Press

23. Pickering, A. (1984). *Constructing quarks*. Edinburgh, Edinburgh University Press. Pickering, A. (1995). *The mangle of practice: time, agency, and science*. Chicago; London, University of Chicago Press

24. Dewey, LW4 p.29

25. Ibid. p.11

26. Dewey, LW4.30

27. Dewey, LW12.108

28. Polanyi, M. (1967). *The tacit dimension*, London: Routledge & Kegan Paul.

29. Dewey, LW12.109

30. Daston, L. and P. Galison (2007). *Objectivity*. New York, Zone; London: MIT.

31. Ibid. p.44

32. Ibid.

33. Shapin, S., S. Schaffer, et al. (1985). *Leviathan and the air-pump: Hobbes, Boyle, and the experimental life: including a translation of Thomas Hobbes, Dialogus physicus de natura aeris by Simon Schaffer*. Princeton, N.J., Princeton University Press.

34. Dewey, LW 12.82-3

35. Dewey, MW8.78

36. Dewey, LW4.204

37. Ibid. p. 34

38. Hempel, C.G. (1994). The function of general laws in history. *Readings in the philosophy of social science*. M. Martin and L. C. McIntyre. Cambridge, MA; London, MIT Press. pp. 43-53

39. Dewey, MW14.135

40. Ibid.

41. Dewey, EW2.328

42. Dewey, MW14. 137

43. Ibid.

44. Ibid. pp.135-6

45. Dorstewitz, P. (2008). Imagination in the Deliberation Process (Technical Report 04/08). *Contingency and Dissent in Science*. D. Fennell, N. Cartwright and S. Hartmann. London, CPNSS at the London School of Economics.

46. Lloyd, G. (1984). *The man of reason: 'male' and 'female' in Western philosophy*. London, Methuen. Frank, R. H. (1988). *Passions Within Reason: the Strategic Role of Emotions*. New York, London, Norton. Damasio, A. R. (1994). *Descartes' error : emotion, reason, and the human brain*. New York, G.P. Putnam. Fricker, M. (2006). "Intuition and Reason." *The Philosophical Quarterly* 45 (179): pp. 181-189

47. Byrne, R. M. J. (2005). *The rational imagination: how people create alternatives to reality*. Cambridge, MA; London, MIT. Frigg, R. (2008). Models and Fiction (Technical Report 05/08). *Contingency and Dissent in Science*. D.Fennell, N.Cartwright and S.Hartmann. London, CPNSS at the London School of Economics. Bronk, R. (2009). *The romantic economist: imagination in economics*. Cambridge, Cambridge University Press.

48. Popper, K. R. (1968). *The Logic of Scientific Discovery. (Revised edition.)*, pp. 479. Hutchinson: London.

49. Dewey, LW2.13

50. Kuhn, T. S. (1996). *The structure of scientific revolutions*. Chicago ; London, University of Chicago Press.

51. Israel Scheffler after Kasser, J. L. (2006). Philosophy of Science, The Teaching Company (www.teach12.com)

52. Kuhn, T. S. (1996). *The structure of scientific revolutions*. Chicago ; London, University of Chicago Press. p. 118

53. Dewey, LW1.132

54. Dewey, LW2.365

55. Stuhr, J. (2002). Power/Inquiry: The Logic of Pragmatism. *Dewey's Logical Theory*. T. F. Burke, D. M. Hester and R. B. Talisse. Nashville, Vanderbilt University Press. pp. 283-4

56. Dewey LW16.260

57. cf. Dewey LW16

58. Callon, M. (1986). Some Elements of a Sociology of Translation: Domestication of the Scallops and the Fishermen of St. Brieuc Bay. *Power, Action and Belief*. J. Law. London, Routledge and Kegan Paul. pp. 196-233. Latour, B. (1987). *Science in action: how to follow scientists and engineers through society*. Milton Keynes, Open University Press. Law, J. (1987). Technology and Heterogeneous Engineering: The Case of Portuguese Expansion. *The Social Construction of Technological Systems: New Directions in the Sociology and History of Technology*. W. E. Bijker, T. P. Hughes and T. J. Pinch. Cambridge, MA, MIT Press. pp. 11-34.

59. Sabatier, P. A. (1988). "An advocacy coalition framework of policy change and the role of policy-oriented learning therein." *Policy Sciences* **21**(2-3). pp.129-68

60. Sismondo, S. (2004). *An introduction to science and technology studies*. Malden, MA; Oxford, Blackwell Pub. p.73

61. Ibid.

62. Boisvert, R. D. (1998). *John Dewey : rethinking our time*. Albany, N.Y., State University of New York Press. p.51

63. Dewey, MW9.7

64. Galison, P. (1997). *Image and logic : a material culture of microphysics*. Chicago, Ill.; London, University of Chicago Press.

Thirteen

A HEDGEHOG WHO THOUGHT HE WAS A FOX: DEWEY BETWIXT THE ONE AND MANY-WORLD PHILOSOPHIES

Ahti-Veikko Pietarinen

1. Introduction

What was Dewey's absolute presupposition upon which he relied in his defense of pragmatism as a sound method of doing philosophy? Did he have one in the first place? Isaiah Berlin famously classified thinkers as Foxes and Hedgehogs: I propose to rephrase the distinction in terms of philosophers who take possibilities as real ('many worlds' philosophers) and those who do not ('one world' philosophers). I will argue that Dewey's philosophy of experimental logic appreciates, to a degree, the value of metasystematic approaches to scientific inquiry which he learned from his teacher Peirce (a Fox), but that he interpreted and applied these approaches in a broadly Jamesian fashion (a Hedgehog). Ultimately, then, Dewey was a universalist about meaning, which implies three tenets: (1) discouraging the falsification of philosophical claims, (2) advocating a naturalistic interpretation of habits, and (3) evaluating the identity of problematic situations with respect to the one, actual world. This paper traces the path by which Dewey comes to subscribe to these three tenets.

We all recall Isaiah Berlin's classic statement in which he suggested grouping the great intellectual personas either as Foxes or Hedgehogs: "The fox knows many things, but the hedgehog knows one big thing."[1] There are those who pursue numerous ends and exploit a variety of methods and experiences to arrive at a world-view, and there are those who see the world through one defining idea, one organizing principle, one single scheme in terms of which, as Berlin tells, they "understand, think and feel."

Such a sweeping literary vision is probably too broad and too vague to satisfy our scientifically trained minds, but some closely related thoughts might well be exemplified in one way or another in dichotomies entertainedby philosophers and historians of ideas time and again. One of them, I might propose at the outset, is found in Peter Strawson's famous classification of metaphysical studies into descriptive and revisionary halves.[2]

A descriptivist believes that there is meaningful structure and organization in the world to be investigated, and though there is no direct access to it, we can well do our metaphysical inquiry to understand the world's inner constituency by studying the structures of the mind and language in correlation with what there is in the world. A revisionist, on the other hand, sees the world as being of a certain kind and satisfying certain criteria: perhaps aesthetic, ethical, religious, intellectual or formal, and he believes that his picture or vision is a product of his or her being in some unmediated contact with the universe at large.

We may furthermore see these two broad approaches as good examples of what another grand historian of ideas, R. G. Collingwood, identified as "absolute presuppositions" characterizing all trustworthy metaphilosophical thinking. Given that pragmatism surely is metaphilosophy, my question can therefore be phrased as: What was John Dewey's own absolute presupposition upon which he came to see pragmatism as a vital and defensible method of doing philosophy? Or, to put the question in a direct way, did he have one in the first place?

To find a concrete ground on which to tackle this question, I advise rephrasing Berlin's loose and metaphorical distinction of foxes and hedgehogs in terms of philosophers who take possibilities as real and those who do not. The former is the 'many worlds' philosophy, or foxes, and the latter the 'one world' philosophy, that is, hedgehogs. I will not argue for the justification of this restatement of the main idea in length here; it is grounded in the notion that accepting real possibilities implies admitting pluralistic methods in philosophical research, such as metatheories of logic or metalinguistic methods of language, and that taking possibilities as real implies a commitment to accept the possibility of stating and assessing conceptual truths in our language of philosophical theorising.

The denial of the reality of possibilities, on the other hand, indicates the sentiment that meaning does not, should not and cannot connect with extra-linguistic factors. Meaning is in the language, or in the very least in the various conceptual or structural relationships construed by the system of language itself. That viewpoint entails several things, including the familiar Wittgensteinian statement of us being prisoners of our colloquial language, despite the fact that we are constantly striving to express what is, strictly speaking, inexpressible. The one who denies the reality of possibilities constantly runs up against the bounds of language in that formidable quest of articulating a metaphysical thought. Among other things, this 'one world' philosophy may lead to relativism, and it can even lead to solipsism, the position that our colloquial language is "the *only* language I understand."

It is appropriate to recall here at the outset Dewey's noteworthy remarks on what he held to be the failures of our common, colloquial language.[3] Dewey is not saying the same thing at all as Wittgenstein with respect to his position concerning the 'unsayables.' By a failure of language, Dewey meant that language is strictly limited to the symbolic means of expression in the one-dimensional, linear media of text or speech and as such is inevitably lacking certain necessary qualitative expressiveness. Such paucity must therefore be counterbalanced with some appropriate ways of extending the overall means and theories of communication. Dewey has in mind what the theory of signs and semiotics suggests: to extend the idea of communication beyond a standard notion and use of language, for example to include iconic, pictorial, visual and metaphoric means among its commonplace resources.

My interpretation of what Dewey in actual fact wanted to mean by his comment on the failure of language goes well together with his lifelong hunt for a better understanding of artistic, rather than linguistic, conceptions and articulations of experience. As such, then, a failure to fulfill an expressive purpose does not imply the stronger claim of the existence of the unsayables, which is a skeptical claim having to do with the inexpressibility of conceptual truths in language. Dewey naturally detested all such skepticism.

My main argument is calculated to conclude that Dewey, in the light of his philosophy of experimental logic, recognizes and appreciates, at least to a certain degree, the value of metasystematic approaches to scientific inquiry—such as the doubt-belief theory, which he learned from his teacher Peirce early on—but that he went on to interpret and apply these methods in a broadly Jamesian, that is to say, individualistic and universalistic fashion. Ultimately, then, Dewey turns out to be a universalist about meaning, though by no means an unqualified or unsophisticated one. Nevertheless, it is characteristic of universalism to ascribe to the metaphysical 'one world' presupposition, and this presupposition is what Dewey appears to accept, although in a unique and distinctive fashion. My main source is thus Dewey's 1916 *Essays in Experimental Logic*, which was a revised and expanded publication with a new introduction of his chapters in *Studies in Logical Theory* he had edited in 1903 and which has recently been republished as a handsome new edition.[4]

I will argue for three implications of universalism and the allied one-world conception that are found in Dewey's thought: namely that they (i) discourage the falsification of philosophical claims, (ii) coerce a naturalistic interpretation of habits, and (iii) presuppose evaluating the identity of problematic situations with respect to the one, actual world. The corollary that remains to be drawn is therefore to point out that these implications are some of the most central theses of Dewey's rendering of pragmatism. My proposal thus boils down to the tracing of the line of thought by which these tenets follow from the one-world conception—the conception that appears by and large to have remained unnoticed and unarticulated by Dewey himself.

2. The Presupposition of Dewey's Pragmatism

We might find some similar suspicions in the following passage by a prominent late philosopher commenting on Dewey and his fellow William James:

> In my opinion, this judgment [by James and Dewey that 'truth' cannot be used to define the basic laws, such as laws of nature] is explained by a failure to distinguish clearly between two different concepts: (1) the degree to which a law is established at a certain time and (2) the semantic concept of the truth of a law.

This weighty opinion was given by Rudolf Carnap.[5] He seems to have been thinking along related lines, though with respect to different subject matter, philosophy of science and the meaning scientists assign to basic laws, rather than metaphysical presuppositions.

It is worth noting that it is indeed a similar idea, the semantic conception of material truth, which Dewey seems to have been missing in his theory of experimental logic. He, correctly, emphasizes the importance of material over logical truth, a fact that shows very well in his disputes with Bertrand Russell, but as far as his implementations of how to tackle the concept of material truth are concerned, the suggestions remain sketchy and fragmentary. For instance, his correspondence with Arthur F. Bentley evidences that he had a hard time digesting Tarski's metalinguistic notion of truth, and he soon leaves aside the entire topic.[7] On the other hand, it is quite understandable that he wished to criticize the notion of logical object in Tarski's theory, because such objects in Tarski's world are devoid of all empirical characteristics.

But these preliminary observations do not compose the main material of the story. Dewey is on the whole very elusive as far as absolute presuppositions are concerned. There are bound to be many hedges, and Dewey's text may, in due course, leave us a little foxed. It is not straightforward to classify Dewey as either-or, a fox or a hedgehog, and so something a little different seems to be going on here.

One portrayal of Dewey's manner of philosophizing is found in a letter Charles Peirce drafted in June 1904, who apparently never sent it to Dewey: "Although I am strongly in favour of your Pragmatistic views, I find the whole volume *[Studies in Logical Theory]* penetrated with this spirit of intellectual licentiousness, that does not see that anything is so very false."[7] This straight-talking judgment passed on Dewey can be seen as illustrating one consequence of the universalist conception: that it is all the actions that we in fact make that contribute to the meaning of philosophical terms. If all actual action contributes to meaning, there can be no false sentences. Such is an example of semantic holism. I have argued elsewhere that such holism was no longer the outcome of Peirce's mature, many-world philosophy of pragmaticism, which is connected with the opposing, many-worlds presupposition.[8] But Dewey's pragmatism seems to come allied with a presupposition radically different from that of Peirce's pragmaticism.

Accordingly, from this version of semantic holism the first tenet is now seen to follow, namely that Dewey's pragmatism discourages the falsification of philosophical claims and propositions. This is my explanation of what Peirce might have had at the back of his mind when commenting on Dewey's "intellectual licentiousness."

At any rate, and not least because he is likely to never have received Peirce's comment, Dewey went on developing a descriptive and non-normative logic, in his terms an experimental logic—a logic that draws from human experience, and not scientific experiments, and which is commonly but slightly misleadingly glossed as psychologistic in certain senses that we need not delve into much here. Peirce was certainly one of those eager to dub Dewey's attempt at the "natural history of thought" as non-normative and psychologistic: recall his misgivings that those in "the Chicago School" led by Dewey are, despite Dewey's own pronouncements to the contrary, in spirit closer to "the German school of logicians" and that they differ sharply from "the English," non-psychologistic school of exact logic. It is the latter according to Peirce's prediction that is inevitably prone to make some significant contributions to "twentieth-century science," whereas the added value of the former is bound to be nil.[9] The case is easily closed, since Boole's, De Morgan's and Venn's works were indeed pivotal to progress in mathematical logic and computer science, Whewell's to the scientific method, and Mill's to linguistics, for example. In contrast, except for perhaps the field of experimental psychology, nothing scientifically relevant has ever arisen from the works of Christoph Sigwart, Vilhelm Wundt, Benno Erdmann or Edmund Husserl.

Like Peirce, Dewey sees pragmatism as not a body of any theses or doctrines but as a method of doing philosophy, but unlike Peirce, as well as unlike F.C.S. Schiller, too (although in quite different respects),[10] Dewey agrees with James's urge to interpret and apply those methods of doing philosophy as widely as possible. This was in sharp contrast with Peirce, who did not want pragmatism to be a philosophy that would "embody the whole man," the notion of which he explains in a letter that this time indeed reached Dewey in good order.[11]

On the other hand, though Dewey's pragmatism is connected with a presupposition quite opposite to that of pragmaticism, he does subscribe to some significant metatheoretic notions such as the control of inquiry. He was keen on reinstituting Peirce's avowed search for the "method of methods" in terms of the "inquiry of inquiry." Though not quite the same as what Peirce's idea was concerning the self-control of rationally behaving agents, Dewey appears to have gained some vital glimpses of the worthiness of metatheoretic methods to the understanding of the workings of scientific theorising.

But it is precisely here that Dewey's tragedy lies. Because of such metasystematic questions around and about in Dewey's work, we apparently would do well to resist the temptation to definitely classify him into one of the two camps. His leanings towards the universalist, one-world conception are quite evident in the prose in which he outlines his philosophy of experimental logic, but the metatheoretic elements here and there that tag along with his philosophy of logic suggest an alternative possibility.

3. Modalities, Existence, and Situations

So what to make of this slight predicament? Where does Dewey truly stand as per the one vs. many worlds position? Or, might we be able to turn the tables here and go on to enquire whether this question is meaningful in the first place? That is, do we really need to make this kind of choice, for fear of losing our argumentative consistency or scholarly integrity? In the hope of clarifying the situation a little more, I will go through a couple of central passages from the *Essays*. These passages concentrate on three key concepts, the precise meaning of which is essential to the resolution of the dilemma: (1) the status of modal notions (especially those of knowledge and time), (2) the nature of existence, and (3) the identity of situations, that is, of Dewey's notion of "problematic situations" in particular.

In Dewey's work, we do find some references to modalities, though they are few and far between. Unlike for Peirce, for instance, modal notions are not the centerpieces of Dewey's overall oeuvre. Yet, for example, knowledge is a thoroughly temporal notion for Dewey: "the hypothesis that knowledge can be rightly understood only in connection with considerations of time and temporal position is a general one," he explains of the logic underlying his epistemology in the introductory chapter he had written anew to the *Essays*.[12] All epistemic notions in Dewey's theory indeed have an unmistakable defeasible and fallible character, resulting from their allegiance to human experience, which according to Dewey is possible only within temporal dimensions.

But knowledge is surely not the ultimate modal conception we are seeking in Dewey's thought. Time and tense themselves are modalities, too. Echoing Peirce's semeiotic, Dewey goes on to discuss, for example, a man who "has acquired certain habits, in other words, in virtue of which certain physical qualities and events are more than physical, in virtue of which they are signs or indications of something else. ... On the other hand, this something else is a [sic] somewhat not physically present at the time: it is a series of events still to happen."[13] Here the temporal nature of habits is well recognized. Habits are those relationships, rules or laws that lie not only in the present but also partly in the future and by which new knowledge can thus be obtained through the proper use and application of signs. This, too, is roughly along the lines of Peirce's account, though Peirce would not have accepted that the general idea of a habit means "acquired" habits, as such habits to him are mere dispositional concepts.

Dewey's instrumentalist version of pragmatism, which he then goes on to develop, demonstrates that, quite unlike Peirce's original formulation of pragmaticism, actions and practices that habits give rise to serve an epistemic rather than interpretative function. To wit, Dewey states later in the *Essays* how "in the logical version of pragmatism termed instrumentalism, action or practice does indeed play a fundamental rôle. But it concerns not the nature of consequences but the nature of knowing."[14]

The divergence is becoming apparent here. Dewey suggests that knowledge is gleaned from actions, whereas for Peirce, signs give information by which we search for the objects of signs that determine their interpretants. The role of interpretants, however, is no longer so important in Dewey's instrumentalist pragmatism. Dewey does not seek to use the operative notion of interpretants as the explication of the nature of different kinds of consequences. He wishes to leave the treatment of consequences surprisingly neutral. Is this a justified move? Is it the result of taking action to serve an epistemic rather than interpretative role and therefore taking the nature of modalities itself as less important a topic in his rendering of pragmatism? I am not sure if we can arrive at a fully pragmatic theory of meaning if we leave the variabilities in the notions of consequences or experimental effects uncharted. Nevertheless, according to Dewey:

> The term "pragmatic" means only the rule of referring all thinking, all reflective considerations, to *consequences* for final meaning and test. Nothing is said about the nature of the consequences; they may be aesthetic, or moral, or political, or religious in quality—anything you please.[15]

For Peirce this would be too vast. Not necessarily impossible, but even the arrival at the theory of the meaning of intellectual signs and concepts, that is, pragmaticism, was a result of a lifetime of work for him. Hence he would have taken Dewey's desiderata (and, for that matter, James's, too) as too expansive and accommodating, and certainly not quite heeding what for Peirce were the mandatory qualities of the "economy of research."[16] For, if the nature of consequences of actions may be "anything we please", there is nothing to be said of the general nature of consequences, and if there is nothing to be said of it, there is nothing to be said of the meaning of intellectual concepts in the first place. Again, Dewey is sliding dangerously close to semantic holism. However, the thing to be noted here is that the reasons are not at all those that Russell had adduced in his criticism on Dewey, since curiously enough, Russell himself was a one-world philosopher par excellence and, it seems, in a much more candid sense than Dewey could ever have been.

It is nonetheless a little peculiar that Dewey is seriously suggesting to call his instrumentalism "the *logical* version of pragmatism." Only in comparison with James and excluding Peirce's pragmaticism, which by that time had already reverted back to the original term 'pragmatism', would such a label be justified.

To mention one historical tidbit relevant to Dewey's instrumentalist restatement, the 1909 *Century Dictionary Supplement Volume* edited by Benjamin Eli Smith bears two important entries, "Pragmatism" and "Pragmaticism." These entries were most likely authored by Dewey. There, pragmaticism is characterised as being:

> a *special* and *limited* form of pragmatism, in which the pragmatism is restricted to the determining of the meaning of concepts … by consideration of the experimental differences in the conduct of life which *would conceivably result from* the affirmation or denial of the meaning in question.[17]

The characterisation is a little deceptive. Sight unseen, the definition reflects Peirce's own traits of thought, but only this much. "Conceivable results" is to the point, provided it is coupled with the 'many-worlds' conception of the modality of the term "would" in the last sentence. But the claim that pragmaticism is a "special" and "limited" form of pragmatism is contestable. The truth of this stronger claim hinges on the exact nature of the presuppositions of the maxim of pragmatism from which Dewey's definition is derived. The maxim itself provides the operational precept for the meaning of intellectual concepts, but it can mean very different things depending on these presuppositions with which the maxim is connected.

In fact, I would venture to suggest that it is the many-world conception that caters for a more general and unlimited form of pragmatism, in the sense of permitting a plurality of methods by which inquiry may be pursued. And likewise, in contrast, it is the one-world conception that requires the experimental differences to be discerned in terms of the outcomes of actual conduct of scientific life and investigation. As such, the result would be a mitigation of the range of those methods of science that may be available as well as conceivable in the light of the inquiry in progress.

Moreover, it is the latter, one-world presupposition that seems to result in the naturalistic interpretation of habits of action and conduct. It does this by necessitating that the evaluation and consideration of "experimental differences" take place in this one, actual world of our investigation. Accordingly, then, here follows the second of Dewey's key tenets of pragmatism: the interpretation of habits in a naturalistic fashion. This interpretation contrasts sharply with Peirce's counterfactual interpretation, in which consequences are the *would-be*s and *may-be*s: they exert counterfactual force to our decisions to act, no matter how improbable the situations may or could be. In other words, in Peirce's account situations may remain irreducibly possible and unactualised. However, habits must suggest the best possible actions in such cases, too, and they cannot do that if they are limited to what Dewey preferred as the acquired, dispositional precepts of individual inquirers.

It is worth remarking that Dewey's naturalistic reinterpretation of habits is very understandable, since it is consistent with taking actions to serve an epistemic instead of interpretative function. In other words, had his interpretation not been naturalistic, it would have remained completely open how the actions we in fact make (and deliberately excluding the possible and conceivable consequences of actions), suffice for Dewey's understanding of the aims of inquiry as knowledge and warranted assertability.

However, if we nevertheless are forced to choose one kind of criteria for consequences over the other, that is, to see the criteria for instance as "aesthetic, or moral, or political, or religious in quality" as Dewey proposes, then we would have already been committed to something analogous to what I mentioned in the beginning of this paper as the revisionary method of metaphysics: we see the world readily as satisfying some such criterion and only then wish to launch the metaphysical inquiry, as if such inquiry could be an afterthought. Dewey would probably have been quite reluctant to accept *this* consequence of his instrumentalist restatement of the original version of pragmatism and its maxim.

Let me take up Dewey's notions of existence and situations here as the final points of discussion. Tom Burke's book *Dewey's New Logic* has done a great service in clarifying the special meanings Dewey assigned to these notions, especially with reference to his 1938 *Logic: The Theory of Inquiry*.[18] The aim of Burke's book was to correct the misconceptions laid on these and related key concepts of Dewey by Russell's polemics. Here I will put the two notions into the perspective of the present discussion with a focus on Dewey's earlier philosophy of experimental logic.

It is commonplace for pragmatists to mark out what is real and what exists, and then go on to exercise caution in order to keep the use of those notions sufficiently separate. Dewey surely does the same. The real is something that exceeds, or goes beyond, that which exists. However, I am inclined to think that for a thinker to be able to successfully operationalize the distinction in his or her actual philosophical output there should be some underlying conceptualization in the philosophical system by which the distinction can be brought into being in the first place. Now in his earlier work of experimental logic, Dewey talks about two "modes of existence"; one is the "given" and the other the "possible" or "inferred" existence: "Correspondence of reference here is as obvious as correlation of existence. In the logical process, the datum is not just external existence, and the idea mere psychical existence. Both are modes of existence—one of *given* existence, the other of *possible, of* inferred existence."[19]

What Dewey is stating here concerning the possible existence appears to suggest a modal nature of reality going beyond actual existence. But it only appears to do so. The modal nature of reality is not what Dewey seems to have had in mind after all. According to him, "the meanings suggested are genuinely and existentially suggested."[20] In the paper *What Pragmatism Means by Practical* he further explains how "the meaning may be the conception or definition of an *object;* it may be the denotative existential reference of an *idea;* it may be actual value or importance."[21] He then clarifies what it means that meanings exist:

> If we stick consistently to this notion of ideas, only *consequences which are actually produced by the working of the idea in co-operation with, or application to, prior existences are good consequences in the specific sense of good which is relevant to establishing the truth of an idea.* This is, at times, unequivocally recognized by Mr. James.[22]

If we consistently stick to *this* understanding of consequences that establish the meaning of signs, ideas and concepts, we are forced to be one-world philosophers, accepting actual consequences produced by applications of ideas as the only kinds of consequences relevant to the semantic notions. Possible or inferred existence is almost a contradiction in terms; from a Peircean perspective it mixes together the categories of secondness and thirdness.

An intermediate conclusion that we can now draw from these remarks is that Dewey surely seems and wants to keep reality and existence as separate categories, recognising the importance of having both to the viability of his experimental logic and the theory of inquiry. But in the end, however, he falls somewhat short of articulating the foremost means, or the clear-cut underlying conceptualisation by which such a separation is to be put into operation and consistently maintained. Namely, he does not derive the distinction between reality and existence in terms of the metaphysics of modalities.

Dewey has much to say about the nature and use of situations in his overall thought. The concept of situations certainly is a predominant feature of his fertile manner of philosophising, which frequently reaches out into new domains. For him, situations are the "universes of experience." As such, they are taken to precede the "universes of discourse," which are the more logical and rigorous renderings of "problematic situations". For Dewey, the former must take priority over the latter: "Every empirical situation has its own organization of a direct, non-logical character," he writes at the outset in the introductory chapter.[23]

These are telling remarks. Dewey seems to be making here, and surprising enough in my opinion, an assumption concerning the distinction between what is logical and what is extra-logical that has a noticeable a priori cling to it. It is quite unexpected to discover such a dichotomy in Dewey's writings. What is there to be said on Dewey's behalf? Perhaps the matter becomes more understandable as soon as we recognize that the assumption of the division of situations into logical and extra-logical may be seen as an instance of the universalist/one-world presupposition concerning the meaning of philosophical terms. What I have in mind is this. The many-world pluralist does not and cannot fix in advance what can and cannot be studied using the tools and methods of logic. The universes of discourse, for instance, are indefinitely expandable so as to possibly accommodate the universes of experiences in the future inquiry, just as they have now managed to constrain the boundaries of what is at issue given some particular systems of languages of logic or the related methods of formal and mathematical sciences. It was one of the major discoveries resulting in the rise of modern logic of the 19th century that there needs to be such a thing as a variable and extendible notion of the universe of discourse. To place a priori limits on such variability or extendibility would be an example of limiting the progress of inquiry. The one-world monist, on the other hand, does crave for a division between what is logical and what is not, since he sees that what the preferred methodologies of viable scientific inquiry are cannot be left to open-ended decision processes to settle in some ideal limit. He needs an appeal to actual human experience to take precedence, to be the ultimate source of guidance in the selection of viable scientific methods.

Another way of making the same point is to say that for Peirce, logic is the theory of formal semeiotic, that is to say the theory of signs in relation to other signs, especially to their objects (logic proper) as well as to their interpretants (methodeutic or speculative rhetoric). For Dewey, logic is a natural history of thought, and a natural history of thought is the theory of inquiry. For Peirce, that is not logic, it is a fixation of beliefs, various processes involving belief revision and doubt. To call it logic would be to keep a tight rein on what can be done with it even in principle.

However, a natural history of *logic* is, on the other hand, a perfectly sensible assumption for any theory of logic, as it concerns the boundary conditions that single out the universes of discourse and provides the background theory and interpretation for the language of logic. I believe that too much stress on the value of thought and thinking per se is inclined to make us think along the lines of one-world philosophers—take Frege as a case in point, though otherwise very different from Dewey. No matter how hard we try, sole thinking is not the source of answers in inquiry; it is the creative use of methods that serve to place the right kinds of questions to sources of information in the hope of gleaning answers from them.

Dewey goes on to remark that, "This doctrine [of experimental logic] implies, moreover, that logical theory in its usual sense is essentially a descriptive study; that it is an account of the processes and tools which have *actually* been found effective in inquiry, comprising in the term 'inquiry' both deliberate discovery and deliberate invention."[24] This passage approves of what I stated above about the differing conceptions of the role of logic and experience in Dewey and Peirce. The other side of the related question eminently worth discussing but which we really must forego here concerns the extent to which Dewey's doctrine, being in his own words "essentially a descriptive study," contrasts to logic as a normative science of the study of possible modes of reasoning of ideally rational agents and to what extent it can also be seen as complementing or partly replacing the normative approach. It seems to me that from the Deweyan standpoint, descriptive and normative should not be thought of being mutually exclusive qualities of inquiry and thus not of experimental logic, either.[25]

In sum, however, Dewey's philosophy of experimental logic presupposes that the evaluation of the identities of problematic situations take place with respect to the one, actual world. He did not follow Peirce's abandonment of his earlier pragmatism, the presupposition of which was, after all, the one-world one. This move became evident by the time of the publication of Peirce's *What Pragmatism Is* in 1905, and Dewey must have been well aware of the manoeuvre. No, he stuck to the original form and content of pragmatism, and seems to have taken the matter of the identification of problematic situations along with it.

This was the third of the three implications of the one-world conception taken up here, the identification of problematic situations with respect to the actual world and excluding unactualised possibilities. Let me now tie the discussion in with some of the recent literature and end with a couple of concluding comments.

4. Wider Vistas

According to R.W. Sleeper, "Dewey was using James's idea of the teleological concept of essences to reconstruct Peirce's conception of logical necessity."[26] In the light of what was said before, it seems that Dewey's reconstruction plan was to make the ontology of problematic situations an actualist one. Sleeper notes how Dewey treats essences and subsistences as "generic features of actual classes or kinds of existences."[27] But if so, logical necessity can no longer mean unactualised possibilities. Sleeper seems to indirectly admit that this indeed is the case with Dewey's theory, since all logical objects there are "empirical characteristics of individual things" and that "logical necessity has existential force only in the context of purposeful behaviour."[28] Insistence on the empirical nature of logical objects was of course the characteristic reason why Dewey could not see the innovative points of Tarski's new theory of logical semantics.

At all events, logical necessity and existence are closely allied notions for Dewey. The concept of logical necessity is not a priori; it is found in the "class of instances of existences." I should note that this characterization Dewey makes is not unlike the Aristotelian theory of universals, in which universals are in some fashion the properties of particular things. For Dewey, extensional, empirical and existential evidence must take precedence over evidence that makes references to intensional notions, unactualized possibilities, or even reality, all of which are to him such types of evidence that are parasitic on the evidence of the former type.

Sleeper concludes that Dewey's ontology of "provisional objects of inquiry" occupies a middle ground between Scholastic realism and James's individualistic nominalism, and that Peirce's doctrine of synechism is not compatible with it. I will not dwell on how well Dewey managed to sort out his reinterpretations of Peirce's synechistic continuum, but seen against the backdrop of the one vs. many-worlds presupposition, the pigeonholing of Dewey's provisional objects of inquiry into the middle ground between Peirce's and James's ontologies is perhaps a relatively apt and mild characterization. I would simply add that the middle ground should be understood in terms analogous to the objects that the theoretical terms of a given law of a theory refer to and which later inquiry may well prove not to exist, but which lack the "semantic concept of a law" to back them up, as Carnap had put it in his comment on the pragmatist misconceptions of the nature of the laws of science.

My final comment moves on from the philosophy of science to society, the realm which nonetheless is seen to carve up the same logical and metaphysical dilemma. Robert B. Talisse accuses in *Pragmatist Philosophy of Democracy* Dewey's theory of democracy of making too many metaphysical commitments: it is no longer normatively neat. He argues that it is not clear what the ontological nature of the "determinate" and "indeterminate" situations in which actual inquirers are posited to find themselves at the end precisely speaking can be. He favours Peirce's theory, which by accepting a trouble-free case of Scholastic realism makes inquiry a practice that can earnestly and solemnly be destined towards a final opinion by unfolding the uncertain continua.[29]

I have considerable sympathy towards stressing the importance of the deeper layers of Dewey's metaphysics, and would therefore like to suggest how to place Talisse's argument, too, within the context of what was suggested in the present paper: to understand Dewey better, and even better than he perhaps understood himself, by observing how he was attempting to relocate modal and intensional concepts to where he thought they must play their explanatory roles: namely as instances or features of the one, actual world of the inquirers' experience. It was this move, which Dewey seems not to have registered in so many words, that I believe gave rise to his notion of "indeterminate" situations. In indeterminate situations it indeed invariably is the "concrete systems" which figure there, and not any unactualized possibilities, and it likewise is the concrete systems and concrete systems only which interact with their presently occupying environment.

As such, Dewey's strategy is interestingly similar to the style in which Richard Rorty—deliberately or accidentally—came to advocate a reduction of the ontology of possibilities to the discourse about some arbitrary locations or situations of the actual world, purging them of the genuinely unrealized, unactualized, synchronic modal status. Admittedly, Rorty in one place states that "the pattern of all inquiry … is deliberation concerning the relative attractions of various concrete alternatives."[30] Here he takes the deliberation concerning concrete alternatives as the constitutive pattern of inquiry. But this surely does not turn him into a many-world philosopher: at no point Rorty takes alternatives to be unactualized real possibilities. He takes them to be alternative constructions of the actual world, all of which are concrete and all of which can be defended and criticized with equal confidence. There is no irreducible modal character to be attributed to such mock alternatives.

And the case is not that dissimilar to Dewey's, whose situations are all concrete though possibly indeterminate instances of experience. They are the non-abstract parts of what exists in the dynamic universe, and bear indexical, and thus causal, relationships with the systems interacting in them.

5. Conclusions

Is it alright to be a hedgehog deep down while at the same time striving towards the fox-like characteristics in one's philosophical outlook? Was it OK for Leo Tolstoy, if we accept Berlin's line of thought, to be a fox that desired to think and act like a hedgehog? Berlin believed that the conflict is liable to result in an intellectual tragedy. We may observe something like this happening in Bertrand Russell's anxieties in finding a proper home for his wit and philosophical accomplishments, for example.

But we should not simplemindedly assume that these two cases, which of course are converses of one another, are of an exactly comparable nature, or that both of them are equally defensible or refutable. It does seem defensible that an earnest many-world philosopher cannot be an earnest one-world one, once and for all, and likewise that one who is committed to be a one-world thinker cannot consistently claim to be advocating the many-world conception. Not even a hardcore Hegelian could accomplish that, it seems to me. But in so far as a certain confusion of the two roles the two very different philosophical temperaments like hedgehogs and foxes can play in one's intellectual life, or an occasional mix-up of broadly two kinds of methodological approaches as to how one is best to fulfill actual philosophical occupations is at issue, maybe a sheer dialethism can be avoided?

It appears that for an indisputable fox to occasionally believe or pretend that he or she is a hedgehog is in a sense a credible option, because in a very straightforward sense what is actual is and must of course be also possible. This is my preferred metaphysical rendering of Tolstoy. However, a genuine hedgehog cannot, I regret, consistently enter the fox realm without coming to annihilate himself, since there is no way an actual world can be carved up into real possibilities at will or whenever the need arises. The logical structure of the metaphysics of modalities dampens any hopes for such a getaway.

So, is there an error in Dewey's philosophizing? What is to be learned from the foregoing remarks, and what is there to be salvaged, if fault it was that Dewey committed? Or, to dig a little deeper, could one, in some yet to be articulated way, have the one-world philosophy and eat the many-world one, too? Dewey's tragedy obviously was the conflict between his aspirations and his natural inclinations. But might his philosophy be to modify a pertinent verse a little, something like an outcome of a 'queer combination' of the brain of an inquiring pragmatist mind with a touch of the soul of a Chinese Taoist? These questions surely cry out for further discussion. One is almost tempted to consult Raymond Smullyan at this point for much more enlightenment and enjoyment.[31]

Supported by the University of Helsinki Chancellor's Travel Grants. My thanks go to the organizers and participants of the "American and European Values" conference on John Dewey in Opole, Poland in June 2009. Larry Hickman challenged my claim of the logical/extra-logical distinction and the interpretation of descriptive/normative in Dewey's experimental logic. I have briefly tackled those challenges and modified some of my claims accordingly. George Pappas pointed out Rorty's use of Berlin that tends to see James as a fox. Seen through the one vs. many-world philosophies, however, which Rorty of course neither applies nor probably would have subscribed to at all, James is indisputably a hedgehog.

NOTES

1. Berlin, Isaiah. 1953. *The Hedgehog and the Fox: An Essay on Tolstoy's View of History*. In Berlin, Isaiah, *Russian Thinkers*. London: Penguin, 436-498, 1994, p. 436.

2. Strawson, Peter. 1959. *Individuals: An Essay in Descriptive Metaphysics*. London: Methuen.

3. Dewey, John. 1930. "Qualitative Thought". In *The Later Works*, 1925-1953, Volume 5, pp. 243-262.

4. Dewey, John. 1916. *Essays in Experimental Logic*. Chicago: University of Chicago Press. Cited as EEL. Dewey, John (ed.). 1903. *Studies in Logical Theory*. With the Coöperation of Members and Fellows of the Department of Philosophy. (The Decennial Publications, Second Series, Volume XI.). Chicago: University of Chicago Press. Dewey, John. 2007. *Essays in Experimental Logic*. Edited by D. Micah Hester and Robert B. Talisse, Carbondale: Southern Illinois University Press.

5. Carnap, Rudolf. 1995. *Philosophical Foundations of Physics: An Introduction to the Philosophy of Science*. Edited by Martin Gardner. New York: Dover, p. 214.

6. Dewey, John and Arthur Bentley. 1954. *John Dewey and Arthur F. Bentley: A Philosophical Correspondence, 1931-1951*. Selected and Edited by Sidney Ratner and Jules Altman. New Brunswick: Rutgers University Press.

7. *Peirce to Dewey*, draft letter, 9 June 1904. Reprinted in Peirce, Charles S. (1931-58). *Collected Papers of Charles Sanders Peirce*. 8 vols., ed. by Charles Hartshorne, Paul Weiss, and Arthur W. Burks. Cambridge, Mass.: Harvard University Press, Volume 8, paragraph 241.

8. Pietarinen, Ahti-Veikko. 2008. "The Place of Logic in Pragmatism", *Cognitio* 9, 247-260.

9. Peirce, Charles S. 1904. Review of *Dewey's Studies in Logical Theory: With the Coöperation of Members and Fellows of the Department of Philosophy*. Published in *The Nation* 79, 1904, pp. 219-220.

10. Pietarinen, Ahti-Veikko. 2009. "Some Remarks on the Peirce-Schiller Correspondence", in W. Oleksy (ed.). *American Studies in the 21st Century*. Peter Lang.

11. *Peirce to Dewey*, 12 May 1905.

12. EEL, p. 26.

13. EEL, p. 34.

14. EEL, p. 331.

15. EEL, p. 330.

16. Peirce, Charles S. 1998. *The Essential Peirce*, Volume 2. Peirce Edition Project. Bloomington: Indiana University Press. Wible, James R. 1998. *The Economics of Science: Methodology and Epistemology as If Economics Really Mattered*. New York: Routledge. Wible, James R. 2008.

"The Economic Mind of Charles Sanders Peirce", *Contemporary Pragmatism* 5, 39-67.

17. Smith, Benjamin Eli. 1909. *The Century Dictionary Supplement*, Volume XII, New York: Century Co., 1909, p. 1050.

18. Burke, Tom. 1994. *Dewey's New Logic: A Reply to Russell*. Chicago: University of Chicago Press. Dewey, John. 1938. *Logic: The Theory of Inquiry*. New York: Henry Holt.

19. EEL, p. 139.

20. EEL, p. 46.

21. EEL, p. 311.

22. EEL, p. 319.

23. EEL, p. 6.

24. EEL, p. 20 (added emphasis).

25. A reaction to Hickman's reaction to Peirce's reaction to Dewey's *Studies in Logical Theory* is Colapietro, Vincent. 2002. "Experimental Logic: Normative Theory or Natural History?", in F. Thomas Burke, D. Micah Hester and Robert B. Talisse (eds.), *Dewey's Logical Theory: New Studies and Interpretations*. Nashville: Vanderbilt University Press, 43-71. Hickman, Larry. 1986. "Why Peirce didn't like Dewey's logic?", *Southwest Philosophy Review* 3: 178-89.

26. Sleeper, R. W. 1986. *The Necessity of Pragmatism: John Dewey's Conception of Philosophy*. New Haven, CT: Yale University Press, p. 76.

27. Ibid., p. 86.

28. Ibid., pp. 86-87.

29. Talisse, Robert B. 2008. *Pragmatist Philosophy of Democracy*. New York: Routledge.

30. Rorty, Richard. 1982. *Consequences of Pragmatism*. Minneapolis: University of Minnesota Press, p. 164.

31. Smullyan, Raymond. *The Tao Is Silent*. New York: Harper & Row.

SOCIETY

Fourteen

TEACHING RELIGION IN PUBLIC SCHOOLS: A CRITICAL APPRAISAL OF DEWEY'S IDEAS ON RELIGION AND EDUCATION[1]

Walter Feinberg

1. Introduction

In this paper I propose to align Dewey's ideas on religion and education with the requirements of a post modern, religiously plural liberal society. By post-modern I mean a society that embraces multiple perspectives and conflicting narratives. By religiously plural, I mean a society where competing religious denominations have the same legal status, and by liberal, I mean a society where the individual person is both the object and the subject of moral action. Although such a society is consistent with Dewey's general philosophy and its emphasis on community, plurality and enriched experience, his writing on religion has a hegemonic side to it and is in tension with his concern to promote community and plurality. In this paper, I draw on recent movements to teach religion courses in the United States to suggest ways in which this tension can be resolved in public schools without rejecting any one religious orientation.

2. Dewey on Religion and Education

Addressing the question of religious education in public schools Dewey wrote:

> "Our [public] schools, in bringing together those of different nationalities, languages, traditions, and creeds, in assimilating them together upon the basis of what is common and public in endeavor and achievement, are performing an infinitely significant religious work. They are promoting the social unit out of which in the end genuine religious unity must grow. Shall we interfere with this work? Shall we run the risk of undoing it by introducing into education a subject which can be taught only by segregating pupils ... ? This would be

> deliberately to adopt a scheme which is predicated upon the maintenance of social divisions in just the matter, religion, which is empty and futile save as it expresses the basic unities of life".[2]

While education is not a significant theme in Dewey's *A Common Faith,* it is not a big stretch to think that Dewey would be comfortable with schools that taught students to eschew the doctrines of a particular religion and to develop what he calls the religious quality of experience. The religious is a quality of experience that, as Dewey puts it: "Is morality touched by emotion[s]. . . that are supported by ends so inclusive that they unify the self."[3] It is the power of an ideal that exists first in imagination but then in thought and action that provides "perspective into the piecemeal and shifting episodes of existence."[4] He seeks to promote a faith that is not shielded by claims about the supernatural and that is open to intelligent inquiry. As Manuel Chambrouty nicely puts it:

> The Religious is more a way of living and behaving, with the help of imagination and experience, in order to convert ideals into action, rather than a special system of beliefs and practices institutionally established and never questioned.[5]

Dewey believed that the religious needed to be emancipated from religion, and from the belief that there is some supernatural realm that is impervious to the scientific method. Hence, throughout *A Common Faith* Dewey sees the religious quality of experience in conflict with religions, and sets his task to freeing the former from the latter.

Dewey denied God existed *as a being* and proposed instead that God served as an aspirational term, an ideal that communicated an imagined harmony that we should strive to achieve. In this sense he is questioning God as an active but transcendent Being, i.e. God as understood by most Christians Jews and Muslims. Moreover, Dewey's view of human nature was largely optimistic and, while not a believer in the necessity of inevitable progress, he did hold the view that the method of intelligence, if applied to the social world, could resolve the problems of men and that progress was a real possibility. He believed that science needed to replace religion as the guide by which people governed their lives:

> The marvelous advance in natural science has come about because of the breaking down of the wall existing in ancient and medieval institutions between "higher" things of a purely intellectual and "spiritual" nature, and the lower things of a "practical" and "material" nature.[61]

3. Dewey's Catholic Problem

Dewey's anti-Catholicism was not always subtle. He singled out the Catholic Church for its divisive potential and anti-scientific bias, and members of the Church often responded in kind, criticizing Dewey for failing to accept the falleness man or for rejecting the supernatural claims of religion. Dewey worried about the influence of conservative Catholic Priests on the Polish American community during the First World War. He praised the 1949 anti-Catholic book by Paul Blanchard as "exemplary scholarship". Blanchard called for a resistance movement to counter the anti-democratic, intolerant and un-American policies of the Catholic hierarchy.[72] Furthermore, in *A Common Faith* he rejected the idea of original sin and proposed that much of organized religion, and especially the Catholic Church, was a rigidified response to old problems. In addressing the difference in value between Protestant and Catholic ideas he writes: "It is better that change be accompanied by the sum total of efforts of men and women who are imbued with personal faith, then they be effected by any wholesale institutional effort that subordinates the individual to an external and ultimately a worldly authority."[7]

It is worth noting that the Church that Dewey was reacting against was quite different from the one that arose some decades later in the aftermath of the Second Vatican Council. This was still the anti-modern, anti-democratic, anti-Protestant and Anti-Jewish Church that Pius IX had promoted in response to the aftermath of the French Revolution and Napoleon. And it was still a time when the proper relationship between religion and science was still intensely debated. In response Dewey allowed for the richness of religious experience while questioning its source in the supernatural.

4. The Religious

For Dewey the religious must thus be liberated from religion. The religious is about imagination and the effort involved in seeking wholeness. It is not about faith as a belief in a body of propositions advanced by particular religions, or as connected to a set of religiously sanctioned moral beliefs. For Dewey the religious involves the recognition of the cooperative relationship between man and nature and an understanding that human nature is "a cooperating part of a larger whole."[8] As Dewey says "Faith in the continued disclosure of truth through direct cooperative human endeavor is more religious in quality than is any faith in a completed revelation."[9] Dewey sees this quality in the struggles of normal people in the pursuit of the aims of everyday life, and that "The claim on the part of religion to possess a monopoly of ideals and of the supernatural means by which alone ... they can be furthered, stands in the way of the realization of distinctly religious values inherent in natural experience."[10] For Dewey, science has largely replaced specific religions as the final arbiter of claims about matters of fact and the universe, and the road to more truths is not through religion, but through patient and scientific inquiry.[11] Given this understanding then "God" as a concept denotes, not a kind of being having an independent existence, but an imagined ideal of unified values that at any given time are granted to have authority over one's volitions and emotions.[12] For Dewey religions get in the way of the religious.

Dewey's ideas on religion are a response to the situation of his time, and, whether he understood it or not, much of his ideas on religion were written in reaction to the rigidity and anti-modernism of a pre-Vatican Two Church, a Church that was still in the grip of Pius 1Xth anti-democratic, totalistic views. It was for this reason that Dewey could express alarm about the influence of the Catholic clergy on the Polish community in Philadelphia during the First World War and it was also for this reason that he could praise without reservation the anti-Catholic book of the theorist Paul Blanchard written shortly after the conclusion of the Second World War. Yet had Dewey emphasized his notions of community and plurality, he might have been able to imagine possibilities for openness and tolerance within Catholic and other religious traditions and he might have also enriched his own understanding of human nature by allowing the value of religion in marking both the joyful and the tragic in the human condition.

Dewey's philosophy of religion emerges out of his conception of experience and instrumentalism while submerging other features of his philosophy such as the value he placed on pluralism and community. Ironically, while Dewey valued plurality he largely failed to scrutinize the differences between religious communities as such just as he failed to understand the plurality within a given religion. This may be because he took American Catholicism as his model of religion and the conservative Church in Rome as the essence of Catholic thought. Given this view, certainly reinforced by the Church hierarchy, he was unable to see the influence that Catholic dissenters, such as John Courtney Murray and Jacques Maritain's efforts at internal reform, efforts that would have significant influence on the Church a few years after Dewey's death.

5. Can Education Recapture Plurality?

Dewey's hope, expressed in *A Common Faith,* that religion would fade as the religious was liberated has certainly proved a hard sell in recent years, as the rise of religious based education in the United States indicates. In what follows I want to suggest a different model, one that I believe is actually more consistent with Dewey's total philosophy than is his common faith because it recognizes pluralism as well as a plurality.

In recent years a few American secondary schools have begun to teach academically oriented courses about religion. As long as schools do not advocate for one religion over another or promote either belief or non-belief this is constitutionally acceptable.

A colleague and I have been studying these efforts for a couple of years, and the results are quite mixed. Here, however, I want to draw on the work of teachers in one school that teaches comparative religions. This school has separate semester long class in both "Eastern" and "Western" religions. The latter covers Christianity, Islam and Judaism, and this is the class I want to describe. Most of the students in the class are Christian, both Catholic and Protestant, including some fundamentalist. There is also a modest minority of Jewish and Muslim students.

The ostensive goal of the course is to provide students with a basic understanding of the beliefs and practices of each religion as well as to inform them of the diversity that can be found within a single religious tradition. In addition there is a consensus among the teachers that while they do not aim to convert students, they do want to provide them with a basic appreciation of religious traditions other than their own, and to make them more knowledgeable about their own faith tradition. To paraphrase one teacher, while we do not want to challenge their beliefs, we would like them to understand why others could find inspiration in a set of religious practices and beliefs different from their own.

One of the important goals of the class is to enable students to reflect on the stereotypes that they carry with them. They approach this indirectly at the very beginning of the term by showing the student pictures of different people and having them match them with their work. The mismatches are quite astounding. The well- dressed white supremacist is identified as a college professor. An African American jurist is seen as a street person and so forth. This then leads to a discussion about religious stereotypes and where they come from. One of the skills the courses seek to develop involves the capacity to dialogue across religious traditions. In the following, the teacher explains to me how this works:

> I don't think that if an Evangelical Christian [student] says that if you don't accept Christ then you're going to hell is disrespectful. I've had students say that, I disagree with that concept because for you to say to me that I'm going to hell because I don't accept your view is both irrational to me and judgmental. That's where I take them. That's allowed, that's respectful. In my class we talk about emotionally charged words and words that have more neutral charge to them. I'll say stick to academic words; crazy is not an academic word because it's open to too many different interpretations. Offensive is a good academic word. Irrational is a good academic word. Vague is a good academic word. So they choose these words, we're able to have this open discussion without people getting too upset.

I believe that these teachers, although never mentioning Dewey, are drawing on a fuller range of Dewey's pragmatism, than we find in Dewey's understanding of religion. They are both recognizing the plurality of religious and non-religious views that are found in their classroom and they are acknowledging the value of pluralism. In overriding this feature of his own philosophy, he also failed to sufficiently interrogate the role that religion plays in individual lives in connecting them to communities that celebrate, mourn and remember together and across generational lines.

To many believers the world is a tragic place that cannot be tolerated without the help of faith. Science may of course help us to live longer, more comfortable lives, but it cannot help us confront the inescapable fact of suffering or death. It can help relieve suffering, but cannot help us to come to terms with a world in which innocent people do suffer. This is not to say that Dewey's understanding of "the religious" could not be taught along with other views about the ultimate meaning of human experience. It is only to say that for schools it should be taught as one of a number of alternative systems of belief and within a concern to preserve plurality.

6. Conclusion

The class that I have just described is not typical of religion courses in the United States and, indeed, most public schools do not teach any courses in religion at all. Hence while the Supreme Court allows schools to teach about religion as long as they do not sponsor worship services or proselytize, most schools choose not to do so at all. This may be changing, however, as more and more schools feel some need to address religious movements, and in many instances Dewey's concerns are still quite legitimate. Many schools, especially in the area called the Bible Belt, teach a form of Bible education that gives priority to a literal reading of the Bible and label their courses Bible history, suggesting that the stories in the Bible had the same epistemic status as, say a historically accurate report on the events of World War II. Since many of these courses dismiss any biblical commentary as "mere opinion" and require that the students read only the King James version of the Bible, the teachers are able to control the information that students are exposed to and, without necessarily violating any Constitutional constraint, privilege one understanding of the Bible over all others. In this sense it is easy to understand Dewey's concern to make the public schools into religion-free-zones. Yet the cost of doing so would be students who are either ignorant of religions other than there own and who are unable to engage others in informed interreligious dialogue. The solution is to provide teachers with sufficient understanding and appreciation of different belief and non-belief systems and with the capacity to teach students to engage them in productive ways.

NOTES

1. Appreciation to the Spencer Foundation for its support of the research on teaching religion in public schools.

2. John Dewey, "Religion and Our Schools", *The Hibbert Journal*, VI (July, 1908), p. 807.

3. John Dewey, *A Common Faith,* New Haven, Yale University Press, 1934/1962, p. 22.

4. Ibid., p. 24.

5. Manuel Chambrouty, "A Common Faith by John Dewey" in *World Religion Watch* March 4, 2009 http://www.world-religion-watch.org/index.php?option=com_content&view=article&id=228&Itemid=66

6. John Dewey, "Challenge to Liberal Thought," *Fortune*, XXX (August, 1944), pp. 154-157.

7. Paul Blanchard, *American Freedom and Catholic Power,* Boston: 1949.

8. Dewey, *A Common Faith,* p. 25.

9. Ibid., p. 26 .

10. Ibid., p. 28.

11. Ibid., p 32.

12. Ibid., p. 42.

Fifteen

JOHN DEWEY AND JOSIAH ROYCE IN DIALOGUE ON THE INDIVIDUAL AND COMMUNITY

Jacquelyn Ann K. Kegley

John Dewey and Josiah Royce are usually viewed as men of very different philosophical temperaments. However, a further look at their work suggests a different story. They, in fact, shared a number of concerns and philosophical positions. Thus, both were concerned about the social conditions of their time and particularly the problem of properly balancing individualism and communal good. They were concerned about the need to redefine "individualism" and to develop conditions for creating a new sense of community. Dewey advocated for the "Great Community" while Royce discussed the "genuine and universal community of interpretation."

In this paper we place the thought of Royce and Dewey in dialogue, focusing on the issues of the individual vis-à-vis the community and the development of a democratic way of life. This dialogue reveals that on these topics Dewey and Royce shared the following beliefs in common: (1) that the social situation of their time was one of insecurity and of "lost" individuals; (2) that this pathological social condition was the result of over-emphasis on self-interest and on pecuniary values; (3) that "individualism" needed redefinition in a manner that facilitated the flourishing of individuals as well as communities; (4) that true individuality is developed in community and communal interaction; (5) that among the conditions for "community-creation" are interaction, shared values, and shared action. Further, both Dewey and Royce argued that "communication" was the key to development of community. However, in my reading of Dewey, he did not develop this notion of communication in any depth while Royce's doctrine of interpretation addresses this aspect of community in some detail. Finally, Dewey and Royce valued the community of science and believed that it, in some manner, provided hope for developing community and a democratic way of life. Dewey saw the scientific mode of inquiry as a model for revitalizing society and democracy while Royce believed the scientific community served as one paradigm for a community of interpretation. Because of limitations of essay length, I will not address this area of concern shared by Royce and Dewey.

While addressing areas of commonality between Dewey and Royce it is important to note that there are also subtle and perhaps significant differences. Thus, for example, Dewey's emphasis is more on social, political, and economic consequences and on economic, political and cultural freedoms. Dewey was concerned to change those conditions that degrade, dominate, or entrap an individual so that these persons become increasingly unable to reach the full freedom befitting a human living in a community. Although Royce was not unconcerned about entrapping social conditions, his focus was more on freeing individuals and communities from self-centeredness and from egoistic blindness to the views and needs of others. Royce believed that a human "lost individual" must be transformed out of a self-centered life "detached" from any genuine community into a person loyal to a form of communal life with others, though always critically aware of the need to continually grow and develop both as an individual and as a community. A person grows and develops through a series of commitments to ever-wider communities. In these the individual takes part by self-aware cooperation, open, sensitive, communication, and coordinated actions designed to build broader community and interactions and to facilitate individual and communal growth.

In focusing more on individual and group self-centeredness as the central problem in the failure of the old Individualism, Royce appears more aligned with the views of critics of Dewey like Reinhold Niebuhr who thought that Dewey was too optimistic about the ability of human beings to raise reason above narrow self-interest. In this regard, Dewey appears to have great confidence in the role of education in producing self-aware citizens who will become an active "public." Royce also honored education and would speak of it in terms of loyalty and building community. He would agree with Dewey that a fundamental task of education in a democracy is learning how to sculpt a genuine individuality while also learning to participate in a democratic community, but his major approach to addressing the problems of his time was to develop a notion of building community from the "bottom-up", developing levels and varieties of "loyal provinces," or "loyal communities." He begins with the family and moves on to discuss other types of human associations such as education, the church, and various social and political units. He argues that all must operate on a principle of "loyalty to loyalty." This principle demands respect for other communities as well as transcendence of one's own narrow concerns. This will be discussed in more detail later in the paper.

Dewey is recognized for his social and political writings, his reconstruction of classical liberalism and his advocacy for democracy as a way of life. In contrast, it is believed that Royce failed to develop a social-political philosophy. My contention is that Royce, in fact, has a viable social-political philosophy, worthy of attention in today's global environment.

1. An Inadequate Individualism and Consequent Social and Political Philosophy

Both Dewey and Royce believed the social conditions of their time had a pathological quality due in large part to the dominating influence of a false individualism that they believed threatened community and even the future of democracy.[1] Dewey's focus is on a reconstruction of classical liberalism, a political view that stresses the value of each individual's pursuit of self-interest or fulfillment. A classical liberal believes that a proper political order would be one that allowed each individual maximum freedom from interference from others. Dewey agreed fully with classical liberalism's belief in the ultimate worth of the individual and a stress on political equality, but he also observed some negative consequences of such an emphasis. In his essay, "Liberalism and Equality," he argues that by insisting upon the minimal role of government, classical liberalism has, as a result, elevated the "wants and endeavors of private individuals seeking personal gain to the place of supreme authority in social life."[2] The problem, as Dewey, saw it, was the dominance of the notion of "economic power."

He argues that this has resulted in a pathological condition of "unrest, impatience, irritation, and hurry." People exhibit, he writes: "a feverish love of anything as long as it is change which is distracting, impatience, unsettlement, nervous discontentment, and desire for excitement."[3] Individuals have become submerged beneath economic, corporate goals and the prevailing mentality is that of a 'business mind,' whose "prevailing standards of value are those derived from pecuniary success and economic prosperity."[4] These kinds of values are not those with which humans can ally nor in which they can invest their imagination and thus these values, rather than providing stability, promote economic insecurity and bewilderment.

In Dewey's view, this type of individualism, advocated by classical liberalism, and based on the idea that property and reward are intrinsically individual, and defining industry and commerce by private pecuniary gain, has not provided the development of full human potential. Dewey cites the philanthropic activity of the capitalist as evidence of the failure of capitalism in this regard. He holds that this activity is really a manifestation of an uneasy conscience. He writes: "As such a manifestation, they [these activities] testify to a realization that a regime of industry carried on for private gain does not satisfy the full human nature of even those who profit by it."[5]

Rather than human development and fulfillment there is the tragedy of the "lost individual." Individuals are caught up in a vast complex of associations which provide no imaginative and emotional outlook on life. Americans, says Dewey, have developed a "joining habit," they seek external contact, often out of fear of solitude," and conformity provides a defense.

Ironically, says Dewey, the "gospel of individualism," becomes conjoined with the suppression of individuality in thought and speech. Dewey writes: "The development of a civilization that is outwardly corporate has been accompanied by a submergence of the individual."[6] Such a statement could well be made about the situation of the present.

Royce makes a similar point about American Society. He argues that focus on a single set of values, namely, economic ones, and an accompanying emphasis on the corporate and on merging of business interests had led to what he calls the "leveling tendency." He describes the situation as follows:

> ...because of the ease of communication amongst distant places, because of the spread of popular education, and because of the consolidation and centralization of industries and social authorities, we tend all over the nation, and in some degree, even throughout the civilized world, to read the same daily news, to share the same external fashions, to discourage individuality, and to approach a dead level of harassed mediocrity.[7]

Royce declares that the result of this leveling is the tendency to crush the individual. Again, Royce's description of the situation seems even more cogent today.

In his book, *Individualism Old and New,* Dewey attacks the older "rugged individualism," of the frontier as no longer viable.[8] He notes that this frontier wilderness and its options no longer existed except in vicarious form in books and films. Royce takes this assessment in another direction. In his history of California,[9] a history still praised today by historians for its insight,[10] Royce portrays the failure of this kind of "rugged " individualism politically and morally. He provides a very realistic picture of the manner in which some Californians ignored the rights of others, engaging in theft of land and actual violence against certain groups, especially those of Hispanic and Native American heritage. Emphasis on property rights and on inalienable rights posited against the common good led to conflict and tragedy, rather than to freedom and benefit for all. Royce provides a concrete example of such a tragedy in his description of the fight over "individual land rights" in the Squatter's Riot of 1850.[11]

Central to this event was the conflict between communal interests and individual interests portrayed in an absolute and idealistic form. The issues, argues Royce, were falsely presented as matters of good versus evil, of manifest destiny and of inalienable rights. The rights of John Sutter to the land, granted by Mexican land grants, were ignored in favor of the squatter's inalienable and American-given rights to their land claims. The riot led to deaths and the destruction of property and land. Royce's discussion of the Squatter's riot seems a nice empirical demonstration of the flaws of a notion of democracy based on an individualism which synthesizes capitalism, natural rights, and morals founded on strictly individual traits.

Both Royce and Dewey objected to the habit of seeing "rights" as inhering in individuals apart from and even in opposition to social claims. Dewey thought that the concepts of "liberty," "freedom," and "rights" needed to be reconstructed to be viewed as conceptual tools functioning in concrete situations. "Liberty, writes Dewey, "in the concrete signifies release from the impact of *particular* oppressive forces at a given time and place."[12] A "right," argues Dewey, "is never a claim to a wholesale, indefinite activity, but to a defined activity; to one carried on, that is, under certain conditions."[13]

Rights and liberties are, in Dewey's view, a type of social power. Royce would agree with this assessment although he would speak about "circumstantial freedom," freedom that enables individuals to accomplish goals and to achieve fulfillment. Various technological tools, for example, provide this kind of freedom. The automobile, birth control, and technologies providing water and energy are good examples in this regard.

How, then, do Dewey and Royce address a situation they perceive as pathological and a society that is dysfunctional? Dewey, as we have noted, sought to reconstruct classical liberalism and Royce would agree with Dewey that both the notion of "individualism" and social and political philosophy need to be re-conceptualized. First, individuals are not to be viewed as atomic, a-social, a-historical, and as existing as individuals prior to social organization. Dewey believed, and Royce would concur, that such an individualistic doctrine has hampered the development of an adequate social and political philosophy. Because of their conviction that individuals are prior to any social/political structure, social/political philosophers are asking the wrong question, namely, "How do individuals become associated?" There is, for Royce and Dewey *no mystery* about association; individuals exist and operate in association. Dewey writes: "Thus man is not merely *de facto associated,* but he *becomes* a social animal in the makeup of his ideas, sentiments, and deliberative behavior. What he believes, hopes for and aims at is the outcome of association and intercourse."[14] Likewise, Royce affirmed the basic social nature of the human self. In fact, he argues that self-consciousness arises out of a social contrast between self and non-self, between what is mine and what is not mine. Royce writes: "I affirm that our empirical self-consciousness, from moment to moment, depends upon a series of contrast effects, whose psychosocial origin lies in our literal social life, and whose continuance in our present conscious life, whenever we are alone, is due to habit, to our memory of literal social relations, and to an imaginative idealization of these relations."[15] Human beings are social by nature; conjoint, combined associated action, for both Royce and Dewey, is a universal trait of all things. The central question for both philosophers is: what types of association make for the conditions needed to develop the full potentialities of all those involved in this relationship?

2. Redefining Individualism and Rebuilding Community

For both Dewey and Royce, true individuality is only achieved in associated living, in community. Dewey writes: "Assured and integrated individuality is the product of definite social relationships and publicly acknowledged functions."[16] Individuality remains for Dewey inexpugnable. He declares that it is so because "it is a manner of distinct sensitivity, selection, choice, response and utilization of conditions."[17] Thus, an individual musician has her style and talents, but, says Dewey, this style and talent happens through her performances before certain audiences and with other musicians. Individuality is of the highest worth, communal living is a necessity. Again, Dewey affirms that "Only in social groups does a person have a chance to develop individuality."[18] Reconstruction of "individuality," for Dewey involves erasing the sharp lines between self/society and self/other. These must be understood as dynamic, interactive, and integral to each other.

For Royce, the supposed tension between individual and community is what can foster genuine individuals and communities which, in turn, foster further creativity and growth for both the individual and the community. For Royce, individuality and community, individual and social interests, are inextricably bound together in a creative, ongoing and infinite process. Individuals, in Royce's view, are both self-made and a social product. Community is the context in which human action and creativity occurs and he understands "genuine community" as providing the enabling conditions for the development of genuine individuality as well as obviously for the creative actions of these individuals. Royce, like Dewey, defines "freedom" not in terms of negative liberty, or non-interference, but in its positive aspect as "circumstantial" providing conditions for individual development and exercise of choice. Civil liberties, Royce would argue, should be seen as enabling conditions for individuals and communities.

Turning back to the interaction between "genuine individuality" and "genuine community," it is clear for Royce that creating a self, although requiring others in a fundamental way, also requires agency and creativity on the part of the individual. Contrast and imitativeness are both necessary, in Royce's view, for adequate self-development. Royce writes: "What the child does instinctively, and without comparison with the deeds of others, may never come to his clear consciousness as his own deeds at all. What he learns imitatively, and then reproduces, perhaps in joyous obstinacy, as an act that enables him to display himself over against others- this constitutes the beginning of his self-conscious life."[19]

Royce argues that without supportive, genuine communities you could not have "genuine individuals," nor, indeed, could "genuine communities develop without the loyalty and work of these "true individuals."[20] Royce asserts that there were parallel sins affecting the human self: (1) the sin of self-loss, becoming part of the crowd, a "they," instead of an "I;" and (2) the sin of self-sufficiency, of the individual who "goes it alone" and believes that genuine selfhood can be achieved in this manner. Royce admired Nietzsche's notion of the individual, noble, courageous self who could transcend narrow interests, mediocrity, and the powerful draw of social conformity in order to live as "captain of one's own soul," possessed of one's unique moral value in the face of the chaos.[21] However, he also asserted that Nietzsche's great limitation was his failure to see that real power for the genuine self lies with the "true life of cooperating individuals."[22] Individuals and communities need each other.

In his Philosophy of Loyalty[23] Royce develops a concept of loyalty in which loyalty helps center a unique self but also helps a self become more than an "isolated, self-centered and detached individual." Royce resolves the conflict between individual and community by arguing that loyalty to a community is the sublime expression of individuality that at the same time satisfies a deep, natural need for meaning and purposes. Individuals need a cause, something to which they can devote their imagination and energies and which will focus their lives. Dewey recognizes this as a crucial need for his time. In describing the lost, bewildered individuals under the sway of economic power, he writes: ". . . .the loyalties that once held individuals, which gave them support, direction and unity of outlook on life, have well-nigh disappeared. In consequence, individuals are confused and bewildered. It would be difficult to find in history an epoch as lacking in solid and assured objects of belief and approved ends of action as in the present."[24] Dewey argues that psychologically speaking, individuals need stable objects to ally themselves with and in which they can invest their imaginations.

Royce argues that individuals need causes. Royce declares: "The great principle of the art of giving sense to life, is the principle—Have a cause, chose it, and having chosen it, be fearlessly and steadfastly faithful to it."[25] Again, for Royce, the stress is both on the individual and on the community. In Royce's view a cause must personally engage an individual, be interesting, and connected to natural temperament and social opportunities. Secondly, a cause must be capable of holding my interest since it defines my life plan and thus must be rich enough to fulfill over a long period my growing and changing desires and needs. It must, in addition, provide the self with indefinite future extension, that is, it must be a cause that can never fully be achieved. Finally, the cause must provide social extension; it must tie me to some "larger whole;" it should unite me with other persons by some social tie, such as personal friendship, family or the state. It ties me to community.

Thus, Dewey and Royce were concerned about providing conditions that developed individuals capable of achieving their full potentialities, rewarding experiences and lives, and yet also capable and concerned to contribute in creative ways to the common life. Both philosophers advocated for the building of a "Great Community." Dewey writes: "Our concern at this time is to state how it is that the machine age in developing the Great Society has invaded and partially disintegrated the small communities of former times without generating a Great Community…Till the Great Society is converted into a Great Community, the Public will remain in eclipse."[26]

How, then, is "The Great Community," created? To search for the great community is, for Dewey, to search for the conditions under which the inchoate public now extant may function democratically. But to function democratically means to function as a community. Dewey declares: "Regarded as an idea, democracy is not an alternative to other principles of associated life. It is the idea of community life itself. . ."[27] Dewey writes: "Wherever there is conjoint activity, whose consequences are appreciated as good by all singular persons who take part in it . . . , there is in so far a community."[28]

Like Dewey, Royce was concerned about the small, vital communities that had been disintegrated by the so-called Great Society. To counter this trend he argued for a rebuilding of community, of "provinces" on the local level. What, then, is a province? Royce writes: "...a province shall mean any one part of a national domain, which is, geographically and socially, sufficiently unified to have a true consciousness of its own unity, to feel a pride in its own ideals and customs, and to possess a sense of its distinction from other part of the country."[29] Further, Royce asserts that the development of a "wholesome provincialism" is a social imperative for the world. A wholesome provincialism, for Royce, is to operate as a genuine community, one that enables its members to become genuine individuals and to engage in creative actions. These provinces are created where there is communication among unique individuals who learn to share a common past and future, where the members of the province seek together to perform common deeds, and to encourage the growth of each member of the province (group, community). This relates to Royce's fourth condition of community which is recognition of cooperative efforts, the recognition that without the contribution of each person the goal would not be achieved, the deed would not be accomplished.

Although Royce focuses on "province," as a social-political unit in his famous essay on provincialism, he, in fact, as indicated earlier, sees various social units in terms of potential loyal communities. As he develops his notion of building community from the "bottom-up,' Royce begins with the family, designated by him as the "original community," a formative crucible for building selves and communities. In Royce's view the family as a "loyal community," has two obligations: (1) to engender genuine individuality; and (2) to follow the principle of loyalty-to-loyalty by building community; urging each self and the family as a unit to expand and transcend its limited, perspective.[30] Using the family's three basic relationships, Royce discusses three species of loyalty. Thus, siblings, as well as the family in general, are called upon to be loyal to the cause of creating adequate free space for each other so that each may enjoy freedom and autonomy and thus achieve adequate physical, social, and psychological development. The loyalty between lovers, friends, or spouses is to the bond of goodness or happiness. This kind of goodness only arises if love grows between them. Royce appears to think of "love" in terms of "mutuality," that is, a sharing of each other, seeking in imaginative, creative, active ways to create a harmony of interests. There is mutual harmony, assistance, support, growth, and shared exaltation. It is most interesting, in fact, to relate this discussion to Dewey's statement on love when he writes: "Love . . . . is the resolute purpose in each to seek the other's good, or rather to seek a common good which can only be attained through a common life."[31] Indeed, I believe Royce and Dewey would see the ideal family as a "unity of interacting personalities," and thus as a democratic community.

As for the parent-child relationship, in Royce's view, the loyalty needed is to the bond of generations. It involves a duty to transmit and receive life, in all its levels. There must be communication and understanding of physical, affective, socio-cultural, intellectual, moral, and religious life, as well as transmission of the refined wisdom of precious generations. Parents, argues Royce, must both transmit and hopefully enhance family, as well as cultural, social wisdom. Such tasks are, of course, crucial to building democratic community.

In addition to building a family community, the family, as a "wise province," must be guided by the principle of loyalty to loyalty. It must seek to create broader community. This, indeed, is the obligation of any loyal community of province. Royce states the principle of loyalty to loyalty as follows: ". . . so choose and so serve your individual cause as to secure thereby the greatest possible increase of loyalty amongst men."[32] This principle demands that individuals, in so far as they serve their individual causes or loyalties, must also secure thereby the greatest possible increase of loyalty amongst men. The individual manifests loyalty to loyalty by fostering the moral growth of others as well as the community. Moral growth, in Royce's view, involves overcoming our natural "narrowness of view," seeing the world only by what is in front of our noses or related to the present, demanding interest. Loyalty to loyalty demands that individuals and communities seek to broaden their encounters with the views of others and correct narrowness of view by confronting and trying to share the views of others.

Above all the "wholesome province," "the loyal community," must combat "false sectionalism," a condition in which the interests of the province or one's community are held above those of others provinces, communities, and humanity as a whole. Royce saw clearly that the sin of pride, of self-sufficiency, can be committed by a community as well as an individual. A community may portray itself as the ultimate community worthy of the ultimate loyalty. In doing so, it betrays the principle of loyalty to loyalty and makes absolute and fixed that which is finite and open to change. Loyalties must always be questioned, broadened and even broken, when a larger loyalty and good demand it. Further, communities must seek to deal with communities who betray the principle of loyalty to loyalty and particularly those communities who bring harm to the loyalties of other persons or causes—communities such as a robber band, or a family engaged in a murderous feud, or, in today's environment, the Mafia or street gangs. Loyal communities, loyal provinces call loyal individuals and communities to creative action to seek to change these harmful loyalties and to build bridges between individuals and communities to allow more community and broader loyalty. Indeed, Royce sees enlightened loyalty and the actions of persons moved by this principle as acting in the "spirit of courageous action" described by James in "The Moral Equivalent of War."[33]

Dewey approaches the same issue from a slightly different perspective. The democratic idea, in its generic social sense, says Dewey, from the standpoint of groups and communities, demands liberation of the potentialities of members of the group in harmony with the interests and goods which are common. Dewey writes: "Since every individual is a member of many groups, this specification cannot be fulfilled except when different groups interact flexibly and fully in connection with other groups...There is a free give-and-take: fullness of integrated personality is therefore possible of achievement, since the pulls and responses of different groups reinforce one another and their values accord."[34] Thus, Dewey advocated for interaction of various groups and for self-reflection of one's loyalties in connection with each other, with the loyalties of other groups and with the common good.

In fact, one of the conditions for building community, for Royce, is the ability to extend self in time and place and this allows the recognition of error, weakness, and the need for others. This also has to do with self-understanding, self-criticism, and self-unification. Self-awareness, in turn, is related to humility and an understanding of limits. Self-extension also relates to creativity and openness to change. The common theme of testing habits and beliefs by the new is stated well by William James when he writes: "The individual has a stock of old opinions already, but he meets a new experience that puts them too strain. He seeks to modify and latch on to a new idea that 'mediates' between the stock and the new experience and runs them into one another most felicitously to make them admit of 'novelty.'[35] A key notion in Dewey's philosophy is the disturbance and changing of habit through encounter with a problem, with the "new."

The ability of individuals and communities to transcend self and to engage in interaction with other individuals and communities are crucial for Royce and Dewey in creating communities that can counter the leveling influence and the mob spirit. Royce describes the "mob spirit" as a state of high emotions, which, when unchecked, result in a "State of irrationality," a "sympathization without thought." In such times, Royce argues, "the social group may be, and generally is, more stupid than any of its members."[36]

Social groups that are wiser than their members tend to be characterized, says Royce, by an emphasis upon the contrasts between various individuals. This emphasis on individual difference is crucial to a genuine community which Royce sees always as a "unity with variety." Further, such a creative individual benefits the community, for it is he who can generate a multiplicity of viable potential strategies for serving a cause or a project or solving a problem, thereby enhancing both individual and communal opportunity for effective action. Royce writes: "the effectiveness of human action at the level of community will be enhanced by a pluralism of ideas and strategies which can be realized only through the cultivation of individual differences."[37]

Individual variety leads to exchange of ideas, as well as to contrast and criticism which tests ideas and make them clearer. These are exactly the skills that Royce argued were needed for a "thoughtful public." Dewey would agree. He writes: "The essential need, in other words, is the improvement of the methods and conditions of debate, discussion and persuasion." [38] He writes also as follows: "To learn to be human is to develop through the give-and-take of communication an effective sense of being an individually distinctive member of a community.....The solution is the perfecting of the means and ways of communication of meanings, so that genuine shared interest in the consequences of interdependent activities may inform desire and effort and thereby direct action.."[39] Finally, he asserts ". . . democracy is a name for a life of free and enriching communication."[40]

Dewey, however, does not develop this notion of communication as a method for developing community. Royce, in contrast, has a rich understanding of such a method, which he calls "interpretation." Community, for him, is a product of interpretation. His second condition for a true community is that there must be communication among its members, communication through attentive listening to the ideas and hopes of others and an attempt to gain some common understanding of goals and ideas. His theory and method of interpretation is significant and valuable for understanding how a community can be developed. Thus, he adds a much needed dimension to the discussion, enriching Dewey's concerns and views.

Interpretation is a triadic process that involves comparison of ideas and/or mediation between two minds, ideas, hopes, or persons. Thus, for example, I am mediating the mind and thought of Royce – accessible through a set of signs contained in his work – to the minds of my audience. Three items are brought into determinate relationship: (1) I, the interpreter, who must seek to understand both Royce and you, my audience; (2) the object, or idea, Royce's thought; and (3) an audience to whom the interpretation is addressed. The relationship is non-symmetrical, that is, unevenly arranged with respect to all three of the terms. If the order of the relationship were reversed, it would change the process. Thus, if you were to write a response to my interpretation of Royce, you would become the interpreter and a new triadic interpretative relationship would result. In fact, a process of interpretation can go on indefinitely; one's ideas and thoughts can be continually interpreted, enriched, and built upon. One is reminded of the remark of Alfred North Whitehead that all philosophy is a footnote to Plato.

Further, Royce sees interpretation as a temporal process; each of the terms of the relation corresponds to the three dimensions of time: past, present, and future. Thus, I am now interpreting what Royce wrote in the past; it becomes present and open to future interpretation as well. The temporal dimensions make a difference to the nature and the value of the communicative process. Thus, once I have spoken, what I have said cannot be revoked. Even if I take it back, it still has been said. This adds caution and care to what I say, to what I communicate. We want to say what we "mean," and to be aware of the impact of our words on others. This adds a moral dimension to the process of communication which is important to developing a democratic way of life that depends on free, open and honest exchange of ideas. Propaganda and distortion of issues has no place in a democratic process. This moral aspect of communication appears missing from Dewey's analysis.

For Royce, interpretation is by its very nature a community-building process. It takes selves outside of themselves and opens up the possibility of creating new relationships and harmonies. Thus, to interpret Royce to you, I had to take my past ideas of Royce, compare them with my present ones, and then achieve a new understanding to convey. In achieving this view, a new union of ideas, I have transcended my past self and I have added to my and the community's stock of ideas. Further, by attending to my interpretation of Royce, you too, open yourself to a transcending of your old self and to a new unity of ideas. To interpret Royce to you, I had to choose to re-immerse myself in Royce's thought, and I committed myself, hopefully, to whatever truth I found therein. I, in opening myself to Royce's ideas, risked having my ideas changed. A further risk, in conveying Royce's or Dewey' thought to you is that I can be wrong and you or the community can rectify my error. By attending to my interpretations, your openness also subjects you to the risk of error in your understanding and to the risk of having your ideas changed. If the attempt at interpretation is successful, new meaning will come forth, for each of us will have achieved a new unity of ideas and hopefully a shared understanding of

Royce and Dewey.

Thus, there are the following crucial elements in interpretation: (1) respect and regard for each self and each idea; (2) the will to interpret; and (3) reciprocity and mutuality. In focusing on respect, interpretation affirms the ethical principle of equality, a commitment to treat each individual as equal, in having an opportunity to contribute to the process. This idea is also essential to Dewey's notion of democracy. This requirement of respect also asks for more active commitment to reach out to others, to seek genuine communication and to engage in attentive, open listening.

The will to interpret involves humility in that one recognizes an narrowness of one's own view and a need to broaden this view through interaction with others (as well as with experience). It also involves a willingness to understand, in the words of William James, what "is significant to the other."[41] Further, as indicated above, engaging in interpretation also involves a willingness to have one's ideas tested and compared against the ideas of others, to risk error and change.

Pursuing Royce's doctrine of interpretation allows us to further probe the conditions of genuine community. A third condition of community, for Royce, is that some unity of meaning is achieved. Through the process of interpretation, each individual in the community comes to share a common past, and/or common future; it becomes a community of memory and of hope. Like the self, the community is a temporal being; it embodies deeds done and ends sought. Royce defines a community as "a being that attempt to accomplish something in and through its members."[42] Like the self, the community is a plan of action. Together individuals determine acts and then act. This notion of community as cooperative activity fits with Dewey's belief that democracy requires that the majority of the population be continuously educated for cooperative inquiry.[43] This Roycean notion of community also relates to the views of Ronald Dworkin in his book, Sovereign Virtue: The Theory and Practice of Equality.[44] In this book, Dworkin identifies four conceptions of a liberal community. The fourth conception involves what Dworkin calls an "integrated community." Such a community, he argues, exists as a collective agency and has a communal life much in the manner of an orchestra. The collective acts are "concerted," that is, these acts are performed self-consciously as contributory to the over all-act or acts of the community. In fact, Dworkin speaks of this kind of community as a "superperson," and says that "its collective life embodies all the features and dimensions of human life."[45] Royce does not use the term "superperson" for his notion of community, but he does use the example of the orchestra and also the jazz combo to illustrate the cooperative action he sees as part of the genuine community.

As a temporal being, a genuine community is always evolving and future oriented. This is also demanded by the principle of loyalty to loyalty. As an interpreting and interpretive entity, a community also weaves a story; it has a narrative. Further, as indicated in our discussion of the province and Royce's principle of loyalty to loyalty, an interpretative community is open-ended and inclusive. This fits well with Dewey's notion of democracy as always being reinvented by each generation. Finally, as a community of loyalty and committed to the principle of "loyalty to loyalty," a community must always seek to extend itself. Exclusivity is, for Royce, a great enemy of community.

In regard to social and political philosophy and the building of community, it is important to recognize that for Royce, interpretation is also a method of overcoming conflict within and between communities. As indicated, an interpreter is one who seeks to build common understanding and community among individuals and communities. Interpretation is a triadic relationship among three agents: an interpreter and two other entities, ideas, individuals, communities. Often the two parties in an interpretative (mediating) relationship are what Royce calls "dangerous pairs." A dangerous pair relationship is one that involves two individuals or groups that approach each other as adversaries; they have or perceive that they have mutually clashing interest, various and mutually estranged motives and activities and seemingly incompatible ideas or views. Each individual or group has its own interpretation of the facts and each is convinced that theirs is the only valid understanding of the situation. The parties essentially see their relationships as a zero-sum game; only one side can win. The interpreter, as a loyal and humble servant of building community, seeks to alter the relationship between the dangerous pairs, allowing a sense of communal relationship to develop. The interpreter represents or interprets "the plans, or purposes, or ideas of one of his two fellows to the other of these two in such wise that the member of the community whom I call the 'interpreter' works to the end that these three shall cooperate as if they were one."[46]

In other words, the relationship between the parties is no longer a zero-sum game but becomes one of a deliberative conversation in which mutually estranged parties, with the help of the interpreter, work to arrive at a kind of consensus, though not necessarily a substantive agreement between the opposing parties. What has been created is a conversational context where the parties no longer seek to trump one another but seek to achieve a mutual understanding of their disagreements as well as their overlapping areas of common understanding.

Further, what is created, is a community in which the dignity of each individual or group is confirmed. At a recent conference in Idaho on Basque studies, I had the opportunity to hear Reverend Alec Reid, who has been instrumental in building a "community of interpretation" among the various parties in the Irish conflict. He cited the "dignity of the human person" as the operating value in his work and argued that seeking "dialogue" confirms each person and gives them respect. His primary method he called "dialogue," allowing each party to defend "their own political cake," but also to listen to others "telling their story." He argued that a key step in Ireland was bringing the Sinn Fèin to the table and allowing them to express their views on the situation. Father Reid spoke of the dialogue as developing a dynamic toward a "common good." This view is, I think, very much in concert with Royce's understanding of the work of interpretation.

Royce's views on "interpretation" and "mediation" provides a solid basis for creating community out of conflict and thus for working to build a global community. As has been argued by others that such a global community will not develop out of political institutions, but rather will be the result of "deliberative dialogues among autonomous international agents acting together" in communities of interpretation.[47] In addition to creative community building, Royce's views are clearly relevant to the need to address the various tragic conflicts so besetting our contemporary world. Thus, in a very fine essay on power and compassion, Shibley Telhami, author of The Stakes: America and the Middle East, [48] argues for the need to build bridges of mutual understanding by establishing dialogue with various stakeholders. He calls this a "compassionate approach that builds coalitions and considers the vital interests and wishes of other states and peoples around the globe."[49] Such a method, he believes, would allow a chance to build moral consistency among various parties about the illegitimacy of terrorist tactics while empowering influential segments of the various Arab and Muslim societies to wage their own struggles for change. It would also reinforce notions of openness and democracy that have been a part of America's greatness.

Royce saw his notion of "communities of interpretation" as promoting those values integral to the stability of a democratic community. Among these values is the principle of equality. The community of interpretation is viewed by Royce as fundamentally egalitarian. No one of the three parties has the power to coerce another into accepting a viewpoint and once the dialogue has begun different members of the community of interpretation will have to assume the role of interpreter as preliminary interpretations and schemes for cooperation are offered. Further, the process of interpretation is open-ended and thus particular interpretations are always open to reconsideration. And, in concert with democratic ideals, the interpretive process provides a model of deliberative decision-making and it promotes dispositions of fairness, open-mindedness and a willingness to alter positions when reason or evidence demands.

Finally, one of the goals of interpretation is to build harmony and a unity within diversity; the diversity of views is not eliminated but enriched. What is achieved, I believe, is something akin to aesthetic unity. Interpretation is a form of art. Given Dewey's emphasis on the aesthetic dimensions of experience, this seems most appropriate to our discussion here.

Royce's doctrine of interpretation and notions of community, I believe, are much in concert with Dewey's vision of the democratic way of life. His discussion of local community as exemplified in a "wholesome provincialism" and his advocacy of building levels of loyal communities adds depth to Dewey's concern to rebuild the communal basis for democracy and the Great Community. His ideas about interpretation adds much to the needed discussion about communication in a democratic process, the nature of community and about building a "Great" and global community.

Thus, we have explored a number of commonalities as well as some differences between Dewey and Royce on the supposed conflict between individualism and community. Among the commonalities of belief we discovered the that both Royce and Dewey perceived that the social situation of their time was one of insecurity and of "lost" individuals, a pathological social condition that they believed was the result of over-emphasis on self-interest and on pecuniary values. To address the situation Dewey and Royce advocated for a new definition of individualism that would honor both the great value of individuality while acknowledging that true individuality is developed in community and communal interaction. Dewey and Royce then explored the conditions for community-creation and identified the need for interaction, shared values and shared action. Further, both Dewey and Royce argued that "communication" was the key to development of community, and Royce developed a doctrine of interpretation which sets out the conditions for "genuine communication" as well as the means for overcoming conflict and building community. In all of the discussion Dewey's emphasis remains on social, economic and political conditions while Royce's promotes a general doctrine of community and a moral model for communication.

Exploring the commonalities and differences between Dewey and Royce on issues of individualism and community, I believe invites fresh exploration of contemporary problems in the democratic vision. Thus, Michael Sandel, in addressing the problems of contemporary democracy, argues that we must overcome the "pathology of uncritical acceptance of rights, fairness, and individual choice and return to notions of responsibility, civic duty, and common good."[50] Both Dewey and Royce address this pathology and propose ways in which a communal, democratic way of life might be re-established. Royce provides us with conditions for genuine community as well as a method of achieving such, namely, through interpretation. Royce also adds an ethical dimension to the task, advocating the virtue of humility and the principle of loyalty to loyalty. Further, Stuart Gerry Brown has argued that the values represented in Royce's "wholesome provincialism," are individual freedom and democracy. Summarizing his assessment of Royce's social theory, Brown writes:

> …The dearest social values are the liberty and dignity of the individual man and the democratic method of arriving at social decisions. In order to preserve and enhance these values, we must re-emphasize, re-dignify, and re-vitalize the smaller units of society which are known as provinces, for within the provincial life individual variety will most surely flourish. A democratic nation requires variety of province just as a democratic province requires a variety of individuals; a world order which is at once free and democratic will require variety of nations.[51]

NOTES

1. See: John Dewey, *The Public and Its Problems,* 1957: New York: Henry Holt & Company and 1965: Athens, Ohio: Ohio University Press; and Josiah Royce, 1908, "On Certain Limitations of the Thoughtful Public in America," in *Race Questions, Provincialism and Other American Problems, 1908,* (New York: Macmillan Company; "Provincialism," (1902); in *Race Questions, Provincialism and Other American Problems, 1908,* (New York: Macmillan Company). See also: *The Philosophy of Loyalty,*1908, (New York: The Macmillan Company); 1889, "reflections After a Wandering Life in Australia," I and II, *Atlantic Monthly* 63, 615-86 and 813-28; 1856, *California, from the Conquest in 1846 to the Second Vigilance Committee in San Francisco [1856]: A Study in American Character,* (Boston and New York: Houghton Mifflin and Company, and 1885, "The Squatter Riot of '50 in Sacramento: Its Causes and Its Significance," *Overland Monthly,* n.s.6 225-46.

2. John Dewey, *The Later Works,* 17 Vols. (Carbondale: Southern Illinois University Press, 1981-91). Reference to this work is LW. "Liberalism and Equality," LW 11:136.

3. John Dewey, "The Lost Individual," LW 5: 68.

4. Ibid. p. 69.

5. John Dewey, "The Lost Individual," in John McDermott, *The Philosophy of John Dewey,* Chicago: University of Chicago Press, Phoenix Edition, 1981, p. 613.

6. Ibid.

7. Josiah Royce, "Provincialism," in *The Basic Writings of Josiah Royce,* edited by John McDermott, Chicago: Chicago University Press, 1969, p. 1069.

8. John Dewey, *Individualism Old and New:* New York: Capricorn Books, 1962.

9. Josiah Royce, *California, from the Conquest in 1846 to the Second Vigilance Committee in San Francisco [1856]: A Study in American Character,* (Boston and New York: Houghton Mifflin and Company, and 1885.

10. Robert V. Hine, *Josiah Royce: From Grass Valley to Harvard, University of Oklahoma Press, 1992.*

11. Josiah Royce, 1885, " The Squatter Riot of '50 in Sacramento: Its Causes and Its Significance," Overland Monthly, n.s. 6, 225-46, reprinted as "An Episode of Early California Life: The Squatter Riot of 1850 in Sacramento," in *Studies of Good and Evil,* 298-347. Citations will be from *Studies,* 301.

12. John Dewey, LW 11:35.

13. John Dewey, "Conduct and Thought," in John Dewey: The Middle Works, 15 vols. (Carbondale: Southern Illinois University Press, 1976-88). This reference will be cited as MW in future citations. MW 5:395.

14. Dewey, The Public, p. 149.

15. Josiah Royce, The World and the Individual, p. 2, p. 260. See also: Royce, "Some Observations on the Anomalies of Self Consciousness," and The Conception of God, pp. 278-289.

16. John Dewey, "The Lost Individual," LW 5:67.

17. Ibid. LW 5:121.

18. John Dewey, "Science, Belief, and the Public," MW 15:176.

19. Josiah Royce, Outlines of Psychology: An Elementary Treatise with Some Practical Application, New York: The Macmillan Company, 1911, p. 326. Italics in the text.

20. See: Jacquelyn Ann K. Kegley, Genuine Individuals and Genuine Communities: A Roycean Public Philosophy. Nashville, Tennessee: Vanderbilt University Press.

21. Josiah Royce, 1917, "Nietzsche," Atlantic Monthly, pp. 119 and 321-31, with an introductory note by W. Fergus Kernan, reprinted in Frank M. Oppenheim, S.J., Editor, Josiah Royce's Late Writings, Volume I, pp. 174-187.

22. Ibid. pp.337-328.

23. Josiah Royce, The Philosophy of Loyalty. New York: Macmillan, 1908 and Nashville: Vanderbilt, University Press, 1995.

24. Dewey, "The Lost Individual," LW 5:66.

25. Josiah Royce to Elizabeth Randolph, November 16, 1910, in John Clendenning, The Letters of Josiah Royce, Chicago: University of Chicago Press, 1954, p. 547.

26. John Dewey, The Public, pp. 126-127.

27. Ibid. p. 148.

28. Ibid. p. 149

29. Josiah Royce, "Provincialism," in Race Questions: Provincialism and Other American Problems, New York: Macmillan Company, 1908, p. 61.

30. For an extensive discussion of Royce's philosophy and family see: Kegley, Genuine Individuals and Genuine Communities, Chapter 5, pp. 106-128. Royce's philosophy as related to education is discussed in Chapter 6, pp.129-157.

31. John Dewey, in John Dewey and James H. Tufts, Ethics, New York: Henry Holt, 1908, pp. 580-581.

32. Josiah Royce, The Philosophy of Loyalty, p. 121.

33. William James, "The Moral Equivalent of War," in William James, The Essential Writings, Edited by Bruce Wilshire, New York: Harper & Row, 1971, pp. 349-61.

34. John Dewey, The Public, p. 208.

35. William James, Pragmatism, (Cleveland: World, 1955), pp. 50-51.

36. Royce, "Provincialism," McDermott, p.1080.

37. Ibid.

38. John Dewey, The Public, p. 208.

39. Ibid. p.154.
40. Ibid., p. 184.
41. William James, "On a Certain Blindness of Moral beings," in The Writings of William James, edited by John J. McDermott, New York: The Modern Library, 1968.
42. Josiah Royce, The Problem of Christianity, New York, Macmillan Company, 1913, Chicago: Henry Regnery Company, 1968, Volume 2, p. 64.
43. See: David L. Hildebrand, Dewey: A Beginner's Guide, New York, Oxford: One World, 2008, pp. 119-123.
44. Ronald Dworkin, Sovereign Virtue: The Theory and Practice of Equality, Cambridge, Massachusetts: Harvard University Press, 2000.
45. Ibid. 227.
46. Josiah Royce, The Hope of the Great Community, New York, The Macmillan Company, 1916; and Freeport, New York: Books for Libraries Press, 1967, p. 64.
47. Jose-Antonio Orosco, Cosmopolitan Loyalty and the Great Global Community: Royce's Globalization, The Journal of Speculative Philosophy, Vol. 17, No. 3, 2003.).
48. Shibley Tellhami, The Stakes: American and the Middle East, Westview Press, 2003 and 2004.
49. Shibley Tellhami, "Of Power and Compassion," The Philosophical Quarterly, September 11, pp. 70-80.
50. Michael Sandel, 2005, Public Philosophy, Cambridge, Massachusetts: Harvard University Press.
51. Stuart Gerry Brown, "From Provincialism to the Great Community: The Social Philosophy of Josiah Royce," Ethics, Vol. 59, No. 1. October, 1948. pp. 14-34 and pp. 19-20.

Sixteen

JOHN DEWEY AND FRIEDRICH VON HAYEK ON INDIVIDUALISM AND FREEDOM

Maciej Kassner

One of the characteristic features of Dewey's political philosophy is his deep awareness of the crisis of the old liberal order and the need for a fundamental redefinition of the basic categories of liberalism. John Dewey developed his late philosophy in an era marked by the crisis of capitalism after the Great Depression, growing disillusionment with the functioning of modern democracy, and an increased feeling of the inadequacy of inherited philosophical beliefs confronted by a new social and political reality. Although Dewey witnessed several social and political upheavals during his lifetime – including the American civil war, the Bolshevik revolution in Russia, the breakdown of democracy in several European countries and two world wars – the crisis of capitalism following the Great Depression was perhaps the single most important event that influenced his political thinking. For Dewey, an economic crash with a subsequent political crisis was an incentive to rethink such issues as our understanding of democracy, the complex relationship between science, technology and social planning, and, last but not least, the prospects of liberal individualism. According to Dewey's *The Quest for Certainty*, the ultimate function of social distress is to force men to address the most basic issues concerning social order:

> We are living in a period of depression. The intellectual function of trouble is to lead men to think. The depression is a small price to pay if it induces us to think about the cause of the disorder, confusion, and insecurity which are the outstanding traits of our social life.[1]

In this paper, I would like to examine Dewey's critique of what he thought to be one of the main causes of disorder, confusion and insecurity, namely the doctrine of *laissez faire* individualism. I am going to confront Dewey's views on individualism and freedom with a position held by one of the leading defenders of the old liberal creed – Friedrich August von Hayek. By comparing Dewey's and Hayek's approaches to individualism, I hope to elucidate the differences in the understanding of individual freedom by proponents of social and classical liberalism. My primary aim will be to show that

Dewey's critique of old *laissez faire* doctrine has not lost its significance and that his notion of individual freedom is superior to Hayek's view. In order to make this case, I will state the controversy between these two authors in terms of Macpherson's opposition of possessive and developmental individualism.[2] I will, then, examine Dewey's polemics against *laissez faire* individualism and Hayek's attempt to resurrect the old liberal doctrine. My thesis is that Hayek's and Dewey's versions of individualism are two different responses to the social and political crisis that Western countries confronted in the interwar period. Although both authors sought to preserve individual freedom in the collectivist age, I will argue that it was John Dewey who offered us a more adequate understanding of personal liberty in a complex society. Therefore, the present paper is not only an exercise in historical reconstruction, but also a case for a social liberal position with regard to the meaning of individualism and freedom. In the conclusion, I will address the problem of the contemporary significance of the Dewey/Hayek debate and discuss the prospects of freedom from a social liberal perspective.

1. Two types of individualism: possessive and developmental

Individualism, as Steven Lukes pointed out, like socialism communism is a 19th-century expression.[3] This very term was introduced to the political language by French conservative writers who condemned the French Revolution for promoting dangerous individualism and therefore destroying traditional social order. Similar to other ideological notions, individualism has been interpreted in various ways depending on the national context and political allegiance of protagonists. In this essay I will confine myself to the discussion of two distinct meanings of this concept that gained prominence within the liberal tradition. On the one hand, the term individualism has been used to describe the doctrine of the spontaneous emergence of social order that is the alleged result of free market exchanges of uncoerced individuals. This view can be traced back to Adam Smith's famous invisible hand thesis and, in a simplified form, it was defended by the proponents of *laissez faire*. Among its contemporary adherents one can find economically oriented liberals such as Friedrich von Hayek, Ludwig von Misses, Milton Freedman, Robert Nozick, James Buchanan, and others. According to the second stream of liberal thought, individualism is a doctrine concerned with the growth and development of an individual personality. The founding text for this new type of thinking on individual freedom is John Stuart Mill's *On Liberty*. Among prominent liberal thinkers, Thomas Hill Green, Leonard Hobhouse, and John Dewey were individualists in this sense. In contemporary debates, this position is represented by Richard Rorty, Amartya Sen and other social liberals. This split in the liberal attitudes towards individual freedom has been aptly described by C. B. Macpherson in terms of the distinction between possessive and developmental individualism. For economically oriented liberals, an individual is primarily a consumer of commodities and possessor of the fruits of one's own labour.

Contrary to that, liberal followers of John Stuart Mill are more concerned with establishing proper social conditions for individual self-creation. Given the fact that developmental individualism is consistent with thinking in terms of the far reaching transformation of the society, it was listed by Peter Lichtenstein as one of the theoretical coordinates of radical liberalism.[4] Needless to say, according to Lichtenstein, John Dewey is one of the leading proponents of this brand of liberalism.

2. Dewey in the 1930s: against *laissez faire*

As John Maynard Keynes observed, "the disposition towards public affairs, which we conveniently sum up as individualism and *laissez-faire*, drew its sustenance from many different rivulets of thought and springs of feeling."[5] The forerunners of economic science such as Adam Smith and David Ricardo, theologians preaching natural harmony between man and society like Paley, utilitarian social reformers in the tradition of Bentham, conservatives such as Burke or Hume, followers of Darwin viewing the market economy as the extension of the laws of evolution into the realm of society, partisans of free trade like Cobden and others – all seem to agree that government should not intervene into the business affairs of individuals. In the United States, the doctrine of *laissez faire* gained a more romantic interpretation through identification with the myth of the frontier and with the notion of endless possibilities open to an adventurous individual. Herbert Hoover, while campaigning for the American presidency, made this concept of individual economic self-reliance central to his political message. In his famous 1928 speech he identified the American political regime with what he called "rugged individualism":

> We were challenged with a peace-time choice between the American system of rugged individualism and a European philosophy of diametrically opposed doctrines – doctrines of paternalism and state socialism. The acceptance of these ideas would have meant the destruction of self-government through centralization of government. It would have meant the undermining of the individual initiative and enterprise through which our people have grown to unparalleled greatness.[6]

The breakdown of the economic system and the coming of the Great Depression made Hoover's speech sound absurd. For several years to come the doctrine of rugged individualism has become a principle target of progressive political argument. During that time John Dewey was working on a book called *Individualism Old and New,* a book that he had started in times of prosperity and continued writing after the stock market crash of October

1929 known as "Black Thursday." Chapters of this book appeared regularly in the *New Republic*, so the public had an opportunity to witness the unfolding of Dewey's argument in connection with the changing economic realities.[7] Although Dewey had criticized the ideology of *laissez faire* long before the Great Depression, most importantly in his book entitled *The Public and Its Problems,* only in the 1930s did the polemics against the old liberal creed became the central theme of his public speeches and political writings. Short essays and articles from that period reveal a lesser known side of Deweyan thinking. During the depression years, John Dewey proved that he was not only America's most respected philosopher, but also a fervent political polemist and partisan on behalf of radical politics. In what follows, I will try to reconstruct the main lines of Dewey's critique of economic liberalism that goes together with the unfolding of his own ideal of individual freedom.

Dewey was convinced that the struggle over the meaning of individualism is a part of a larger battle that would determine the course of the liberal political argument. The split between old and new liberalism was aptly characterized in Dewey's programmatic statement called *A Liberal Speaks Out for Liberalism*. In that article Dewey wrote:

> In spite of the extreme clash, both schools of liberalism profess devotion to the same ultimate ideal and goal. The slogan of both schools is the utmost possible liberty of the individual. The difference between them concerns the province in which liberty and individuality are most important and the means by which they are to be realized. [8]

Arguments concerning the identification of the right province of an individual stem naturally from the opposition of developmental and possessive individualism. For proponents of *laissez faire* the sphere of individual liberty is limited to economic matters as they tend to believe that economic freedom will automatically lead to all other kinds of freedom. As Dewey noted, this conception of liberty can be traced back to struggles against government encroachment in private property and limitations on economic activity imposed by a feudal system. Nonetheless, the movement towards greater economic liberty was accompanied by another stream of liberalism that Dewey labelled humanitarian. This breach of the liberal movement was concerned primarily with alleviating unnecessary misery. It culminated in an emergence of a doctrine of social liberalism advocating the use of government intervention in the name of social reform. According to Dewey, liberals of this stripe assumed a much wider conception of the sphere of individual freedom:

> Their conception of what I called the province of liberty and individuality is broader and more generous than of those who come forward as self-appointed champions of liberty. They think that

> liberty is something that affects every aspect and phase of human life, liberty of thought and expression, of cultural opportunity, and that is not to be had even in the economic sphere, without a degree of security that is denied to millions by the present economic system.[9]

A wider conception of the right province of the individual was accompanied by an equally altered notion of freedom. Dewey understood liberty as an "effective power to do specific things."[10] Therefore, it was misleading to speak of freedom in general. Rather, one has to examine what people can do and what they cannot do at any given time and place. Freedom, as understood by Dewey, was a function of social environment and required some measure of equality, security and education to fulfil its promise. More than that, it needed individual participation in the shared culture and institutions of the society. Having this in mind, Dewey recommended breaking the old patterns of domination and destroying all "the external arrangements of status, birth, wealth, sex, etc., which restrict the opportunity of each individual for full development of himself."[11] Participatory ideals, Dewey believed, could help unite the demands of liberty and equality and the goals of individual fulfilment with the promotion of a common good. The theory of individualism was for Dewey a necessary supplement to his theory of democracy,[12] and it lies therefore in the heart of his political philosophy.

Given these two different views of individual freedom, we can better understand the policies advocated by proponents of classical and social liberalism. Judging from the perspective of effective liberty, the net effect of policies supported by economic liberalism was securing power and opportunity only for those at the top of the social ladder. In Dewey's own words, "the liberty of concentrated wealth to do as it pleases is that which practically reduces so many to the condition of serfdom."[13] Conversely, policies aiming at the redistribution of wealth and substantial social capabilities are increasing the actual freedom of large number of people and helping to restore the genuine equality of opportunity. Indeed, it was Dewey's account on the nature of individual liberty that committed him to radical politics, a politics exemplified by the program of the League of Independent Political Action, where John Dewey served as a chairman.[14]

Second, Dewey's argument against *laissez faire* was rooted in his observation that the notion of the individual exemplified in that philosophy was wholly detached from social and economic realities. This line of reasoning was pursued by Dewey in *Individualism Old and New.* In the opening chapters of the book, Dewey argued that the doctrine of rugged individualism, which once had been a living faith capable of inspiring human conduct, now became an empty formula. Following this further, Dewey observed a profound split between people's actions and their professed goals. In his own words, "our old creeds have become ingrowing; the more we depart from them in fact, the more loudly we proclaim them."[15] For Dewey *laissez faire* individualism alongside with some religious doctrines was a primary example of such an outworn creed. The bulk of Dewey's argument is that the process of industrialization and a growing role of large corporations in American public life left no room for the old kind of individualism. Within the scope of this essay I can only list the major social changes that made the *laissez faire* doctrine look preposterous.

Firstly, one needs to take into account the transition from free labour to wage labour that altered the situation of working people in the United States. Part of the ideological appeal of *laissez faire* consisted in emphasizing the capacity of an entrepreneurial individual to set up his own business and make a living. However, as Michael Sandel pointed out, by the year 1870, two thirds of productively engaged Americans were simply wage earners.[16] The ideal of America as a land of self-employed independent individuals was in decline and a significant part of *laissez faire* mythology died with it. Secondly, an important advancement was an emergence of what John Dewey called "United States Incorporated," that is the domination of corporations in America's social and economic life. The consequences of these processes for individuals were tremendous. In the social realities of industrial capitalism, individuals confronted with corporate organization found themselves powerless. Traditional liberal arguments stating that private property is a necessary prerequisite for individual autonomy presupposes the social system of dispersed property. However, as Dewey realized, the concentration of wealth and power in the hands of a few great corporations made the old line or reasoning outdated. Under changed circumstances, the old liberal defence of private property and economic freedom looked as nothing more than shameless justification of corporate profits. All social trends mentioned above, together with the experience of severe economic depression, led liberals to question the American version of the *laissez faire* mythology. Having proclaimed the collapse of the romance of business enterprise, Dewey wrote: "We should at least get rid of the vast load of nonsense which now adds to our depression if we were coolly to take business for just what it is, (…) the purse for profit, and cease glorifying it for what it is not." [17]

Despite tremendous social change, old rhetoric managed to persist and play a major part in Hoover's campaign discourse. On behalf of progressivism, Charles Beard gave the doctrine of rugged individualism the treatment it deserved. Bringing examples of government interference with economy, such as the establishment of several federal bodies like the Federal Farm Board or government regulation on railways, tariffs, highways, waterways etc that somehow always turned out to benefit big business, he argued that rugged individualism is pure delusion.[18] Beard's critique contained a degree of irony as he pointed out that businessmen and farmers, both regarded as paradigmatic rugged individuals, appeared to receive more support from the government than any other professional group. Dewey's own account on the role of the *laissez faire* rhetoric was far more sophisticated. In *Individualism Old and New* he diagnosed what he called a "seeming hypocrisy" of modern American life, which is an unabridged gap between words and deeds, ideas and the reality.[19] However, he stressed the need for the psychological readjustment of individuals to altered social conditions and a call for a new ideal of individualism suitable for a technological age.

Another one of Dewey's arguments against economic liberalism is concerned with the alleged metaphysical fallacy inherited in that philosophy. In *Liberalism and the Social Action* John Dewey offered a short version of the genealogy of *laissez faire* individualism that was meant to disclose a dubious presupposition upon which the whole doctrine was based. According to Dewey, early economic liberalism can be interpreted as an attempt to give new meaning to the Lockean doctrine of natural law. In this view, natural laws are equated with the workings of the market economy, whereas man-made laws are regarded as less important and artificial. As Dewey observed, the first scholars who gave the doctrine of natural law economic interpretation were the French physicrats. They believed that land is the source of all wealth and therefore the organization of society should be modelled on the pattern set by nature. The basic idea of natural order had been taken over by proponents of economic liberalism with the emphasis shifted from land to labour. Another metaphysical ingredient of the old liberal creed was the idea that people's actions driven by an invisible hand would lead to a pre-ordered harmony of some sort. This alleged harmony can in turn be disclosed by human beings endowed with the natural light of reason. Needless to say, Dewey distrusted these aspects of liberal doctrine, which he regarded as semi-theological and semi-metaphysical illusions. From the practical point of view, the whole doctrine of the "natural order of liberty" represented but another obstacle for intelligent collective action and much needed social reforms.[20]

The next charge against *laissez faire* liberalism is concerned with authoritarian traits present in that doctrine and can be seen as a logical extension of the "metaphysical fallacy" argument. This line of reasoning is developed in an article called "The Future of Liberalism", where Dewey challenged what he recognized as "*laissez faire* doctrine [that] was held by a degenerate school of liberalism to express the very order of nature itself" with the accusation of absolutism.[21] A defining feature of this absolutism is an attempt to deny the historical relativity of liberal views, an attempt to create a "philosophy in which the particular ideas of individuality and freedom where asserted to be absolute and eternal truths; good for all times and all places."[22] The roots of these authoritarian traits within liberal philosophy, according to Dewey, can be traced back to the 18th century, when liberalism was born in the struggle against the absolutist state and the feudal elite. Fighting against a particular kind of state, namely a monarchic form of tyranny, liberals proclaimed that the state in general is the chief enemy of human liberty. In Dewey's view, that "authoritarian inheritance" made the previous form of liberalism unnecessarily dogmatic and therefore appearing dangerously similar to the rejected doctrine of absolutism.[23] Contrary to that, advanced liberalism learned the lesson of historic relativity and "knows that the content of individual and freedom change with time."[24] In the end, Dewey was far more willing than the partisans of *laissez faire* to utilize sociological insights into the condition of individuals in a given society to determine what forms individualist argument should take.

The assertion of individualism as an absolute dogma also troubled Dewey for another reason. He sought problematic similarities between the doctrine *laissez faire* and Marxian socialism. As Dewey pointed out in his *Freedom and Culture,* both philosophies were equally dogmatic in their admiration of overall principles and equally hostile to the spirit of trial and error experimentalism. From the point of view of pragmatism, class struggle and *laissez faire* individualism were ready-made explanations of the nature of social phenomena, explanations that could not be tested against empirical evidence. However, as Dewey warned, "two opposite errors do not constitute one truth, especially when both errors have the same root."[25] This common root was a metaphysical and absolutist attitude prevailing in both philosophies. In tracing the similarities between economic liberalism and Marxism, Dewey went as far as to suggest that "Marxist doctrine might even be regarded as a generalized version of that aspect of classic economic theory which held that completely free society in the open market would automatically produce universal harmony of persons and nations."[26] According to Dewey the only thing that Marx changed was "converting the competition of individuals into the war of classes."[27] In the end, it seemed to Dewey that Marx turned the "optimistic psychology of l*aissez faire* liberalism" on its head, pretty much the same way as he did with Hegelian idealism.[28] From Dewey's point of view, common features were more important than differences, for he asserted that "all absolutisms tend to assume theological form" and therefore create obstacles for free inquiry into the nature of social problems.[29]

Besides ethical objections against authoritarian features of old liberal philosophy, Dewey frequently complained about the lack of clarity and methodological scrutiny of the *laissez faire* argument. More specifically, he deplored the crude opposition between the individual and the social that was central to the rhetoric of economic liberalism. He regarded the collectivism/individualism distinction as deeply flawed for it was based on a myth of the abstract individual existing prior to society. More than that, he deplored terms like "the individual" and "the society" as too wholesale of notions to mean anything.[30] Ultimately, the only way to make the distinction between collectivism and individualism meaningful was to distinguish between the realm of private and public action. Having accepted that distinction, Dewey argued that liberalism has to come to terms with organized social action. Even more so, he claimed that intelligent public action is necessary for preserving and enhancing the freedom of individuals. Clearly, the division here is not between friends and foes of individual liberty. The real debate is rather about what policies are most conducive for individual flourishing. Furthermore, a closer look at that issue would reveal that liberals of both camps are equally against any attempt to submerge an individual to some higher collectivist order.

Reflecting on liberal confusion concerning the relation of an individual to the society, Dewey traced back the sources of the present misunderstanding to the philosophy of John Locke. Locke's contribution was to formulate a liberal political program in terms of the protection of the rights of the individual against intrusion by government. However, it was not the only possible formulation of the early liberal creed. "It is now easy for the imagination," wrote Dewey "to conceive circumstances under which the revolts against prior governmental forms would have found its theoretical formulation in an assertion of the rights of groups, of other association that those of a political nature."[31] The stark opposition between the individual and society was therefore historically contingent and open to revision. Had liberals refused to follow Locke's lead, the political language of the West would have been utterly different. Liberalism in particular would have assumed more communal and less individualistic forms. This line of reasoning was pursued by Dewey in *The Public and Its Problems* and it is now widely recognized for foreshadowing communitarian argument.[32] It is interesting to note that if this alternative path had been followed, there would not have been any need for inventing the distinction between individualism and collectivism. For the purpose of this paper, it is important to note that the empowering of communities can be a third possibility beyond the expansion of the state and the market, a possibility that was overlooked by the proponents of economic liberalism.

3. From rugged individualism to the road to serfdom

The outcome of political and ideological struggles of the 1930s was a profound redefinition of basic liberal categories. The term "liberty" was stretched to encompass freedom from want (including economic security) and freedom from fear, to use the vocabulary from Franklin Delano Roosevelt's famous speech.[33] New Deal policies in the United States and Lord Beveridge's report in Great Britain were the most prominent attempts to give the new liberal philosophy a practical application. However, this new liberal consensus soon was challenged by those who tried to resurrect the old creed.

Friedrich von Hayek was an economist and political philosopher who made the old doctrine of "rugged individualism" respectable again. Having experienced grave defeats as an economist in the debates on the business cycle theory, Hayek turned to liberal philosophy and political matters.[34] He gave his views on the topic of individualism and freedom in his 1944 bestselling book *The Road to Serfdom* and – in a more sophisticated form – in an important essay called "Individualism True or False", first delivered as a lecture at University College Dublin in 1945. In *The Road to Serfdom* Hayek presented himself as a self-styled defender of Western Civilization against what he identified as totalitarian tendencies inherited in any attempt to intervene in a free market economy. The bulk of Hayek's argument was that central planning lies at the heart of totalitarianism both in Nazi Germany and Soviet Russia. Having realized that fundamental truth, Hayek felt obliged to warn his fellow citizens of Western democracies against the proponents of social planning, the socialists of all stripes, as he called them. Only going back to the firm principles of classical liberalism and *laissez faire* policy could rescue Western countries from sharing the sad fate of Germany or Russia. Among the doctrines primarily responsible for present confusion, he listed the idea of freedom that can be found in the works of "the leading philosopher of American left-wingism," John Dewey.[35] Hayek accused Dewey of confusing freedom with power by maintaining that liberty is the effective capacity to do certain things. This mistake, Hayek thought, is the typical error of collectivism and, as any other collectivist invention, has to lead to disastrous consequences. It is worth noting that Hayek repeated that charge against Dewey later on in his philosophical *opus magnum* – *The Constitution of Liberty.* [36]

Notwithstanding the enormous popularity of *The Road to Serfdom,* Hayek's most important arguments in defence of the old individualistic creed were laid down in various articles published in scholarly journals. In "Individualism True or False" Hayek distinguished two main streams of individualist thinking.[37] True individualists are those who defend the principles of limited government, private property and the spontaneous order of a free market economy. The names of John Locke, Bernard Mandeville, Adam Smith, Adam Ferguson, Edmund Burke, Lord Acton, and Alexis de Tocqueville among others stand for the grand tradition of liberty in the spirit of classical liberalism. False individualism, on the contrary, was born from the original sin of Cartesian rationalism. It is exemplified by the work of Encyclopaedists, Jean Jacques Rousseau, and others who tried to base social order on the study of the nature of individual reason. In Hayek's view, this false individualism always turns to its opposite, namely collectivism and socialism. That is because the followers of Descartes adhered to the design theory of social institutions. According to this theory, individual reason plays an important role in drawing social and political institutions. This belief is best exemplified by Descartes dictum that ancient Sparta is politically superior to Athens because Spartan laws were created by one man. Hayek believed that he can trace the development of the tradition of Cartesian rationalism from the early social contract theories to the contemporary Marxist attempts to build social order on alleged scientific grounds. In opposition to that, true individualists maintained that social order is a product of human action but not of human design. A free market economy, where actions taken by individuals can lead to socially beneficial outcomes, is a primary example of such spontaneous ordering.[38] From Hayek's perspective, developmental individualism, as exemplified by the work of John Stuart Mill, was situated somewhere between the opposing camps of classical and Cartesian individualists. However, Hayek had little doubt that Mill's social philosophy, affected by the deadly germs of German Romanticism, would ultimately lead to some form of collectivist serfdom.

In spite of the fact that Hayek presented himself as the last defender of classical liberalism in a collectivist age, it is worth stressing that he made a conscious effort to revise the old creed. Most importantly, he reformulated a central argument for free market in terms of the economy of information. He ridiculed what he called "the bogey of the 'economic man' and all similar attempts to identify the free market with the single psychological motif of self-interest.[39] For Hayek, the moral worth of the market lies in the fact that market mechanism allows all participants in economic exchanges to make free use of the unique knowledge they possess. In other words, a market economy can solve the problem of processing dispersed information better than any form of central planning. Apart from these major changes, Hayek also dropped the label *laissez faire* as a discredited name for the liberal principles he stood for. Furthermore, he allowed for a slightly wider sphere of government interference than previous economic liberals were ready to accept. Despite all the improvements, Hayek remained a firm believer in the old creed. He concluded his *Road to Serfdom* with a reassertion of what he thought to be a guiding principle of *laissez faire*, namely the view "that a policy of freedom for the individual is the only truly progressive policy, remains as true today as it was in the nineteenth century."[40]

4. Rejoinder to Hayek from a Deweyan perspective

In assessing arguments brought about by Dewey and Hayek, I will divide the discussion into two sections. First, I will address the controversy concerning the proper response to the political crisis in the West and then turn to epistemological and meta-philosophical claims supporting each position.

John Dewey was by no means unaware of the theoretical challenge represented by Friedrich von Hayek. Dewey chose to respond to that challenge in a rather indirect way through the discussion of the work of Karl Polanyi, whose book *The Great Transformation* he regarded as "the most enlightening account of the most important historical events in the last century-and-a-half that I ever read."[41] In a letter to Sidney Hook, Dewey expressed his firm belief that Polanyi's work is the best cure for recent trends to resurrect the doctrine *of laissez faire*: "if Hayek has ever read one chapter of [*The Great Transformation*]," he wrote, "I think that he would have been ashamed to write his book [*The Road to Serfdom*] – for it is convincing proof that all the evils and objectionable problems Hayek builds on are the products of the necessity of social protection against a market economy..."[42]

In the book that made such a tremendous impression on John Dewey Karl Polanyi questioned several assertions central to Hayek's social theory. To start with, Polanyi claimed that the free market itself cannot guarantee a stable social order. "Our thesis is that the idea of self-adjusting market implied a stark utopia," wrote Polanyi on the very first page of his book.[43] "Such an institution could not exist for any length of time" – he went to say – "without annihilating the human and natural substance of society; it would have physically destroyed man and transformed his surroundings into wilderness."[44] However strange it may sound, it is important to note that the central thesis of *The Great Transformation* was a commonplace among American liberals of that time. Certainly, it was shared by Dewey, who frequently expressed his doubt whether the "profit system" can serve as a sole basis for social organization.[45] To be sure, what is asserted here is not that the market mechanism has to be abolished altogether. Rather, it is argued that the institution of the market has to be embedded in society that is supplemented by other social institutions, which by necessity interfere with the workings of the market economy.

The attempt to dismantle the market from the larger social fabric and make it the only principle of social organization lies at the heart of economic liberalism. However, for reasons given by Polanyi and others such social order was impossible to achieve. In the words of another admirer of Polanyi's work, Joseph Stiglitz, who wrote the foreword to the second edition of *The Great Transformation*, "Polanyi exposes the *myth* of the free market: there never was a truly free, self-regulating market system."[46] Consequently, Polanyi rejected the doctrine of the spontaneous emergence of the market order. He sought the free market project as a constructivist attempt to impose the rules of *laissez faire* on the society. Contrary to that, the movement towards the protection of society aroused spontaneously and took many different forms varying from Bismarck's authoritarianism to Roosevelt's liberalism. "Laissez faire was planned, planning was not," Polanyi paradoxically quipped.[47] Last, but not least, Polanyi claimed that both fascism and socialism represented two different reactions to the collapse of the old liberal order. The difference between them is moral. Under a social democratic regime freedom was preserved and enhanced, while under fascist rule it was doomed. Contrary to Hayek, who warned that "those who preach New Order" in fact "can think of nothing better than to imitate Hitler," Polanyi greatly admired New Deal policies as consistent with freedom in a complex society.[48]

In an essay tellingly called "Crisis in Human History. The Danger of Retreat to Individualism" Dewey gave his own account of the transition to serfdom that was based on the critique of the *laissez faire* philosophy offered by Karl Polanyi. This essay is extremely important for the purpose of this paper for it contains the only direct allusion to Hayek's philosophy in general and to *The Road to Serfdom* in particular that can be found in Dewey's works.

I will quote the relevant paragraph at length:

> So called "free enterprise" in business and finance was taken to be identical with the very essence of freedom. When it become apparent as to such Freedom that its net result was recurrent industrial depressions, each one more severe than its predecessor, in which the mass of human beings were reduced to a state of insecurity and fear, it is hardly surprising that peoples who were already habituated to dependence upon superior political authority preferred "the road to serfdom" which promised them some stability of life.[49]

According to Dewey it was important to bear in mind that the outbreak of totalitarianism had been preceded by an underlying phase of *laissez faire*. The movement that goes under the name of individualism, Dewey argued, was largely responsible for creating conditions leading to social disorder. He repeated the diagnosis laid down by Polanyi when he contended that "totalitarian socialism is without paradox a legitimate consequence of and reaction to the *laissez faire* Liberalism that proclaimed the subordination of the political to the economic."[50] More than that, Dewey was also convinced that the structure of totalitarian thought was shaped by a false distinction between an abstract notion of an individual and equally abstract concept of the society, a distinction first made by the proponents of *laissez faire*. As Dewey reminded us, "placing the socialistic in stark opposition to individualistic was not the creation of Fascism and Totalitarianism."[51] It was, in fact, "direct inheritance from laissez faire Liberalism."[52] The only difference is that while liberals claimed the absolute sovereignty of individual, proponents of totalitarianism professed equally absolute subordination of the individual to totality of some sort. However, in order to stop the pendulum going back and forth between those two utterly unattractive extremes, some sort of more balanced position was needed. The true solution, according to Dewey, lies in the recognition of the unitary character of the human being, which pays due attention both to social and individual aspects of a person. A more accurate view of the human condition would result in more adequate – and more balanced – politics, which could overcome the "crisis in human history" and heal the wounds.

A closer look at the historical evidence will show that both liberal theory critical towards modern forms of tyranny and the political practice of several countries that manage to evade totalitarianism were far from what Hayek imagined as classical liberalism. As Jan Werner Müller pointed out, cold war liberals such as Isaiah Berlin, Karl Popper and Raymond Aron, professed sympathy for the welfare state and took pains to distance themselves from Friedrich von Hayek and the libertarian movement.[53] All three authors believed that without some measure of security, equality and social compromise liberty cannot be preserved. John Dewey himself took a more radical stance as he argued that a proper response to totalitarianism should involve the radicalization of democracy and a further move towards the left of the political spectrum. Still, mainstream liberal opposition against totalitarian menace for most of the time was closer to the position of John Dewey than to that of Friedrich von Hayek. More importantly, as even a brief survey of the types of European political regimes in the interwar period will disclose, the alternative to fascism was social democracy (or traditional dictatorship) and not classical liberalism.[54] In that time most of the countries that managed to preserve a democratic regime adopted substantial social democratic policies. Czechoslovakia, Sweden, Norway, Denmark and, to a lesser extent, Great Britain may serve as examples. One could hardly find a single country whose political order or adopted policy would meet Hayekan recommendations. In light of empirical evidence it seems that social democracy rather than classical liberalism proved to be the real saviour of freedom in the West.

It would not be difficult for Dewey to defend himself from epistemological charges brought by Hayek against the proponents of false individualism. He can hardly be considered guilty of falling under the spell of Cartesian rationalism. From Dewey's perspective this kind of rationalism was but another example of failure to recognize that intelligence is a social asset and can be fulfilled only in public debate. Since Dewey's own case for state interference with economy was based on stressing the need of public deliberation, he cannot be called a Cartesian rationalist. As a good pragmatist, Dewey was always keen to find dubious metaphysics in the views of his political opponents. From the pragmatist perspective Hayek's radical critique of design theory of social institutions amounts to the denial of the role of human creativity in the shaping of the course of human affairs. In *The Quest for Certainty* Dewey identified that attitude with a spectator theory of knowledge, a theory that radically separates knowing and doing. Dewey regarded the doctrine of *laissez faire* as a "logical conclusion" of the spectator theory of knowledge and deplored it for the "paralyzing effect on human action" it exercised.[55] Contrary to Hayek, Dewey did not believe that market left alone will somehow generate a stable social order. By arguing that liberalism has to come to terms with organized social action he left more room for human invention in political affairs.

More importantly, Dewey would find several of Hayek's views concerning ethical theory to be highly objectionable. According to the author of *The Road to Serfdom,* an individual participating in social affairs has to "submit to conventions which are not the result of intelligent design, whose justification in the particular instance may not be recognizable, and which to him will often appear unintelligible and irrational."[56] In Hayek's view, free market can work without friction only if people accept inherited codes of behaviour. From Dewey's perspective, statements like that are tantamount to the denial of individual freedom in the sphere where it is most valuable, namely the sphere of ethical consideration and personal development. In *Ethics,* a book written together with James Tufts, Dewey appraised transition from group ethics that rested on customs and tradition to reflexive morality based on critical reasoning and conscious consideration. Part of that latter view was strong admiration for the "intense moral individualism like that of Rousseau and that of Tolstoy," an individualism that would question established social convictions in the name of new moral intuitions.[57] Contrary to that, Hayek regarded all pleas for individuality and originality as exemplified by J. S. Mill's *On Liberty* as but another example of false individualism. "It must remain an open question whether a free or individualistic society can be worked successfully if people are too 'individualistic' in the false sense," warned Hayek.[58] In fact, the author of "Individualism True and False" fell short from retreating to an old kind of morality based exclusively on custom and convention, which was rightly criticized by Dewey and Tufts. Judging from the perspective of developmental individualism, despite all his critique of alleged collectivism in economics, Hayek himself proved to be a collectivist in ethics.

To be fair to Hayek, one has to admit that his contention that replacing market mechanisms with central planning lies at the heart of totalitarianism can be rendered plausible, especially with regards to its Marxist version. He was on the right track when he pointed out that many of the left wing authors of that period were uncritical admirers of the Soviet experiment. One can also acknowledge Hayek's insights concerning the role of information in the market economy that earned him recognition from the Nobel prize committee. However, as it is often the case with Hayek's writings, one can appreciate his best arguments without accepting the radical political posture of the author of *The Road to Serfdom.* Notwithstanding all the possible merits of Hayek's work as an economist and political theorist, it would be wiser for us to derive our understanding of individualism and freedom from the work of John Dewey. To my mind, the merit of Dewey's position is that it takes into account threats to human freedom caused by vast inequality and insecurity that are likely to exist in unrestricted markets. Concentrating on the effective freedom to achieve valuable goals, Dewey managed to avoid tautological equitation between liberty and the market. Social liberalism – as I will try to show later in my discussion of the work of Amartya Sen – does not need to be hostile to the market as such. However, from the standpoint of social liberalism, the freedom to participate in market exchanges is neither the only form of liberty nor the most fundamental one. More than that, sometimes the doctrine of economic liberalism itself can pose a serious threat to individual freedom, as it was indicated in John Dewey's "depression writings" and in the work of Karl Polanyi. In the end, John Dewey was able to offer a broader view on liberty, a view that would encompass not only political and economic freedom but also important values of free communication, self-development and participation.

5. Contemporary relevance of the Dewey/Hayek debate

Recently, the doctrine of Hayek has received some attention due to the several breakthroughs in social and economic policy that were associated with his teachings. Margaret Thatcher's government in Great Britain and Ronald Regan's presidency in the United States are primary examples of what has become known as a neoliberal agenda. Neoliberalism, as it is often argued, can be seen as an effort to remake *laissez faire* and apply it to contemporary social conditions.[59] The rebirth of the old liberal creed is proof that an important lesson of the 20th century has been forgotten. As Paul Samuelson reminded us, not only the false god of Marxian socialism has failed in the long 20th century.[60] In fact, during the years following the Great Depression his archetypical rival, the ancient god of *laissez faire,* did not fare any better. For the purpose of this paper, it is important to note that with the rebirth of *laissez faire,* the old notions of individualism and freedom regain prominence.

Nevertheless, in times of neoliberal domination the doctrine of developmental individualism has not altogether perished. Let me here just mention the sophisticated social theory of Amartya Sen, Indian philosopher and noble prize winner in economics, whose work seems to me to be the best restatement of the social liberal position with respect to the meaning of individual freedom. It is not without significance that Sen presented one of his first papers on that subject during the *Dewey Lectures* that he delivered at Columbia.[61] Similar to John Dewey, Sen holds that freedom is a complex notion that involves among other elements several positive liberties such as freedom from hunger, freedom to avoid premature morbidity, freedom to obtain proper shelter, freedom of movement, etc. Sen uses the term capabilities for these different forms of positive liberty. As Sen explains, "a person's capability set can be defined as a set of functioning vectors within his or her reach."[62] Putting it simply, capabilities tell us what people can do or be. Sen's capability approach to freedom can be seen as a contemporary version of the theory of effective liberty held by Dewey and other social liberals.

In my view, Sen made at least two important advancements in the theory of effective liberty. To start with, he made it clear that any consistent defence of positive freedom has to face the charge of mistaking freedom with security, power, equality or something else. In other words, proponents of effective freedom have to find an answer to the accusation that they are overextending the meaning of the word "liberty." Trying to meet that charge Sen devised the idea of formulating a list of basic capabilities that should be made a citizens concern. Having admitted that any such index has to be the subject of public debate, he nevertheless discussed several important capabilities, including right to participate in public deliberation, the right to basic education and health assistance, and other positive freedoms mentioned above. From the standpoint of effective liberty, the market mechanism has to be evaluated from the point of view of enhancing human capabilities. Far from being hostile to the market itself, Sen argued that it has to be supplemented by the public provision of some goods, most importantly health care and basic education.[63] Last but not least, Amartya Sen sought to elaborate concrete policy proposals based on his theory. As he demonstrated in his book entitled *Development as Freedom*, an approach focused on maximizing human capabilities, including basic education and universal health care, can be far more beneficial for developing countries than simple policies aiming exclusively at the "freeing of markets."[64] I cannot discuss Sen's work at greater length here. Suffice to say, his powerful restatement of social liberal position proves that the idea of effective liberty can be an important instrument in coping with the problems of contemporary world.

Another important point of reference for the contemporary relevance of the struggles over the meaning of liberalism and liberty are the former communist countries. As Jerzy Szacki, the leading Polish social scientist, famously paraphrased Marx: "after the fall of Communism the spectre was haunting Eastern Europe – the spectre of liberalism."[65] Reflecting on the unexpected popularity of liberal philosophy among countries seeking for a way out of communism, he remarked that "liberalism today seems to offer the most consistent blueprint for a political-social-economic order."[66] Regrettably, all too often people in Eastern Europe chose to learn their lesson of freedom from neoliberals like Milton Friedman or Friedrich von Hayek. In consequence, the rich and complex tradition of liberal political philosophy was obscured by the identification with rather simple minded economic liberalism. As Szacki observed, liberals in the East "have shown an exaggerated tendency to associate it [liberal philosophy] chiefly with the really not so miraculous economic policies of Ronald Regan and Margaret Thatcher."[67] Had they derived their views on individualism and freedom from John Stuart Mill and John Dewey rather than from Hayek and Friedman, they would have arrived at a more sophisticated understanding of liberalism and liberty.

Last but not least, the economic crisis that accelerated after the fall of the Lehman Brothers led many people all around the world to question the credibility of the neoliberal doctrine. In the words of Joseph Stiglitz, "the fall of Wall Street is for market fundamentalism what the fall of the Berlin Wall was for communism – it tells the world that this way of economic organization turns out not to be sustainable."[68] Our situation is in some respects similar to that of the Great Depression. We have indulged in the "orgy of so-called economic liberty," to borrow Dewey's apt phrase.[69] Now we have to face the consequences. The real function of trouble, as Dewey taught us a long time ago, is to lead men to think about the causes of confusion, insecurity and disorder. Admittedly, some of the thinking has already been done and the need for reforms is widely recognized. However, the economic crisis can serve as an opportunity not only to establish a new regulatory framework, but also to re-evaluate much broader issues concerning our understanding of individual freedom. If we are going to learn our lesson from the present economic crises, we should question not only several neoliberal policies, but also the ethical foundations of this very doctrine. Looking for new ideas we can find ourselves not far away from what Dewey advocated as "advanced liberalism" or "new individualism."

6. Conclusion. Overcoming rugged individualism as a precondition for true individualism

Admittedly, economic individualism of a Hayekan type is only one factor among many reasons responsible for individuals feeling "lost." Nevertheless, it is rather important one. As Dewey, observed, "the loss of individuality is conspicuous in the economic region because our civilization is so predominantly a business civilization."[70] However, breaking with the false creed is only the first step towards new individualism. After all, as Paul Kurtz in his introduction to *Individualism Old and New* observed, "one cannot hope to develop integrated individuality by an all-embracing social program alone."[71]Overcoming rugged individualism, in all its historical and contemporary forms, is only a precondition for a more ambitious project of developing a new kind of individuality suitable for the technological age. "It is difficult for us to conceive of individualism except in terms of stereotypes derived from former centuries," complained Dewey.[72] "A stable recovery of individuality" – he went on to say – "waits upon an elimination of the older economic and political individualism, an elimination which will liberate imagination and endeavour for the task of making corporate society contribute to the free culture of its members."[73] That is, no doubt, a noble and tempting vision. The chances of fulfilling its promise, I shall conclude, depend crucially on whether we manage to break with all the false individualistic doctrines that obscure our understanding of individuality and freedom.

NOTES

1. John Dewey, "Science and Society *[Philosophy and Civilization],"* in *The Later Works of John Dewey*, Vol. 6, ed. by. Jo Ann Boydston, Carbondale, 2008 p. 63. If not stated otherwise, all quotations from Dewey's works refer to this edition of *The Collected Works of John Dewey.*

2. C. B. Macpherson, "Pluralism, Individualism, and Participation," in *The Rise and Fall of Economic Justice and Other Essays*, Oxford, New York, Toronto, 1987, pp. 92-101.

3. On the rich semantic history of the term see, Steven Lukes, *Individualism*, Oxford, 1985, pp. 1-42.

4. See Peter Lichtenstein, "Some Theoretical Coordinates of Radical Liberalism," American Journal of Economics & Sociology, July 1984, Vol. 43 Issue 3.

5. John Maynard Keynes, *The End of Laissez Faire,* Second Impression, London, 1926, p. 6.

6. Herbert Hoover, "Rugged Individualism Speech" October 22, 1928 in Richard Hofstadter and Beatrice K. Hofstadter, *The Great Issues in American History, Vol. III: From Reconstruction to the Present Day, 1864-1981,* New York,1982, p. 331.

7. Paul Kurtz, "Introduction," in John Dewey, *The Latter Works of John Dewey*, Vol. 5, p. xvii.

8. John Dewey, "A Liberal Speaks Out for Liberalism," in *The Latter Works of John Dewey,* Vol. 11, p. 285.

9. Ibid., p. 286.

10. John Dewey, "Liberty and Social Control," in *The Latter Works of John Dewey,* Vol. 11, p. 361.

11. John Dewey and James Hayden Tufts, "Ethics," 1932 Edition, in *The Latter Works of John Dewey,* Vol. 7, p. 349.

12. Ibid., pp. 344-350.

13. John Dewey, "American Ideas (I). The Theory of Liberty vs. the Fact of Regimentation," in *The Latter Works of John Dewey,* Vol. 9, p. 89.

14. On Dewey's political activity during that time see, Edward. J. Bordeau, "John Dewey's Ideas about the Great Depression," Journal of the History of Ideas, Vol. 32, No. 1, 1971, pp. 67-84.

15. John Dewey, "Individualism, Old and New," in *The Latter Works of John Dewey*, Vol. 5, p. 48.

16. Michael J. Sandel, *Democracy's Discontent. America in Search of a Public Philosophy*, Cambridge, Massachusetts, London, 1998, pp. 168-200.

17. John Dewey, "The Collapse of a Romance," *The Latter Works of John Dewey,* Vol. 6, p. 73.

18. Charles A. Beard, "The Myth of Rugged American Individualism" in *New Deal Thought,* ed. by Hoard Zinn, Indianapolis, New

York, Kansas City, 1966, pp.1-10.

19. John Dewey, "Individualism, Old and New," p. 68.

20. John Dewey, "Liberalism and Social Action," in *The Latter Works of John Dewey*, Vol. 11, pp. 9-11.

21. John Dewey, "The Future of Liberalism" in, *The Latter Works of John Dewey*, Vol. 11, p. 291.

22. Ibid., p. 290.

23. Ibid., pp. 290-291.

24. Ibid., p. 292.

25. John Dewey, "Freedom and Culture," in *The Latter Works of John Dewey*, Vol. 13, p. 125.

26. Ibid., p. 125.

27. Ibid., p. 125.

28. Ibid., p. 134.

29. Ibid., p. 122.

30. John Dewey, "Liberalism and Social Action," p. 44.

31. John Dewey, "The Public and Its Problems," in *The Latter Works of John Dewey*, Vol. 2, p. 289.

32. See Michael Walzer, "The Communitarian Critique of Liberalism," Political Theory, Vol. 18, No. 1, 1990, pp. 6-23.

33. Franklin D. Roosevelt, "Four Freedoms Speech," January 6, 1942, in *Great Issues in American History*, Vol. 3, pp. 386-91.

34. Paul Samuelson, "A few remembrances of Friedrich von Hayek (1899-1992)," Journal of Economic Behavior & Organization, No. 69, 2009, pp. 1-4.

35. Friedrich A. von Hayek, *The Road to Serfdom,* Abingdon and New York 2006, p. 26 (Footnote 1).

36. Friedrich A. von Hayek, *The Constitution of Liberty,* Chicago, Illinois, 1960, p. 17.

37. See Friedrich A. von Hayek, "Individualism: True and False," in *Individualism and Economic Order*, Chicago, London, 1958, pp. 1-32.

38. See Steven Horowitz, "From Smith to Menger to Hayek. Liberalism in the Spontaneous-Order Tradition," The Independent Review, No. 6, Summer 2001, pp. 81-97.

39. Friedrich A. von Hayek, "Individualism: True and False," p. 11.

40. Friedrich A von Hayek, *The Road to Serfdom,* p. 246.

41. John Dewey, "Comment on Bell and Polanyi," in *The Later Works of John Dewey*, Vol. 15, p. 361.

42. John Dewey, J.D to Sidney Hook, 28 June, 1945, quoted in Robert Westbrook, *John Dewey and American Democracy*, Ithaca and London, 1991, p. 461.

43. Karl Polanyi, *The Great Transformation. The Political and Economic Origins of Our Time*, Boston 2001, p. 3.

44. Ibid., p. 3.

45. See for example John Dewey, "Individualism, Old and New," pp. 99-110.

46. Joseph Stiglitz, Foreword, in K. Polanyi, *The Great Transformation,* p. xiii.

47. Karl Polanyi, *The Great Transformation*, p. 147.

48. See Friedrich von Hayek, *Road to Serfdom,* p. 246; Karl Polanyi, *The Great Transformation,* pp. 245-256.

49. John Dewey, "The Crisis in Human History. The Danger of the Retreat to Individualism," in *The Latter Works of John Dewey,* Vol. 15, p. 216.

50.Ibid., p. 218.

51.Ibid., p. 215.

52. Ibid., p. 215.

53. Jan W. Müller, "Fear and Freedom. On '*Cold War Liberalism',* European Journal of Political Theory, Vol. 7, No. 1, 2008, pp. 45-64.

54. Gregory M. Luebbert, "Social Foundations of Political Order in Interwar Europe," World Politics, Vol. 39, No. 4, 1987, pp. 449-478.

55.John Dewey, *The Quest for Certainty,* in *The Latter Works of John Dewey,* Vol. 4, p. 169.

56. Friedrich A. von Hayek, "Individualism: True and False," p. 22.

57. John Dewy and James Tufts, *Ethics,* in *The Latter Works of John Dewey*, Vol. 7, p. 326.

58. Friedrich A. von Hayek, "Individualism: True and False," p. 26.

59. See Jamie Peck, "Remaking laissez-faire," Progress in Human Geography, No. 32, 2008, pp. 3-43.

60. Paul Samuelson, "Two Gods that Fail," Challenge, Vol. 42, Issue 5, 1999, pp. 29-33.

61. Amartya Sen, "Well-Being, Agency and Freedom: The Dewey Lectures 1984," Journal of Philosophy, Vol. 82, No. 4, 1985, pp. 169-221.

62. Ibid., pp. 200-201.

63. On market and freedom see Amartya Sen, "Economic Development and Capability Expansion in Historical Perspective," Pacific Economic Review, No. 6, 2001, pp. 179-191; Amartya Sen, "Moral Standing of the Market" in Dieter Helm (ed.), *The Economic Borders of the State,* 1992, pp. 92-109; Amartya Sen, "Markets and Freedoms: Achievements and Limitations of the Market Mechanism in Promoting Individual Freedoms," Oxford Economic Papers, Vol. 45, No. 4, pp. 519-541.

64. Amartya Sen, *Development as Freedom*, Oxford University Press, 1999.

65. Jerzy Szacki, *Liberalism After Communism*, Budapest 1995, p. 1.

66. Ibid., p. 10.

67. Ibid., p. 6.

68. Joseph Stiglitz, "The Fall of Wall Street is to Market Fundamentalism What the Fall of Berlin Wall Was to Communism,"

interview with Nathan Gardels, [at] www.huffingtonpost.com

69. John Dewey, "Freedom," in *The Latter Works of John Dewey*, Vol. 11, p. 251.

70. John Dewey, "Individualism, Old and New," p. 70.

71. Paul Kurtz, "Introduction", in The Latter Works of John Dewey, Vol. 5, p. xx.

72. John Dewey, "Individualism, Old and New," p.75.

73. Ibid., pp. 75-76.

Seventeen

DEWEY'S ETHICAL-POLITICAL PHILOSOPHY AS RESOURCE IN TODAY'S GLOBAL CRISES AND AS A GUIDE TO A POST-IDEOLOGICAL POLITICS FOR THE 21ST CENTURY

Gregory Fernando Pappas

> "Better it is for philosophy to err in active participation in the living struggles and issues of its own age and times, than to maintain an immune monastic impeccability."
>
> —John Dewey, "Does Reality Possess Practical Character?"

Infrequently, the term 'Pragmatism' is both popular and in circulation. In political discourse currently (on the web and in newspapers), 'Pragmatism' is both in use and in fashion. For this we need to thank the recent debates surrounding the "pragmatism" of President Obama as well as the fact that 'Pragmatism' is often assumed to be the alternative to ideological ways of thinking in the political arena. Hence, today may be a unique opportunity for scholars on pragmatism to contribute to public dialogue, and at the very least to rectify some common misconceptions. More importantly, now is a unique opportunity to show that Dewey's particular form of Pragmatism is worth reconsidering. Dewey's ethical-political philosophy is a present resource to reflect about the possibility of post-ideological politics. Dewey made two important contributions that I wish to highlight because they bear directly on today's problems. First, in Dewey's philosophy, we find tools of diagnosis and understanding of how ideological oppositions may have contributed to our present economic-political crisis. Secondly, there is in Dewey a positive proposal in regard to what sort of politician is needed if we are ever to transcend how opposing ideologies have dominated politics. Whether Obama may be a Pragmatist is not my concern here. Instead, I want to consider what could Obama's alleged pragmatism mean and address the critics of Pragmatism, *i.e.*, those that worry that Obama is a Pragmatist or exemplifies the practices recommended by Pragmatism.[1]

1. Ideological Oppositions in Socio-Political Discourse

For Dewey, ideological oppositions in socio-political matters have had a long history. The oppositions are based on untenable dualism that, in practice, has led to a dogmatic and narrow-minded approach to problems. Ideological oppositions are an obstacle to the sort of intelligent and context-sensitive deliberation/communication that Dewey thought was needed in a democracy. He thought that, throughout history, one of the most common ideological oppositions has been between forms of individualism versus collectivism. The opposition between these types of political theories is of consequence because they each prescribe a particular form of inquiry about social problems.

Political theories that center on individual natural rights and the notion of negative freedom are suspicious of all collective action because such collective action tends towards regimentation, mechanical and mass uniformity, censorship, and suppression. Collectivism, for its part, considers organized action as the source of all that is good and civil in nature. Both views assume an untenable dualism between the individual and the social but, more importantly, they share the same dogmatic approach to problems. For the individualist, social organization (government) is at best a necessary evil, and it tends to be oppressive. Therefore, social problems are analyzed in terms of how individual initiative, freedom, incentives, and independence have been suppressed by some collective action or organization. The collectivist, on the other hand, tends to analyze problems in terms of the disintegration and instability created by a rampant individualism that has undermined social order or communal bonds. Depending on who represents the status quo in this debate, one accuses the other of the present problems, but neither one cares to examine situations on their own merits. Hence, Dewey referred to them as dangerous opposing doctrines that seem to exempt us from the responsibility and the hard work of ameliorating social problems in light of their contextual uniqueness. "The person who holds the doctrine of 'individualism' or 'collectivism' has his program determined for him in advance."[2]

The existence of these opposed schools of thought would be inconsequential if it were only a historical artifact. Unfortunately, the Libertarians and Communitarians hold sway in our society, and continue to apply their peculiar way analyzing to try to solve our *current* problems. This is not the place to go in depth about this ongoing debate, but I trust that what has been said already is sufficient to make anyone suspicious of any efforts to interpret Dewey as belonging to either camp. Instead, Dewey was simply a "contextualist" who, however much he cherished community avoided prejudicing the communal over the individual good.

It seems plausible to believe that recently (since the second half of the 20th century), parts of the world have been living under the assumption of a related but more significant opposition. Here is some recent commentary on this matter:

> The irony of the post-Cold war period is that the fall of Communism was followed by the rise of another utopian ideology, which created the laissez-faire financial system that has now disintegrated. In America, a type of market fundamentalism became the guiding philosophy. [3]
>
> For the better part of half a century after World War II, democratic capitalism built its modern framework against the backdrop of its death match with totalitarian Communism. In the two decades since the fall of the Berlin Wall, the American model of capitalism, largely unchallenged by ideological alternatives and increasingly dominant around the world, drifted toward what conservatives viewed as a more pure form of economic liberty and what liberals came to view as misguided free-market fundamentalism… But now, as the United States and other nations look for lessons in the wreckage from the excesses of that period, political leaders are confronting uncertainty about what economic structures and values should define capitalism's next chapter. [4]

One common assessment of our current situation (implicit in the quotes above) is that going too far in the direction of capitalism has caused our present global crisis. As Dewey feared, the argument between socialists and capitalists has been used to describe our situation, and that argument dominates our current political discourse which forms an excuse for precluding constructive action. Most political analysis and commentary today assume that, in the map of possible political positions, there is a left and there is a right; each one defined according to opposing policies and values. These labels are applied to countries, political leaders, and Supreme Court Justices. One obvious objection to this way of approaching problems is that reality is very different, or that we are relying on "a map" that oversimplifies the actual political landscape. Perhaps, but I will set this issue aside and instead consider how Dewey's philosophy can be useful in addressing this last common assessment of our recent problems in terms of ideological oppositions.

Dewey warned us about the dangers of historical shifts between opposing ideologies, and provided a philosophical framework to move beyond those ideologies. Dewey was concerned with how oppositions tend to invade our thinking and lead people somehow to believe that they must choose between them. This myopic way of thinking is what leads to extremism, and to entire societies moving in a pendulum between one extreme to another. For some reason, humans fall prey to fixing a one-sided extremism with a craving for the opposite extreme. This is often called a "backlash," but Dewey thought of it in terms of a non-reflective pattern of compensation. We tend to fall easily into this pattern because within the experience of one kind of excess, its contrary seems desirable. The desired excess might represent the direction towards which we should aim, but we must not confuse direction with final aim. Extremism is appealing because it is easily noticed and it seems to represent a position of strength. But Dewey argued otherwise. It is when we live within an extreme that we are the most vulnerable, because we are more likely to compensate by shifting to the opposite extreme. Extremism is also appealing because it offers simple answers to complex problems, and the quest for simple answers is part of the quest for certainty.

In his capacity as a philosopher, Dewey sought to undermine the philosophical views that propounded dualisms and thereby sanctioned the idea that we are stuck between extreme options. He said, "[t]he modern world has suffered because in so many matters philosophy has offered only an arbitrary choice between hard and fast opposites."[5] Philosophy, as criticism, is needed to unearth these prejudices and to suggest better ways to live. This is what Dewey tried to do when he proposed his democratic contextualism.

2. Dewey's Democratic Contextualism

Dewey thought about the important questions that are the basis of ideological debates among politicians such as: What is role of government and markets in the ideal society? Should we have more taxes? More regulation? More public investment? Where and how much is too much? Should we rely on the market to fix the problem of health care? How do we make these decisions? Dewey's personal stand on these and other questions was often misunderstood. He was in fact subject to criticism from both sides of the political spectrum. People at contrary extremes perceived Dewey's position as a disguised threat to their desired extreme. At the very least, it was hard to figure out what Dewey advocated. Was he on the left or was he on the right? The same thing is happening now with President Obama (more on this latter).

Dewey's politico-philosophical stand is, however, clear once one makes a genuine effort to read what he wrote. Dewey stood for the values and virtues of democracy as an ideal. For Dewey, every question regarding what should be socialized or otherwise left to market forces must be considered on its own merits. On this point, Dewey could not have been clearer:

> We have to consider the probable consequences of any proposed measure with reference to the situation, as it exists at some definite time and place in which it is to apply. There cannot be any universal rule laid down, for example, regarding the respective scope of private and public action. [6]

Even if Dewey — in his time — may have favored a more socialized economy to remedy particular problems, it would be a mistake to call him or his philosophy "socialist" or left. His heart was with democracy *as a way of life*, and his approach to means and problem solving is a radical, non-partisan contextualism.

Contextualism is a view about good judgment and decision-making. Dewey's contextualism can be expressed in a negative or positive way. His contextualism refutes the notion that judgments which do not follow rules are necessarily arbitrary. Dewey is against the notion in philosophy that one can or should set up criteria or rules in advance to determine good judgment in a situation. Context matters in a profound way in all areas of life where judgment is needed. Whether a feature of an action counts as a reason to do it (or not) is dependent on the context. Whether a stroke of green in a painting contributes to the beauty of a painting depends on what other colors and lines are present. The green that helps with the beauty of one panting may be what ruins another.

In today's political climate, it is not easy to be a contextualist about the appropriate means to democraticize our experience (Dewey's normative stance). Consider the implications. A contextualist may require that we resolve one problem by market means and another through government intervention, or even with that we treat the same problem differently at different times. Such an approach by a politician today would probably be perceived as a sign of indecision, a character flaw, or what has been called "flip-flopping." For many, political integrity is personified by standing fast to one's ideological (preconceived) way of dealing with a problem, no matter what. Granted, a genuine contextualist politician may at times be difficult to distinguish from an opportunist politician, in so far as both seem flexible. For Dewey, however, there is a profound difference between these characters in both their methods and the result. Now, all of this strikes me as particularly relevant today since people are trying to figure out where Obama stands on many issues and, as with Dewey, he has been criticized on all fronts.

Neither the left nor the right can make sense of Obama's economic philosophy or lack of one. Is he center-left, right-center, or is he just another politician capable of reaching across the ideological spectrum for convenient reasons, but without a moral backbone? Could it be that Obama is instead a contextualist, and hence he is a Deweyean Pragmatist? The label 'Pragmatism' has been used to describe Obama's view. "Mr. Obama is stepping into the debate characteristically intent on avoiding polarizing labels, and his advisers describe his philosophy in terms of pragmatism rather than ideology."[7]

Ever since the publication of Chris Hayes' article entitled "The Pragmatist" in the December 29, 2008 edition of *The Nation*, there has been much discussion, especially in Internet blogs, about the extent to which Barack Obama's thought and action can be located in the tradition of American Pragmatism. Hayes and Mitchell Aboufalia have even made us aware of the indirect historical and biographical similarities between Obama and John Dewey.[8] Others have considered these speculations a "stretch," or find them unhelpful in the attempt to decipher where Obama stands in relation to his contemporaries. In my view, it is too early to offer a judgment about's pragmatism. I am only interested in the debate insofar as it is revelatory of how many conceive of Pragmatism today, and how the debate presents an opportunity to address some objections that have been raised about it. More importantly, my goal is to introduce Dewey's Pragmatism as an alternative approach to deliberation and problem solving in the political arena, and contrast it with recent attempts to articulate what 'Pragmatism' entails in politics.

3. The Varieties of Pragmatism in Contemporary Politics

Pragmatism is appealing today because it points to a new style of politics that transcends the partisan and ideological divisions that have marred the history of political deliberation. However, there are different senses of 'Pragmatism' that are used today when referring to politicians. To be a 'pragmatists' sometimes means simply to be "expedient" and "efficacious" in resolving problems and pursuing one's interests. At first blush, there seems nothing inherently bad about this; in fact, we wish many of our leaders had this quality. The problem is that, in politics, "expedience" is an excuse to achieve self-serving aims (power, fame, re-election) or to acquiesce to the status quo that benefits the interests of the few. Hence, a "pragmatist" in politics could well refer to the politician who is just a practical opportunist and is usually contrasted with the politician that bases his decisions upon theory, ideas, or some fundamental principles and values. Richard Posner has, however, given a positive spin to this type of pragmatism. He characterizes his "everyday pragmatism" as "the mind-set denoted by the popular usage of the word "pragmatic", meaning "practical and business-like, 'non-sense,' disdainful of abstract theory and intellectual pretension, contemptuous of moralizers and utopian dreamers."[9] For Posner, there is nothing inherently problematic about the "self-serving" aspect of a "pragmatist" politician. This is only problematic for those who hold unreasonable ideals (utopias) about politics and democracy. Posner thinks we need to be neo-Schumpeterian "realist" about what to expect from democracy. Democracy is merely a procedure, "a kind of market,"[10] a "competitive power struggle among members of a political elite … for the electoral support of the masses."[11] It should be clear that Posner is using "pragmatism" in a way that, while common, is far from "Pragmatism" as a philosophy that includes John Dewey.

In his *Nation* article, Chris Hayes is aware of the ambiguity of the use "pragmatism" in politics and is hopeful that Obama will be a "pragmatist" in the more philosophical and Deweyean sense. He writes:

> Pragmatism in common usage may mean simply a practical approach to problems and affairs. But it's also the name of the uniquely American school of philosophy… This tradition is a worthy

> inheritance for any president, particularly in times as manifestly uncertain as these. And if there's a silver thread woven into the pragmatist mantle Obama claims, it has its origins in this school of thought. Obama could do worse than to look to John Dewey, another onetime resident of Hyde Park and the founder of the University of Chicago Laboratory Schools, which Obama's daughters attend[ed].

There are however, others that are not so optimistic about Obama, precisely because they are worried that he may be a Pragmatist in the philosophical sense. They have raised criticisms of Pragmatism as a philosophy because they feel it falls short of what is needed today in the political arena. While many of these critics are seeking a post-ideological way to do politics, they find Pragmatism as an alternative to be shallow, empty, without a moral backbone, and ultimately a view that sanctions politics as usual. For instance, a recent editorial in the *Christian Science Monitor* asserts a critic's concern about relying on Obama's or anyone else's Pragmatism as a guide to lead a nation. "Policy devoid of clear ethical theory creates a nation without principle, and a nation without principle is a nation on stilts." "An analysis of what pragmatism really means explains why Mr.'s plan has not (and cannot) work. It also reveals the emptiness of pragmatism as national principle."[12] Hayes voices a similar concern when he says, "Obama may have told Steve Kroft that he's solely interested in 'what works,' but what constitutes 'working' is not self-evident and, indeed, is impossible to detach from some worldview and set of principles." [13]

The concern behind the remarks critical of Obama is clear: unless Pragmatism in the philosophical sense can convince us that it has enough substance beyond repeating the vague assertions that it is committed to "what works" instead of ideology, it is hard to see a significant difference between it and pragmatism in the narrow expedient sense. Obama may be a Pragmatist in the philosophical sense, but that makes him susceptible to the same problem of a pragmatist as a self-serving opportunist—not enough of a moral backbone to avoid relativism and to change the pessimistic or cynical view that many have about politicians.

I am sympathetic to these recent critics of Pragmatism. Part of the problem is that, although pragmatism is often mentioned as the alternative to ideological politics, outside of the academic world there has been very little effort to articulate what Pragmatism means. I have already suggested the importance of "contextualism," but more is needed to answer the critics. It is not enough to reply, as many scholars in Pragmatism have, that Pragmatism in politics is more than mere political expediency because Deweyan Pragmatism is committed to deliberation, experimentation, fallibilism, and social intelligence. None of these broad prescriptions address moral values, and in fact seem perfectly compatible with a politician that has these commitments and traits but lacks the sort of moral character needed in an ideal democracy.

The recent "normative modesty" of Robert Talisse and Cheryl Misak[14] in restricting the understanding of Pragmatism in politics to an epistemology is also insufficient. Their argument does not address the concerns of the objectors who are searching for more moral substance. These recent epistemologists of Pragmatism seem to put great faith in the inculcation of proper epistemic habits (*e.g.*, concern for truth, reasoning well, etc.) but are these habits sufficient to ensure we have democratic political leaders? A politician may be a fallibilist and abide by epistemic principles. He/she may base policy decisions on reasons, arguments, listening to others, and evidence. Yet this seems compatible with someone driven by money and power instead of care for the people that he or she represents. Nor does being epistemically responsible guarantee moral integrity. Seemingly most people who are cynical about politics think of politicians as "moral" failures and not just epistemic failures. We wish to hold politicians morally responsible and not just epistemically responsible.

Granted, to have excellent epistemic habits avoids ideologues, wishful thinking, and dogmatism, but doesn't the pragmatist politician need some sort of moral commitment in order to avoid becoming the opportunist-self-serving politician that "flip-flops" on positions and is willing to compromise on the basis of mere expediency? I think this is the intuition behind those who want to defend Obama by ascribing onto him some ideology, at least vaguely conceived as some conceptual framework or set of principles to stand on. Otherwise, Obama would remain susceptible to becoming just another "pragmatist" politician in the narrow-opportunistic sense. Here is how Hayes puts it: "while ideology can lead decision-makers to ignore facts, it is also what sets the limiting conditions for any pragmatic calculation of interests… Principle is often pragmatism's guardian."[15]

Dewey's pragmatism is, however, not susceptible to the objection of lacking a moral backbone because moral values, virtues, and commitments are integral to his conception of the ideal politician. How is this consistent with the well-known notion that a pragmatist is committed to "what works"?

4. The Pragmatist' Commitment to "What Works": The Ideal Contextualist Politician

The vague commitment to "what works," often cited as key to being a pragmatist politician, has not helped with the moral reputation of Pragmatism. For many, this vague commitment appears as either spineless, as an empty assertion, or as leading to an unwanted relativism. Let us examine some of the most common reasons for these impressions and concerns, and how Dewey's pragmatism would answer those critics.

For some critics, to state "what works" does not answer the important questions. For them, "what works" raises the questions: "For whom?" "For what?" "What works" is always relative to some ends that people may disagree with. People may derive different judgments about "what works" because they have disagreement about ultimate ends-values. Consequently, we must first determine the desired ends.[16]

Dewey scholars reply to the "ends" objection by noting the objector's impoverished understanding of Pragmatism. For Dewey, the means and ends distinction is functional. Dewey argues that ends can be subjected to evaluation by its means and other contextual factors. This is an unsatisfying reply to the objectors because it does not address their implicit need for an ultimate standard of evaluation of *both* means and ends. The objector finds "what works" empty because he/she cannot conceive how "what works" in a particular case can be determined without presupposing some general standard or criterion of "what works." But the objector is asking for precisely that which Dewey denies and cannot provide given his contextualism: in matters of evaluation and judgment, there is no ultimate single criterion for a right judgment across all situations. For Dewey, "what works" is not what meets a standard but "what fits or resolves *this* present problematic situation." Dewey's approach can be called a type of "relativism" but it is relativism to each situation as the ultimate context that grounds our judgments. "What works" has to remain a general, empty, and vague assertion because "what works" is different in each situation. Saying more tends to defeat the purpose of the expression.

The ideal Pragmatist politician is sensitive to context. Sensitivity to context means, among other things, that one does not impose a single end or a universal criterion across problematic situations. A politician that tries to deal with contentious issues by always seeking answers that reach consensus or bipartisan support is not a pragmatist in Dewey's ideal sense. Sure, sometimes the problematic situation is such that preventing legislative gridlock trumps everything else, but we rely on the judgment of the ideal politician to figure out what the situation calls for, and other goals, ends, obligations, and values, are to be given consideration in that deliberation. One implication of the notion of the ideal contextualist politician is that he/she is more unpredictable than those who are ideologues or those who adhere to one single end or criteria. One may argue that politicians should be predictable, but it is not obvious why predictability is so desirable.

For other objectors, the problem with Pragmatism in politics is not that it lacks a criterion of "what works," but that the presupposed criterion is narrow and leaves out moral values and principles. In the political arena for example, it is assumed that the politician who is committed to "what works" believes that deliberation, decisions, or judgments should be based solely on some type of scientific or cost-benefit calculation. Bronsther says:

> For James, things are "true" – and thus we ought to accept them and act upon them – insofar as they "work" for people, making their lives

> more satisfactory. Through philosophical [P]ragmatism, the ethical questions of politics became scientific questions. A lawmaker was to assess a policy's impact on the real world."[17]

Why is the opposition to dogma and ideology and the Pragmatist's corresponding emphasis on inquiry so often understood as implying scientism or consequentialism?[18] Granted, to inquire about policies "that work" may require science and deliberating about possible consequences. It does not follow, however, that consequences (of any particular kind) or science determines the ultimate criteria of judging "what works" in a particular situation.

Bronsther and others are suspicious of the Pragmatist's rhetoric in appealing to "what works." They see the appeal as naïve, and the rhetoric to be a way of ignoring or avoiding moral values, moral discussion, moral ideals, and moral disagreement:

> He [Obama] presumes that policies forged by reason, evidence, and "unbiased" expertise – those policies that "work" – will garner the support of all reasonable members of Congress and thus bridge partisan divides. He bases his belief in the possibility of national and political consensus on this faulty argument. Consensus has not emerged in Washington because disagreement exists.... What "works" for liberals doesn't work for conservatives... The deeper partisan disagreements are ethical and philosophical.[19]

The assumption in Bronsther's criticism of Pragmatism seems to be: "what works" requires deliberation about value-neutral facts or that "what works" entails the primacy of an impartial or value-free standpoint in deciding policy. First, this interpretation assumes a dualism between facts and values, something that any good philosophical Pragmatists would deny. Moreover, the Pragmatist does not assume *a priori* (prior to a situation) and naïvely that there is more agreement about facts than there is about values. Ideologues often cling to ideals, theories, and values in such a way as to ignore facts. A Pragmatist emphasizes facts, concrete circumstances, present situational needs and possibilities, but she does not ignore the important function of ideals, theories, and values in deliberation and good judgment. Let's make this clear.

Pragmatists are utterly opposed to decisions and judgments guided by ideology, as implied by the contextualism already explained. We must be careful, however, not to conclude that Pragmatists therefore hold the contrary view: that one must remain neutral or empty in regard to principles, ideas, ideals, concepts, and values or to any of the content that is associated with "ideology." For the Pragmatist, there may be here a false dichotomy. We are not caught up with an either/or situation, e.g., either there are politicians whose decisions are ideologically driven or they are purely focused on "what works" in a situation (in some value-neutral way). The truth of the matter is that even the most ideological person or politician has some degree of situational-flexibibility, and even the person most willing to compromise and be flexible is usually committed to something. Finding the proper balance between openness and commitment is in fact part of what is ideal, according to Dewey.[20] If having an "ideology" merely means the inclusion of principles, ideals, and values in political deliberation, then Pragmatists (at least Dewey) had one: it is called "democracy." I suspect, however, that most people would find this last sense of "ideology" too broad to be useful, especially in politics where "ideology" is used to describe the "ideologue," someone who is contrary to Pragmatism.

Dewey does not think that his contextualism, and the openness and flexibility that it requires of us, to be incompatible with wholehearted commitment to ideals, principles, and values. The latter are, in fact, part of the contextual resources (tools) in a situation that are needed to determine relevance. Sometimes these tools provide useful standpoints that can help in deliberation, but they do not specify the specific course of action. In some cases, principles may be what keeps good political leaders "anchored," and prevents them from going down the "slippery slope" in their flexibility and compromise. Hence, a general demand for some principles or values to guide a politician's deliberation seems reasonable. What is not reasonable, for Dewey, is the expectation that these principles and values need to operate in deliberation as criteria or standards that determine judgment. "What works" does not mean not guided by principles. Instead, it means that what is a justified or proper course of action cannot be determined in advance of the problematic situation that calls for judgment. In facing social problems, one should try to do "what works" and not what some absolute principle or antecedent criteria tells him/her. The Pragmatist is very flexible. She is even willing to "test" principles in light of present problems. Yet, she does not enter the situation empty-handed or "neutral" among values, nor should she try to do so.

Far from an empty or value-free assertion, "what works" in Dewey's version of Pragmatism presupposes a "thick" or substantive normative standpoint. "What works" means that which ameliorates or resolves the particular problematic situation calling for our judgment. However, Dewey has a substantial prescription about how this resolution should be accomplished. According to Dewey, we must try to come to judgment by being sensitive and open to the present and unique problematic context, while also being guided by democratic values and principles. This is a different prescription from doing what is most expedient (as in the "pragmatist opportunist") or what safeguards the validity or subsistence of an ideology. It is also a different prescription than to pursue truth or any of the epistemic goals of the recent epistemic pragmatists.

Dewey's contextualism is part of a normative ethical philosophy, with substantial recommendations about character, dialogue, and community.

In a democracy, reasonable judgments require characters that transact with others who are equipped with certain habits, such as sensitivity and openness to context. Dewey also mentions virtues regarding our treatment of others, such as trust, tolerance, and empathy. This is not the place to examine all the character traits of the ideal "Pragmatist" politician that is needed for Dewey's ideal of democracy *as a way of life*. I must, however, say something about empathy since it refutes the notion that Dewey's Pragmatism prescribes an impartial, value-free standpoint in the political arena. Furthermore, Dewey's view on empathy could be considered as another reason in favor of the view that Obama is a Deweyan Pragmatist.

5. Empathy as a Democratic Virtue

From the standpoint of Dewey's philosophy, impartiality has been overrated in the history of ethical and political theory. "Impartiality" can be a virtue but only in some contexts—and there is no such thing as absolute impartiality. Selectivity and partiality are at bottom unavoidable traits of any inquiry:

> It is not concern which is objectionable even when it takes the form of bias. It is certain kinds of bias that are obnoxious. Bias for impartiality is as much a bias as is partisan prejudice, though it is a radically different quality of bias. To be "objective" in thinking is to have certain sort of selective interest operative. One can only see from certain standpoint but this fact does not make all standpoints of equal value….") [21]

Strict objectivity or impartiality is a myth. Pragmatists are suspicious of the separation of inquiry and facts from value. Inquiry starts with, and cannot totally divest itself from, values and the normative to achieve a neutral "objectivity" from a "god's-eye view". For the classical Pragmatists, values are unavoidable. Where there are always partialities, the only reasonable question is: which partiality is best. There are better and worse "biases" relative to their capacity to disclose the scope and depth of the situation.

> A standpoint which is nowhere in particular is an absurdity. But one may have an affection for a standpoint which gives a rich and ordered landscape rather than for one from which things are seen confusedly and meagerly. [22]

Which "habits of desire and imagination"[23] one brings to a situation will determine ones selectivity, and hence the scope and depth of the present situation that is being immediately appreciated (had) and reflectively examined. There are habits that provide the agent a rich, broad landscape upon which to deliberate and to act, and there are others that in comparison are constraining, one sided, and narrow. Dewey prescribed "the fostering of those habits and impulses which lead to a broad, just, sympathetic survey of situations." [24] Empathy and openness are two valuable "undergoing" habits in this regard.

For Dewey, whether our experience is wide or narrow depends also on the extent and depth of what he called sympathy, but today is known as empathy. To exchange places—emotionally and imaginatively—with another is the only way to widen our intellectual horizon in situations and to determine effectively what ought to be done. As Dewey said, this is the most "generous thought." On the other hand, "a person of narrow sympathy is of necessity a person of confined outlook upon the scene of human good." [25]

The rationalist tradition in moral and political theory has assumed that the avoidance of a distorting partiality in deliberation, and the guarantee for considerate and just treatment of others, requires the adoption of the universal and objective standpoint provided by reason. Dewey cannot make sense of an impartial and universal standpoint. However, Dewey holds that the closest we could get to a similarly broad intellectual standpoint—that might be useful for deliberation—is with the aid of sympathy. "Sympathy, … furnishes the most reliable and efficacious intellectual standpoint. It supplies the tool, par excellence, for analyzing and resolving complex cases ... for an effective, broad, and objective survey of desires, projects, resolves, and deeds."[26] The concern implied by a genuine sense of justice cannot be accounted for by the intellectual capacities associated with the impartiality of reason, it requires sympathy. "A person entirely lacking in sympathetic responses might have a keen calculating intellect, but he would have no *spontaneous sense* of the claims of others."[27] Failure to treat others (beyond ones' social group) as equals is not a deficiency in the ability to perform a reasoning process from a formal principle of justice, nor is it a result of failing to grasp intellectually and abstractly the respect in which all people are equal. Justice ceases to be a virtue if it is not fused with sympathy. It is through sympathy that one can appreciate what justice demands because it makes possible for us to understand what each person needs in a vivid and more reliable fashion. Hence Dewey claimed that:

> [T]o put ourselves in the place of another, to see things from the standpoint of his aims and values ... is the surest way to appreciate what justice demands in concrete cases. [28]

Let us return to the topic of Pragmatism and the notion of an ideal politician.

For Dewey, there is nothing problematic about a politician being partial or biased with regard to moral values and ideals. Dewey would admit that not all biases are the same. A bias towards the values of democracy in our political leaders is important to avoid such practices as imposition and oppression. For Dewey, avoiding narrow partiality, and therefore respect for pluralism and inclusivity of understanding, require what Obama has recently called "empathy" in his stated criteria for a Supreme Court Justice. On the campaign trail, Obama said that he would want his Supreme Court nominees to have "empathy." Notice in his remarks to Planned Parenthood (July 17, 2007) how Obama seems to be contrasting the impartiality of an umpire with empathy:

> You know, Justice Roberts said he saw himself just as an umpire. But the issues that come before the court are not sport. They're life and death. And we need somebody who's got the heart to recognize— the empathy to recognize what it's like to be a young, teenaged mom; the empathy to understand what it's like to be poor or African-American or gay or disabled or old. And that's the criteria by which I'm going to be selecting my judges.

Critics of Obama's criteria consider his view a threat to the "objectivity" and "impartiality" needed to be a good judge. Empathy seems, to them, to be irreconcilable with decision-making, which to them must be based solely on law and facts. Everyone agrees that ideology should not guide a judge or a politician. The disagreement is about the role of empathy. For Dewey, and perhaps for Obama, strict impartiality or neutrality is impossible and in any case undesirable. When empathy is added to other important qualities, such as intellect, integrity, and respect for the law, those qualities broaden deliberation, and counteract narrow partiality to achieve justice.

6. Conclusion

Dewey's ethical-political thought is a resource worth considering as a guide to a post-ideological politics. A commitment, even a partisan commitment, to certain values and principles, while also being open and sensitive to the uniqueness and complexity of each problematic context, is a virtue for Dewey. Refusing to pre-define "what works" ahead of problematic situations gives the ideal pragmatist politician the flexibility that is not present in the ideologue. It is true that there is always the danger that this flexibility could turn eventually into the undesirable flexibility of the self-serving "flip-flopper." Yet this "slippery slope" problem is not inevitable, and it is a danger worth taking in comparison to the alternatives, especially the rigidity of the ideological political leader. Moreover, the best way to prevent this danger is by making sure that virtues (such as good habits) are not turned into vices, and not by strict adherence to rules or absolute values. This is the importance of character in Dewey's ethics. For Dewey, good judgment and deliberation depend on the cultivation of virtuous habits. Encouraging in a community certain virtues is the best way to prepare for particular collective decision-making, and the best way to counteract the seductions that politicians have to face (*e.g.*, power, money, status, and wishful thinking). Instilling virtuous traits is the alternative to the imposition of proper rules or restrictions on public discourse, and the reason why Dewey puts so much faith in education.

The sort of intelligent and context-sensitive deliberation and communication that Dewey thought was needed in a democracy requires characters with certain dispositions: sensitivity to context, open-mindedness, tolerance, and empathy. These virtues, however, cannot be cultivated unless this last type of deliberation and communication is practiced, not just in the White House, but also within society as a whole. The obstacles needed to cultivate the virtues required for a post-ideological age in politics are too numerous to mention, but this is an inquiry worth having.

Pragmatists have been severe critics of abstract moral arguments, fixed or absolute values, and the quest for some ultimate criteria of all judgments. This has somehow been interpreted as implying the denial of the important role of values, principles, and morality. Dewey's normative view may still be considered too "thin" from the point of view of those who believe that the ideal politician must guide himself by absolute moral values. In other words, objectors may still insist that only adherence to absolute values would help us avoid moral anarchy or apathy. Under that assumption, it seems impossible to convince them that a Pragmatist has a strong enough "moral backbone" to make judgments and lead a country.

Do we need to worry about President Obama being a Pragmatist in the philosophical sense? I have shown that these worries are based on misconceptions about the ethics of Pragmatism. A careful reading of this philosophical tradition, especially of Dewey, reveals an ethics that, while not centered on an ultimate criterion of right conduct, is substantive enough to offer normative moral guidance centered on democratic values and virtues. What we really need to worry about is putting too much hope on one single individual (*e.g.*, Obama) to break the shared habits and the environment that continue to perpetuate ideological thinking and practical opportunism in politics. Dewey's ethics of democracy puts the emphasis not on particular politicians transcending ideology, but on having more communal inquires and deliberations where people can rightfully and wisely choose from both sides of the political spectrum if the context calls for it. The recent brawls in town hall meetings about health care policy[29] are evidence that we are failing at all levels of society to have the sort of discussion about broad policy questions that democracy requires. We need to inquire into what it takes to break with ideological ways of thinking that have dominated discourse at all levels. To leave things as they are will perpetuate pessimistic and cynical views about politicians *and* people. As we all know, losing faith in the people is the beginning of the end of a democracy.

NOTES

1. I agree with Michael Eldridge when he writes, "It is not necessary that Obama knows this intellectual movement and considers himself to be a part of it. It is sufficient that he practices what they recommended. He exemplifies the pragmatism they articulated and defended. By placing Obama in this context we are able to make use of the resources of this tradition to explain and justify his practice. Some of these charges are that his pragmatism is a matter of mere expediency or that he is anti-ideological and thus devoid of principle." http://www.obamaspragmatism.info/MEonOP.htm

2. John Dewey (*LW* 2:361) Citations of the works of John Dewey in this article refer to the critical edition published by Southern Illinois University Press. In the citations the initials of the series are followed by volume and page numbers. Abbreviations for the critical edition are:

EW The Early Works (1982-1898)
MW The Middle Works (1899-1924)
LW The Latter Works (1925-1952)

3. John Gray, "Utopia Falls" Harpers Magazine, Dec 2008, p, 17.

4. Richard W. Stevenson "Obamanomics: Capitalism after the Fall" New York Times April 19, 2009.

5. John Dewey (*MW* 12:137).

6. John Dewey (*LW* 7:336)

7. John Gray "Utopia Falls", p, 17.

8. See http://www.obamaspragmatism.info/LandS.htm for links to the online contributions of Mitchel Aboufalia and others in blogs.

9. Richard Posner *Law, Pragmatism, and Democracy* (Cambridge, Mass.: Harvard University Press, 2003), 50.

10. Ibid., p. 166.

11. Ibid., p. 130.

12. Jacob Bronsther "The emptiness of Obama's pragmatism" from the May 26, 2009 edition (http://www.csmonitor.com/2009/0526/p09s02-coop.html)

13. Chris Hayes "The Pragmatist" in December 29, 2008 edition of *The Nation* http://www.thenation.com/doc/20081229/hayes/single

14. See Robert B. Talisse *A Pragmatist Philosophy of Democracy* (New York and London: Routledge, 2007) and Cheryl Misak *Truth, Politics, and Morality: Pragmatism and Deliberation* (New York: Routledge, 2000).

15. Chris Hayes "The Pragmatist"

16. See, for example, Robert Reich, "Obama and Pragmatism: Thinking Through Values" Robert Reich's Blog (5 May 2009).

17. Jacob Bronsther, "The emptiness of Obama's pragmatism"

18. In Ethical theory, consequentialism is a view about the ultimate or universal criterion of correct judgment.

19. Jacob Bronsther, "The emptiness of Obama's pragmatism"

20. See Gregory Pappas "Openmindedness and Courage: the Virtues of the Pragmatist Ideal Believer," *Transactions of the Charles Peirce Society* Vol. XXXII, No.2 (Spring 1996) pp.316 335.

21. John Dewey (*LW* 6:14-15, emphasis added).

22. John Dewey (*LW* 6:14-15, my emphasis).

23. John Dewey (*LW* 7:187).

24. John Dewey (*MW* 14:144).

25. John Dewey (*LW* 7:270).

26. John Dewey (*MW* 5:303).

27. John Dewey (*LW* 7:270, emphasis added)

28. John Dewey (*LW* 7:251)

29. See http://www.nytimes.com/2009/08/09/weekinreview/09stolberg.html?hp

LIST OF CONTRIBUTORS

James Campbell was educated at Temple University (B.A., 1972) and SUNY/Stony Brook (M.A., 1978; Ph.D., 1979). He teaches at the University of Toledo, Ohio, where he is Distinguished University Professor of Philosophy. He is the author of numerous articles and reviews, as well as four volumes: *The Community Reconstructs: The Meaning of Pragmatic Social Thought* (1992), *Understanding John Dewey: Nature and Cooperative Intelligence* (1995), *Recovering Benjamin Franklin: An Exploration of a Life of Science and Service* (1999), and *A Thoughtful Profession: The Early Years of the American Philosophical Association* (2006). He is also editor of *Selected Writings of James Hayden Tufts* (1992), and co-editor of *Experience as Philosophy: On the Work of John J. McDermott* (2006).

Ramón del Castillo. Professor of Philosophy, Universidad Nacional de Educación a Distancia (UNED), has written on American philosophy and culture. Among his works: *Conocimiento y acción* (1995) on American and European pragmatism, and several essays on Emerson, James and Santayana: "Varieties of American Ecstasy", "The Glass Prison", "Portrait of an Anxiety". He has also published two long introductory studies to the Spanish editions of Dewey's *Old and New Individualism* and *The Public and Its Problems*.

Philipp Dorstewitz, joined the Faculty of Arts and Social Sciences of Maastricht University in 2007. He teaches courses on political philosophy, philosophy of science and diverse courses in the Arts & Culture and European Studies programs. Philipp holds degrees in Philosophy and Business Administration from LSE, Hogeschool van Utrecht and FOM, Essen. Prior to joining Maastricht University he held teaching appointments at the London School of Economics and at the University College Dublin.

Ángel M. Faerna, is professor of philosophy at Universidad de Castilla-La Mancha, Toledo (Spain). He is the author of *Introducción a la teoría pragmatista del conocimiento* (1996), and co-editor of *Identidad, individuo e historia* (2003), with M. Torrevejano, and *Caminos de la hermenéutica* (2006), with J. Muñoz. He is also the editor of a selection of Dewey's essays on epistemology, *La miseria de la epistemología* (2000), and the translator of Dewey's *Teoría de la valoración* (2008). He is presently working on a new translation into Spanish of George Santayana's *Scepticism and Animal Faith*.

Walter Feinberg, Professor of Philosophy of Education at University of Illinois at Urbana-Champaign; authored: Common Schools/Uncommon Identities: *a Philosophy of Education in a Multicultural Society* (1998); *On Higher Ground: Education and the Case for Affirmative Action* (1998); *Japan and the Pursuit of a New American Identity: Work and Education in a Multicultural Age* (1993); *Understanding Education: Towards a Reconstruction of Educational Inquiry* (1983); *Reason and Rhetoric: The Intellectual Foundations of Twentieth Century Educational Policy* (1975).

Ángel M. Faerna, is professor of philosophy at Universidad de Castilla-La Mancha, Toledo (Spain). He is the author of *Introducción a la teoría pragmatista del conocimiento* (1996), and co-editor of *Identidad, individuo e historia* (2003), with M. Torrevejano, and *Caminos de la hermenéutica* (2006), with J. Muñoz. He is also the editor of a selection of Dewey's essays on epistemology, *La miseria de la epistemología* (2000), and the translator of Dewey's *Teoría de la valoración* (2008). He is presently working on a new translation into Spanish of George Santayana's *Scepticism and Animal Faith*.

Matthew Caleb Flamm is Associate Professor of Philosophy at Rockford College, in Rockford Illinois. He has written articles on George Santayana and John Dewey and has co-authored and edited (with Krzysztof Piotr Skowroński) the book anthologies, *Under Any Sky: Contemporary Readings of George Santayana* (Cambridge Scholars Publishing, 2007), and *American and European Values: Contemporary Philosophical Perspectives* (also with John Lacks, Cambridge Scholars Publishing, 2008).

Richard A. S. Hall received the B.A. in philosophy from Boston University, the M.A. in philosophy from Dalhousie University, and the Ph.D. in philosophy from the University of Toronto. He currently is an assistant professor of philosophy at Fayetteville State University, a constituent institution of the University of North Carolina. He has authored two books, *The Ethical Foundations of Criminal Justice* (Boca Raton, Florida: CRC Press/LLC Publication, 2000), and *The Neglected Northampton Texts of Jonathan Edwards: Edwards on Society and Politics* (Lewiston, New York: The Edwin Mellen Press, 1990), and both edited and contributed to *The Contribution of Jonathan Edwards to American Culture and Society (The Northampton Tercentenary Celebration 1703-2003)* (Lewiston, New York: The Edwin Mellen Press, 2008). Hall has also published articles and essays in the areas of the history of American philosophy, aesthetics and the philosophy of art, ethics, and the philosophy of religion.

Larry Hickman is director of the Center for Dewey Studies and professor of philosophy at Southern Illinois University Carbondale. He is the author of

Modern Theories of Higher Level Predicates (1980), *John Dewey's Pragmatic Technology* (1990), *Philosophical Tools for Technological Culture* (2001), and *Pragmatism as Post-Postmodernism* (2007). He is the editor or co-editor of more than a dozen volumes, including *Technology as a Human Affair* (1990), *Reading Dewey* (1998), *The Essential Dewey* (with Thomas Alexander, 1998), *The Correspondence of John Dewey,* (1999, 2001, 2005, and 2008), and *John Dewey: Between Pragmatism and Constructivism* (with Stefan Neubert and Kersten Reich, 2008). In 2007 he received the Phi Kappa Phi National Scholar Award, and he holds honorary doctorates from Soka University (Japan) and the University of Cologne (Germany).

Robert E. Innis is professor of philosophy at the University of Massachusetts Lowell. He has been Fulbright Professor at the University of Copenhagen and Humboldt Fellow at the University of Cologne. His books include /Karl Bühler: Semiotic Foundations of Language Theory /(1981), /Semiotics: An Introductory Anthology/ (1985), /Consciousness and the Play of Signs /(1994), /Pragmatism and the Forms of Sense: Language, Perception, Technics /(2002), and, most recently, /Susanne Langer in Focus: The Symbolic Mind/ (2009).

Maciej Kassner, PhD candidate at University of Warsaw. He is currently involved in a research project that aims to investigate the doctrine of American liberalism through the study of the political works of John Dewey.

Jacquelyn Ann K. Kegley is CSU Outstanding Professor of Philosophy and Wang Family awardee for outstanding teaching, research, and service. She is Professor of Philosophy and Chair of the Department of Philosohy and Religious Studies at California State University, Bakersfield. She is author of *Josiah Roxce in Focus* and *Genuine Individuals and Genuine Communities: A Roycean Public Philosophy*. She is an author and editor of Genetic Knowledge as well of numerous articles on American Philosophy, Genetic technology, and contributed to the volume, Pragmatic Bioetchics. She is Vice President of the Society for the Advancement of American Philosophy and a recipient of the Herbert Schnieder Award for outstanding contributions of American Philosophy.

Hugh McDonald is an Assistant Professor of philosophy at the CUNY New York City College of Technology, is the author of *Radical Axiology: A First Philosophy of Values* and *John Dewey and Environmental Ethics.*

Gregory Fernando Pappas works within the American Pragmatist and Latin American traditions in ethics and social-political philosophy. He is the author of numerous articles on the philosophy of William James and John Dewey. His most recent publication is the book with Indiana University Press titled "*John Dewey's Ethics: Democracy as Experience*" It is the first comprehensive interpretation of Dewey's ethics. He has been the recipient of a Ford Foundation Postdoctoral Fellowship as well as the William James and the Latin American Thought prizes by the American Philosophical Association. He is working on the philosophical connections between American Pragmatism and Latin American Philosophy. Pappas is the editor of Pragmatism and the Hispanic/Latino World (Fordham University Press, forthcoming).

Ahti-Veikko J. Pietarinen is a Professor of Philosophy at Department of Philosophy at University of Helsinki. He authored *Signs of Logic*: *Peircean Themes on the Philosophy of Language, Games, and Communication* (2006). He edited some books and published over 100 papers on Philosophy of Science, Pragmatism, History of Analytic Philosophy and others.

John Ryder is Professor of Philosophy and the Director of International Programs, State University of New York. His publications include *Interpreting America: Russian and Soviet Studies of the History of American Thought* (1999), *American Philosophic Naturalism in the 20th Century* (editor) (1994), *The Philosophical Writings of Cadwallader Colden* (2002), co-edited with Scott Pratt, and *The Blackwell Guide to American Philosophy* (2002), co-edited with Armen Marsoobian. He has also co-edited four volumes (2004-2009) of papers from the Central European Pragmatist Forum, of which he is co-founder and co-chair.

Krzysztof (Chris) Piotr Skowroński currently teaches Contemporary Philosophy, Aesthetics, Cultural Anthropology, Polish Philosophy, and American Philosophy at the Institute of Philosophy, Opole University, Poland. He authored the book *Santayana and America. Values, Liberties, Responsibility* (Cambridge Scholars 2007), co-edited, with Matthew Flamm, *Under Any Sky. Contemporary Readings of George Santayana* (Cambridge Scholars 2007) also, coedited, with Matthew Flamm and John Lachs, *American and European Values: Contemporary Philosophical Perspectives*, (Cambridge Scholars 2008). His latest book, *Values and Powers. Re-reading the Philosophy of American Pragmatism*, was published by Rodopi in 2009.

Sebastian Stankiewicz is secretary of John Dewey Research Center at Jagiellonian University in Cracow. In 2009 he achieved his PhD. degree with a thesis on John Dewey's philosophy and pragmatist aesthetics. His current work deals with problems of art theory and pragmatist approach to philosophy of perception including its relevance to contemporary cognitive sciences.

INDEX

VIBS

The **Value Inquiry Book Series** is co-sponsored by:

Titles Published

Volumes 1 - 189 see www.rodopi.nl

190. Oscar Vilarroya and Francesc Forn I Argimon, Editors, *Social Brain Matters: Stances on the Neurobiology of Social Cognition*. A volume in **Cognitive Science**

191. Eugenio Garin, *History of Italian Philosophy*. Translated from Italian and Edited by Giorgio Pinton. A volume in **Values in Italian Philosophy**

192. Michael Taylor, Helmut Schreier, and Paulo Ghiraldelli, Jr., Editors, *Pragmatism, Education, and Children: International Philosophical Perspectives*. A volume in **Pragmatism and Values**

193. Brendan Sweetman, *The Vision of Gabriel Marcel: Epistemology, Human Person, the Transcendent*. A volume in **Philosophy and Religion**

194. Danielle Poe and Eddy Souffrant, Editors, *Parceling the Globe: Philosophical Explorations in Globalization, Global Behavior, and Peace*. A volume in **Philosophy of Peace**

195. Josef Šmajs, *Evolutionary Ontology: Reclaiming the Value of Nature by Transforming Culture*. A volume in **Central-European Value Studies**

196. Giuseppe Vicari, *Beyond Conceptual Dualism: Ontology of Consciousness, Mental Causation, and Holism in John R. Searle's Philosophy of Mind*. A volume in **Cognitive Science**

197. Avi Sagi, *Tradition vs. Traditionalism: Contemporary Perspectives in Jewish Thought*. Translated from Hebrew by Batya Stein. A volume in **Philosophy and Religion**

198. Randall E. Osborne and Paul Kriese, Editors, *Global Community: Global Security*. A volume in **Studies in Jurisprudence**

199. Craig Clifford, *Learned Ignorance in the Medicine Bow Mountains: A Reflection on Intellectual Prejudice*. A volume in **Lived Values: Valued Lives**

200. Mark Letteri, *Heidegger and the Question of Psychology: Zollikon and Beyond*. A volume in **Philosophy and Psychology**

201. Carmen R. Lugo-Lugo and Mary K. Bloodsworth-Lugo, Editors, *A New Kind of Containment: "The War on Terror," Race, and Sexuality*. A volume in **Philosophy of Peace**

202. Amihud Gilead, *Necessity and Truthful Fictions: Panenmentalist Observations*. A volume in **Philosophy and Psychology**

203. Fernand Vial, *The Unconscious in Philosophy, and French and European Literature: Nineteenth and Early Twentieth Century*. A volume in **Philosophy and Psychology**

204. Adam C. Scarfe, Editor, *The Adventure of Education: Process Philosophers on Learning, Teaching, and Research*. A volume in **Philosophy of Education**

205. King-Tak Ip, Editor, *Environmental Ethics: Intercultural Perspectives*. A volume in **Studies in Applied Ethics**

206. Evgenia Cherkasova, *Dostoevsky and Kant: Dialogues on Ethics*. A volume in **Social Philosophy**

207. Alexander Kremer and John Ryder, Editors, *Self and Society: Central European Pragmatist Forum*, Volume Four. A volume in **Central European Value Studies**

208. Terence O'Connell, *Dialogue on Grief and Consolation*. A volume in **Lived Values, Valued Lives**

209. Craig Hanson, *Thinking about Addiction: Hyperbolic Discounting and Responsible Agency*. A volume in **Social Philosophy**

210. Gary G. Gallopin, *Beyond Perestroika: Axiology and the New Russian Entrepreneurs*. A volume in **Hartman Institute Axiology Studies**

211. Tuija Takala, Peter Herissone-Kelly, and Søren Holm, Editors, *Cutting Through the Surface: Philosophical Approaches to Bioethics*. A volume in **Values in Bioethics**

212. Neena Schwartz: *A Lab of My Own*. A volume in **Lived Values, Valued Lives**

213. Krzysztof Piotr Skowroński, *Values and Powers: Re-reading the Philosophical Tradition of American Pragmatism*. A volume in **Central European Value Studies**

214. Matti Häyry, Tuija Takala, Peter Herissone-Kelly and Gardar Árnason, Editors, *Arguments and Analysis in Bioethics*. A volume in **Values in Bioethics**

215. Anders Nordgren, *For Our Children: The Ethics of Animal Experimentation in the Age of Genetic Engineering*. A volume in **Values in Bioethics**

216. James R. Watson, Editor, *Metacide: In the Pursuit of Excellence*. A volume in **Holocaust and Genocide Studies**

217. Andrew Fitz-Gibbon, Editor, *Positive Peace: Reflections on Peace Education, Nonviolence, and Social Change*. A volume in **Philosophy of Peace**

218. Christopher Berry Gray, *The Methodology of Maurice Hauriou: Legal, Sociological, Philosophical*. A volume in **Studies in Jurisprudence**

219. Mary K. Bloodsworth-Lugo and Carmen R. Lugo-Lugo, *Containing (Un)American Bodies: Race, Sexuality, and Post-9/11 Constructions of Citizenship*. A volume in **Philosophy of Peace**

220. Roland Faber, Brian G. Henning, Clinton Combs, Editors, *Beyond Metaphysics? Explorations in Alfred North Whitehead's Late Thought*. A volume in **Contemporary Whitehead Studies**

221. John G. McGraw, *Intimacy and Isolation (Intimacy and Aloneness: A Multi-Volume Study in Philosophical Psychology, Volume One)*, A volume in **Philosophy and Psychology**

222. Janice L. Schultz-Aldrich, Introduction and Edition, *"Truth" is a Divine Name, Hitherto Unpublished Papers of Edward A. Synan, 1918-1997*. A volume in **Gilson Studies**

223. Larry A. Hickman, Matthew Caleb Flamm, Krzysztof Piotr Skowroński and Jennifer A. Rea, Editors, *The Continuing Relevance of John Dewey: Reflections on Aesthetics, Morality, Science, and Society*. A volume in **Central European Value Studies**